Ius Comparatum – Global Studies in Comparative Law

Volume 40

As globalization proceeds, the significance of the comparative approach in legal scholarship increases. The IACL / AIDC with almost 800 members is the major universal organization promoting comparative research in law and organizing congresses with hundreds of participants in all parts of the world. The results of those congresses should be disseminated and be available for legal scholars in a single book series which would make both the Academy and its contribution to comparative law more visible. The series aims to publish the scholarship emerging from the congresses of IACL / AIDC, including: 1. of the General Congresses of Comparative Law, which take place every 4 years (Brisbane 2002; Utrecht 2006, Washington 2010, Vienna 2014, Fukuoka 2018 etc.) and which generate (a) one volume of General Reports edited by the local organizers of the Congress; (b) up to 30 volumes of selected thematic reports dealing with the topics of the single sections of the congress and containing the General Report as well as the National Reports of that section; these volumes would be edited by the General Reporters of the respective sections; 2. the volumes containing selected contributions to the smaller (2-3 days) thematic congresses which take place between the International Congresses (Mexico 2008; Taipei 2012; Montevideo 2016 etc.); these congresses have a general theme such as "Codification" or "The Enforcement of Law" and will be edited by the local organizers of the respective Congress. All publications may contain contributions in English and French, the official languages of the Academy.

More information about this series at http://www.springer.com/series/11943

Académie Internationale de Droit Comparé
International Academy of Comparative Law

Franz Werro
Editor

The Right To Be Forgotten

A Comparative Study of the Emergent Right's Evolution and Application in Europe, the Americas, and Asia

Springer

Editor
Franz Werro
Georgetown University Law Center
Washington, DC, USA

ISSN 2214-6881 ISSN 2214-689X (electronic)
Ius Comparatum – Global Studies in Comparative Law
ISBN 978-3-030-33511-3 ISBN 978-3-030-33512-0 (eBook)
https://doi.org/10.1007/978-3-030-33512-0

This Springer imprint is published by the registered company Springer Nature Switzerland AG.
The registered company address is: Gewerbestrasse 11, 6330 Cham, Switzerland

Contents

The Right to Be Forgotten: The General Report—Congress of the International Society of Comparative Law, Fukuoka, July 2018

Franz Werro

Abstract The present general report is based on the work of fifteen national rapporteurs. It finds that jurisdictions embrace the right to be forgotten mostly where the right to privacy imposes limits on the right to free expression. Regardless of labels or formal legal recognition, the right to be forgotten takes various forms. In its most traditional form, this right has existed in some parts of Europe for over two centuries. It gives individuals the right to preclude the media from revealing true facts about their private life where no public interest prevails. In today's world, the right to be forgotten has a more multifaceted meaning. With respect to personal data, this right can involve the right to access, control, and erase these data. The access and the control in turn will depend on various elements, including the roles of data processors, technological devices, competing interests, and the interest of the state. As the world is still assessing the roles of these elements, the right to be forgotten, at least in some of its current manifestations, will gain importance.

1 Introduction

The question that this book addresses is whether and to what extent an individual has the right to preclude anyone from publicizing a particular true fact or event relating to his or her private life, which has lost its newsworthiness or public pertinence. This right to be forgotten has found express recognition in some places, such as France, since the middle of the nineteenth century. In today's world, this right also comprises a person's entitlement to access, control and, sometimes, erase personal data held by others. In practice, the right to be forgotten serves as a shield against media platforms that would otherwise enjoy the right to publicize this fact or event. On the Internet

The publication of this piece considers materials up until September of 2019.

F. Werro (✉)
Fribourg University, Fribourg, Switzerland

Georgetown University Law Center, Washington, DC, USA
e-mail: fgw@law.georgetown.edu

F. Werro (ed.), *The Right To Be Forgotten*, Ius Comparatum – Global Studies in Comparative Law 40, https://doi.org/10.1007/978-3-030-33512-0_1

and in e-commerce, it can serve as a means to know and control information about one's own personal data, and possibly to erase them.

In this European understanding, the right to be forgotten in all its forms finds its roots in fundamental precepts of human dignity. As such, it is part of the right to privacy, understood both as an entitlement between private individuals, as well as a constitutional one against the state. When recognized, the right to be forgotten necessarily comes into conflict with other private or constitutional entitlements, such as the right to freely speak and inform, the right to property or the right to engage in commercial activity, domestically or across borders. The limits of the right to be forgotten thus result from a balance between these rights. In that sense, it is never unconditional. To enjoy recognition, the right to be forgotten depends on the perceived legitimacy of its limits on other fundamental rights.

Although the right to privacy has a clear public dimension, this report will focus on the right to be forgotten as the prerogative of an individual against other individuals or private corporations and leave aside other possible entitlements regarding information held by states. The international or transnational aspects of the questions will only be briefly referred to, namely with respect to enforcement issues. The focus will be on a comparison between the various national approaches to the right to be forgotten in its private law dimension.

Our report finds its basis in the work of fifteen national rapporteurs. These rapporteurs are scholars from Argentina, Belgium, Brazil, Canada, Czechia, Denmark, Finland, Germany, Ireland, Italy, Japan, Romania, Taiwan, Turkey, and the United Kingdom. I am very thankful for the work of these scholars, who answered a questionnaire I had drafted.[1] Their work helped create a contemporary image of the right to be forgotten and deepen the meaning one can give to it, particularly with respect to its multifaceted content. As the work of the rapporteurs shows, the right to be forgotten has acquired global notoriety in recent times, particularly in the application of the right to de-indexation on the Internet, as decided by the Court of Justice of the European Union in *Google Spain*.[2] The unexpected, and at times scandalous, news in recent times also showed the ways in which information held by social networks and other Internet players trigger many troublesome questions about the private life of Internet users.

The work of the rapporteurs further suggests that the right to privacy—understood as the right to keep one's private life free from private and/or public infringements—prevails more strongly in some states than in others. These legal differences surely cannot be separated from cultural ones, which in turn bear on historical and social values. At the same time, I find that a state's treatment of the right to be forgotten directly correlates with the extent to which a state is willing to recognize and guarantee the right to privacy. Conversely, I found a correlation between a state's treatment of the right to be forgotten and the importance its legal systems give to capitalism and free market ideas. The more a system values free entrepreneurship,

[1]The questionnaire is an annex to the present report.

[2]Case C-131/12, Google Spain SL Google Inc. v Agencia Española de Protección de Datos (AEPD) Mario Costeja González [2014] ECR I-000, at pt. 35 ff, 80 ff, 86 [Hereinafter *Google Spain*].

the less it appears willing to compromise free speech with the protection of private life.

The report is a survey of what the national reports have exposed, rather than the outcome of our individual research and critical analysis. At the same time, it does not do full justice to the wealth of information provided by the national reports nor to the depth of the work of the individual rapporteurs, some of whom have written books and produced major contributions on the subject.[3] In its present form, the general report only touches the surface of many issues. It merely tries to suggest an assessment of the current state of affairs and possibly a *grille de lecture* that will help sort things out. Accordingly, I will first explore how the general right to privacy comes into opposition with freedom of expression (2). I will then assess the implications of this opposition for the right to be forgotten (3). Subsequently, I will identify the core feature of the right to be forgotten (4) and examine, beyond media and information, its implications with respect to individuals' ability to control their personal data (5).

2 Balancing the Freedom of Expression Against the General Right to Privacy

In this section, I will compare the European and American[4] approaches to privacy as balanced against free speech and illustrate their distinct features. Subsequently, I will contrast our findings with the picture that emerges in Latin America and Asia.

2.1 *The European Approach vs. the United States' Approach*

In continental European states, the right to privacy often finds its express basis in the constitution.[5] In Ireland, the right found its expression in judicial recognition under the doctrine of un-enumerated rights.[6] Despite the absence of a written Constitution,

[3]A quick look into the bibliography of the national reports will give an idea of the wealth of the research work of their authors and of what is missing in the present report.

[4]We did not receive a United States report, but we analyzed various contributions published in this country. For a recent analysis of the right to be forgotten in the United States' jurisprudence, *see* Gajda (2018), p. 201, which interestingly and somewhat unexpectedly claims that the right to be forgotten finds some acceptance in the United States, at least to a larger extent than what has been traditionally acknowledged. *See also* Post (2018), pp. 1059–61.

[5]*See, e.g.*, Convention for the Protection of Human Rights and Fundamental Freedoms, art. 8, Nov. 4, 1950, 213 U.N.T.S. 222 [hereinafter Eur. Conv. on H.R.]; Wildemeersch, Belgium Report, p. 3; Kühling, Germany Report, p. 2; D'Antonio, Pollicino, Italy Report, p. 1–2; Şandru, Romania Report, p. 11.

[6]*See* O'Callaghan, Ireland Report, p. 5.

the situation appears very similar in the United Kingdom, especially under the influence of the European Convention on Human Rights.[7] In Turkey, the constitution expressly proclaims a constitutional right to privacy.[8]

Free speech finds protection in all national constitutions, as well as in the European Convention on human rights (Art. 10). This right is considered essential to insure the functioning of democracy. The French 1789 « *Déclaration des Droits de l'Homme et du Citoyen* » was the first one to consecrate this fundamental right in Europe. It found its inspiration in the 1776 United States Declaration of Independence.[9] It was then included in the First Amendment of the US Constitution in 1791.[10] While they originate from the same source, these European and United States rights did not define the limits to these rights in the same way.[11]

Interestingly, and the point deserves attention at the outset, these constitutional entitlements, both privacy and free speech, have implications on the correlative rights between private individuals. Indeed, because of what one calls the horizontal effect of constitutional rights, private life protected by the constitution also determines the limits of private life or speech, in private law in general and tort law in particular. As we will see, the constitution does indeed not only oblige the state to refrain from infringing on individuals' rights. It also obliges the state to preclude private individuals and corporations from interfering with other private individuals' rights.

This so-called "positive obligation" of the state gives individuals public and private protection, and consequently shapes the law of tort in general, as well as the law dealing with the protection of personality and personal data. Moving away from its original *raison d'être*, the constitution is not just a tool of protection against state action jeopardizing individual's liberties, but one that entitles actors to protection against other private individuals. This development reinforces what one sometimes refers to as the "constitutionalization of private law,"[12] applying constitutional rights between individuals.[13]

By contrast, in the United States, a right to the respect of one's private life finds no explicit direct protection in the U.S. Federal Constitution.[14] Instead, at the federal level, the right to privacy developed as part of a larger movement finding

---

[7]Jacques, UK Report.

[8]Article 20 of the Turkish Constitution. *See* Kapanci B, Paksoy S, Turkey Report, p. 2.

[9]Heyman (2008), pp. 7–22.

[10]McLean (2004).

[11]Whitman (2004), p. 1180.

[12]See for example, Brüggemeier et al. (2010), p. 31.

[13]On the effect of fundamental rights in private relations, *see* Clapham (2006). *See also* Alston (2005), p. 2. For a recent Swiss perspective, see Müller (2018).

[14]*See* Werro (2009), pp. 285, 291, 299.

unenumerated rights in the penumbras of the Bill of Rights.[15] In practice, especially under the conservative lens of the current U.S. Supreme Court, the U.S. will, rarely if ever, favor an unenumerated right over an enumerated one—thereby relegating the right to privacy to a second tier right behind those explicitly enumerated in the Bill of Rights, such as the right to freedom of speech (First Amendment). Unless the speech in question implicates one of the few narrow First Amendment carveouts, the U.S. presumption in favor of free speech almost universally prevails.[16]

The lack of counterweight to this strong recognition of free speech at the constitutional level has had private law implications. In contrast to Warren and Brandeis' scholarship,[17] in large part inspired by European conceptions of privacy, the protection of privacy in the law of tort in the United States has predominantly yielded to free speech. Accordingly, other than in very specific situations,[18] individuals have generally been precluded from successfully claiming infringements of their private life against private media reporting about them either with respect to their contemporary or past life.[19]

2.2 An Illustration

The European Court of Human Rights' ("ECtHR") ruling in *von Hannover v. Germany* illustrates the European approach to privacy in its relation to free speech in general.[20] This case neatly demonstrates the European approach to balancing the right to privacy of an individual against the freedom of expression of another. The balance struck between these constitutional entitlements reflects the ways in which private law plays out between individuals—be it in the realm of general tort law or in the private right to protect one's personality as measured against the right of the press to disseminate and sell information.

The plaintiff in *von Hannover v. Germany* was the eldest daughter of Prince Rainier III of Monaco. She claimed that several media outlets had taken pictures of her in a number of places where the Princess asserted she had a legitimate

---

[15] *E.g.*, Griswold v. Connecticut, 381 U.S. 479, 484–85 (1965) (discussing constitutionally-derived "zones of privacy"); Mapp v. Ohio, 367 U.S. 643, 656 (1961) (holding that the Fourth Amendment creates a "right to privacy"). Note, though, that the right to privacy is found in the Constitutions of ten states: Alaska, Arizona, California, Florida, Hawaii, Illinois, Louisiana, Montana, South Carolina, Washington. National Conf. of State Legislatures, *Privacy Protections in State Constitutions* (May 5, 2017), http://www.ncsl.org/research/telecommunications-and-information-technology/privacy-protections-in-state-constitutions.aspx.

[16] *See* Werro (2009), pp. 296, 300.

[17] Warren and Brandeis (1890), pp. 193–220.

[18] For a discussion of the convergence of EU and US privacy regulations, and more specifically the recent California Consumer Protection Act (CCPA, June 2019), see Büyüksagis (2019).

[19] For a detailed account, see Page (2010), p. 38.

[20] *Von Hannover v. Germany* (59320/00), [2004] E.M.L.R. 21.

expectation of privacy. She claimed that the media's publication of candid photos of her and her family in these places violated her right to privacy under Article 8 of the European Convention on Human Rights.[21] Relying on the Princess' position as a semi-public figure, the German courts found her right to privacy inherently diminished by virtue of her social status, and authorized the publication of these pictures, based on the freedom of the press and of expression.[22] Claiming that the German tribunals had failed to adequately protect her privacy rights against infringement by mass media, thereby violating their positive obligations under the constitution and the European Human Rights Convention, the Princess sued the state of Germany before the European Court of Human Rights.

The ECtHR reversed the decision of the German courts. Despite the Princess's status as a semi-public figure, the European Court "reiterate[d] that the concept of private life extends to aspects relating to personal identity such as a person's name or a person's picture."[23] The Court held that the concept of "private life. . .includes a person's physical and psychological integrity," thereby interpreting Article 8 "to ensure the development, without outside interference, of the personality of each individual in his relations with other human beings."[24] This, in turn, allowed the Court to conclude that there exists "a zone of interaction with others, even in a public context, which may fall within the scope of 'private life.'"[25]

In so concluding, the Court was keen to recognize that the freedom of expression guaranteed by Article 10 of the Convention has to be balanced against the protection of private life.[26] The Court considered the publication of the photos a violation of the Princess' right to privacy because it "cannot be deemed to contribute to any debate of general interest to society despite the applicant being known to the public."[27] Note that because of its constitutional dimension, a mere tort conflict ended up involving a private person against the State. Again, in the European approach, constitutional rights not only protect the individual against State infringements. They also oblige the State to take adequate and *positive* measures to insure the protection of individuals against other individuals.[28] Thus, the *von Hannover* decision effectively analyzes a claim of a violation of the right to privacy through the lens of the individual's rights rather than limiting the right to privacy to its public law dimension.[29]

---

[21]In relevant part, Article 8 of the European Convention on Human Rights provides that "Everyone has the right to respect for his private and family life, his home and his correspondence." 213 U.N.T. S 221.

[22]*See von Hannover v. Germany* (59320/00), [2004] E.M.L.R. 21at ¶ ¶18–42.

[23]*Von Hannover v. Germany* (59320/00), [2004] E.M.L.R. 21 at ¶ 50 (citations omitted).

[24]*Von Hannover v. Germany* (59320/00), [2004] E.M.L.R. 21 at ¶ 50 (citations omitted).

[25]*Von Hannover v. Germany* (59320/00), [2004] E.M.L.R. 21 at ¶ 50 (citations omitted).

[26]*Von Hannover v. Germany* (59320/00), [2004] E.M.L.R. 21 at ¶ 58.

[27]*Von Hannover v. Germany* (59320/00), [2004] E.M.L.R. 21 at ¶ 65 (citation omitted).

[28]For a further discussion of these positive duties from a Swiss law perspective, see Müller (2018).

[29]The *von Hannover* case was deliberately styled as a grievance against the Federal Republic of Germany. Per Articles 32, 34 of the Convention, individual applicants may only petition the European Court of Human Rights for grievances they claim to have suffered at the hands of a

The ECtHR's ruling in *von Hannover* is no outlier in the European approach to privacy rights. Many examples could be provided here. Much in line with the Court's viewpoint in *von Hannover*, Romania, for example, recognized "broadcasting news or other written or audio-video materials [related to an individual's] private life, without the consent of the interested person" as an "infringement. . .of. . .privacy."[30] Unsurprisingly, this approach has shaped the understanding and the limits of the right to be forgotten in ECtHR jurisprudence, as we will see below. Respect for one's private life, autonomy, and dignity imposes limits on what one can publicize when the information has lost its newsworthiness due to the passage of time.

Because the aforementioned values are the same, this approach holds essentially true when it comes to the processing of personal data and the right of individuals to control and even possibly erase them when they have lost their public or private relevance.[31] The law dealing with data protection is a modern extension of the traditional law protecting the personality of individuals. In part because of the European Directive of 1995, this is true in all countries of the European Union.[32] Indeed, all European Union countries follow the same approach. This approach has been reinforced under the General Data Protection Regulation ("GDPR"). Turkey goes a step further. Article 20 of its constitution, which provides for the right to private life, states that "everyone has the right to request the protection of his/her personal data. This right includes ... requesting the deletion of his/her personal data."[33]

Through very different mechanisms of adjudication, the United States' approach yields an opposite outcome. *Smith v. Daily Mail Publishing Company* provides an illustration of the limits of tort law and the primacy of the constitutional right to free speech. In *Smith* a juvenile murder suspect brought suit against a local newspaper for publishing his full name in violation of a West Virginia statute that disallowed newspapers from publicly divulging the names of juvenile criminal defendants.[34] The United States Supreme Court struck the West Virginia statute down, finding it contrary to the newspaper's free speech. The Court considered that because the

---

Party to the Convention (i.e. States). Applicants to the European Court of Human Rights may not style their claims in the form of private grievances against another individual. *See* Eur. Conv. on H. R., 213 U.N.T.S 221, art. 34. Thus, although the *Hannover* case was presented as a claim against the Federal Republic of Germany, the decision carries direct implications for private entitlements for all Parties to the European Convention of Human Rights.

[30] Şandru, Romania Report, p. 10.

[31] For a contrary view as to the divisibility of the right to be forgotten, see Post (2018), pp. 993–994, who argues that the traditional individual right to be forgotten protecting dignitary privacy is distinguishable from the RTBF, "the distinct bureaucratic version of the right to be forgotten created by the Directive to protect data privacy. . . ."

[32] *See* D'Antonio, Pollicino, Italy Report at 2; O'Callaghan, Ireland Report, p. 9 (indicating that Article 17 of the GDPR improves adds to data protection currently afforded by national law by making data subjects' consent the touchstone of the data's use); Kühling, Germany Report, p. 7; Wildemeersch, Belgium Report, pp. 12–13.

[33] Kapanci B, Paksoy S, Turkey Report, p. 2.

[34] 443 U.S. 308 (1977).

published information was "lawfully acquired and in the public interest," any attempt to restrict the newspaper's free speech "must be necessary to advance a state interest 'of the highest order.'"[35]

Following *Smith*, further Supreme Court caselaw suggests that there is little to be done "to prevent the media from disseminating sensitive information so long as that information is legally obtained."[36]

That reasoning sits in stark contrast to the ECtHR's ruling in *von Hannover*, which held that "the decisive factor in balancing the protection of private life against freedom of expression should lie in the contribution that the published photos and articles make to a debate of general interest."[37] As we will see below, claims brought under the right to be forgotten have, unsurprisingly, found little success in the United States.

2.3 Variations in Latin America, South East Asia and the Far East

Approaches to the relationship between privacy and free speech in Latin America, Japan, and Taiwan appear to be both different from that in Europe and that in the United States. Examining them as variations on the dichotomous Euro-American approaches, I found that the balance between freedom of expression and a person's right to private life shifts across Latin America and Asia in ways that do not resemble either approach.[38]

2.3.1 Latin America

On the basis of the national reports, the scope of the right to privacy in Latin America does not appear to be quite as expansive as it is in Europe. For example, although Article 5 of the Brazilian Constitution codifies a right to privacy, Brazilian case law has favored the public dissemination of information even when an individual's privacy is infringed. As the Brazilian rapporteur explains in *Xuxa v. Google Brasil* with respect to the right to be forgotten, a children's television show host sued to remove search results connected with the terms "Xuxa pedophile."[39] The Brazilian Superior Court of Justice denied the action and said that the balance weighed in favor of the *public's right* to information. Even more directly, in the *Candelária's Slaughter* case, a Brazilian court distinguished between the right to be forgotten's

[35]Werro (2009), p. 295.

[36]Werro (2009), p. 296.

[37]*Von Hannover v. Germany* (59320/00), [2004] E.M.L.R. 21 ¶ 76.

[38]*See* Alfonsín ML, Argentina Report, p. 3; Gonçalves R, Brazil Report, pp. 4–10.

[39]Gonçalves R, Brazil Report, p. 8.

applicability in cases involving mass media versus its applicability within the context of an internet de-indexing request.[40]

Argentina, which constitutionally guarantees a right to privacy, also readily limits it when there is a conflict with the freedom of expression. In *Rodriguez v. Google*, a professional model sued Google for defamation because it rendered search results that would link users to pages containing sexual content that depicted her. In denying her claim, the Argentina Supreme Court held that Article 13 of the American Convention on Human Rights places a limit on the right to privacy.[41] Article 13 consecrates the freedom of expression as free from "abuse of government or private controls...."[42] The Court anchored its decision in Law No. 26.032, which places digital search, reception, and dissemination of information within the purview of the constitutionally protected freedom of expression,[43] establishing clear limits on the right to be forgotten in Argentina.

These cases reflect a pattern of judicial reluctance to restrict mass publications based on the sole reason that the news published trends closer to sensationalism than pure informational facts. These cases exhibit an alignment with the U.S. approach.[44] The Argentina rapporteur also speaks of an "Inter-American" approach to the balancing between the freedom of expression and personal data privacy. Such commonality is further substantiated by the Canadian rapporteur's account of the strong protection granted to the right of expression in a unanimous Canadian Supreme Court decision against the Province of Alberta's Information and Privacy Commissioner.[45]

2.3.2 Japan and Taiwan

Japan does not explicitly proclaim the right to privacy in its Constitution. Japan's courts have derived the right to privacy from the guarantee of the pursuit of happiness, found in Article 13 of Japan's Constitution.[46] Yet, the balancing of that right against the freedom of expression remains unsettled.[47] In a defamation case, Japan's Supreme Court has ruled that the "right to personality" is akin to a "property

---

[40] *See* Gonçalves R, Brazil Report, p. 5. The Brazilian high court seems to have implied that information erasure requests are more actionable in mass media cases than in internet de-indexation requests.

[41] Alfonsín ML, Argentina Report, p. 3.

[42] Organization of American States, Am. Conv. on Human Rights, Nov. 22, 1969, O.A.S.T.S. No. 36, 1144 U.N.T.S. 123.

[43] *See* Alfonsín ML, Argentina Report, p. 3.

[44] *See* Alfonsín ML, Argentina Report, p. 7 (noting "the differences between the European framework for the protection of personal data and the strong emphasis on the right to freedom of expression in the Inter-American system.").

[45] *See* Eltis, Trudel, Canada Report, pp. 5–7.

[46] Yamaguchi, Japan Report, p. 7.

[47] Yamaguchi, Japan Report, p. 7.

right," in which the exclusivity of ownership is a defining feature.[48] Nevertheless, the Court has also stated that restraints on "expressive conduct should be allowed 'only and under strict and definite requirements' as an 'exception' to the purpose of Article 21 of the Constitution which guarantees the freedom of expression and prohibits censorship."[49]

Taiwan recognizes a constitutional "right of reputation and privacy."[50] Interestingly, the Taiwanese rapporteur indicates conflicting case law on the balance of the freedom of expression and the right to privacy.[51] In one case, Taiwan's High Court refused to require that Yahoo Taiwan remove a purportedly defamatory article posted on a website Yahoo Taiwan ran.[52] The High Court "fear[ed] that the freedom of speech [would] be overly suppressed if an ISP is to play the role of the speech police on the internet."[53] However, in another case, involving criminal slander and a de-indexation request, a district court ordered Google "to remove all search results from the domains of google.tw."[54]

As previously mentioned, the contrast between these two jurisdictions' balancing of the right to privacy vis-à-vis the freedom of expression spills over into these jurisdictions' considerations relating specifically to the right to be forgotten. These points require further development in the next section.

3 The Right to Privacy's Implications for the Right to Be Forgotten

Because the right to be forgotten derives from the precepts described with respect to privacy, it always found recognition as part of the right of privacy.[55] As the Italian rapporteur aptly puts it: "[an] individual's request to be forgotten. . .[is] an expression of the right to privacy."[56] The same is true for French law and Swiss law.[57] This explains why the decisions of the ECtHR regarding the right to be forgotten are in line with the rational scheme used in the *von Hannover* case, even when dismissing a

---

48 Yamaguchi, Japan Report, p. 8.

49 Yamaguchi, Japan Report, p. 8.

50 Chiou, Taiwan Report, p. 5.

51 Chiou, Taiwan Report, p. 4.

52 Chiou, Taiwan Report, p. 4.

53 Chiou, Taiwan Report, p. 4.

54 Chiou, Taiwan Report, p. 4.

55 *See* Gajda (2018), pp. 203–204. Wildemeersch, Belgium Report, pp. 2–3 ("La seconde sous-catégorie du « droit à l'oubli » est une expression du droit à la vie privée. . .D'abord utilisé dans le cadre de la presse traditionnelle, il connaît de nouveaux développements à l'aire des archives numériques au travers du droit à l'anonymisation.").

56 D'Antonio, Pollicino, Italy Report, p. 2.

57 For Swiss law, see Werro (2009).

claim.[58] This is also in part why the right to be forgotten, as proclaimed by the CJEU in *Google Spain*,[59] has been met with so much surprise in the United States and has found practically no recognition, if any.[60] Some U.S. commentators, like Robert Post, challenged the CJEU decision, in part, because in their view, the court conflated data privacy—purporting to ensure fair information practices and the use of personal information—with dignitary privacy—purporting to restrict inappropriate communication that threatens to degrade, humiliate, or mortify individuals.[61] Their surprise was amplified by the fact that the court held Google liable for the diffusion of information that had lost its newsworthiness, while allowing the original newspaper to maintain the information on its own website.[62]

A complete response to this critique is outside the scope of this report. However, one should note that the distinction made between the two types of privacy ignores the fact that data protection is derivative from the general right to privacy, and that its ultimate justification lies in the same foundational values.[63] It is because the court found the information sensitive, harmful and outdated that it found it appropriate to have it removed from Google's listings and the kind of public access thereby granted. In a persuasive way, it considered the power of a search engine to disseminate information to be incomparably larger than that of a single local news outlet, even when that local outlet has a website,[64] and accepted that impact-oriented distinctions can be made between linked defendants. Under this rationale, it accepted

[58]*See, e.g.*, ML and WW v. Germany, Nos. 60798/10 and 65599/10 (Eur. Ct. HR. 2018) (upholding German constitutional court's decision to quash application by two convicted murderers for the anonymization of stories concerning their conviction, finding, under the *Axel Springer* criteria, that Article 10 rights outweighed Article 8 rights in this case; Satakunnan Markinaporssi Oy and Satamedia Oy v. Finland, *Application no. 931/13,* (Eur. Ct. HR. Jun. 27, 2017) (upholding Finish court's decision to enjoin the dissemination of tax information (lawfully received and published) via sms message); Furst-Pfeifer v. Austria, Application nos. 33677/10 and 52340/10 (Eur. Ct. HR May 17, 2017) (upholding 4-3 the Austrian courts' judgment that Article 8 was not infringed by the publication of truthful medical information about a registered psychological expert for court proceedings in custody and contact-rights-related disputes on public care and child abuse); Axel Springer AG v. Germany, App. No. 39954/08 (Eur. Ct. H.R. Feb. 7, 2012) (striking down 12-5, following the application of a 6-part balancing test, as a violation of Article 10, German courts' decision to fine and enjoin German media companies from publishing the details of a prominent television actor's arrest for cocaine possession).

Post (2018), pp. 1058–1059.

[59]Case C-131/12, Google Spain SL v. Agencia Espanola de Proteccion de Datos, 2014 E.C.R. 317.

[60]For more details, *see* Werro (2009); for suggestions as to how to accommodate the European right to be forgotten in the US environment, see Bennett (2012), p. 161.

[61]Post (2018), pp. 1059–1061.

[62]Post (2018), p. 1010.

[63]The Directive itself mentions the protection of these values. It will not grant the right to erasure that could come into conflict with a public interest. Gratuitous harmful information is not protected. Art. 94 GDPR is a repeal of 95/46/EC.

[64]Case C-131/12, Google Spain SL v. Agencia Espanola de Proteccion de Datos, 2014 E.C.R. 317 pt. 35, 80, 86.

that the nature of Google's communication could justify restrictions to its free speech that would not apply to other media outlets.[65]

Before we shed more light on the implication of privacy for the right to be forgotten, this section will explore competing background principles of liberty, dignity, and capitalism as animating factors beneath the application of the right to be forgotten (3.1), discuss the unavoidable conflict between free expression and privacy in the application of the right to be forgotten (3.2), and catalog the limits on the right to be forgotten (3.3).

3.1 Liberty, Dignity, Capitalism and the Right to Be Forgotten

Differing cultural and social values give rise to different laws that either promote or preclude the recognition of privacy.[66] The same is true for the right to be forgotten. A number of considerations prove relevant to the right's statutory or judicial recognition, including the relationship between individuals and the state in the relevant jurisdiction, that state's protection of freedoms or rights that may directly compete with the right to be forgotten (i.e. freedom of expression)[67] and localized public policy arguments.

Because these considerations are also relevant to evaluating jurisdictions' treatment of the right to privacy, one can generally use a jurisdiction's treatment of the right to privacy as an indicator of the jurisdiction's openness toward recognizing the right to be forgotten. Indeed, the right to be forgotten operates in the shadow of the right to privacy, and where this right yields to other constitutional entitlements, the same is true for the right to be forgotten.[68] In his work on the right to be forgotten, Post states that "the difference between the right to be forgotten in the United States and the right to be forgotten in other legal systems is that American courts adopt an exceptionally strong presumption in favor of allowing publication."[69] Post's statement surprises me. Indeed I do not know of any case in which an American court would have recognized a right to be forgotten as it is understood in the European

[65]Case C-131/12, Google Spain SL v. Agencia Espanola de Proteccion de Datos, 2014 E.C.R. 317 pt. 17.

[66]For a recent in-depth account on the cultural dimension of privacy, see Legrand (2017), p. 1 (involving a comparison between the work of James Gordley and that of James Whitman). For another fundamental analysis, see also, Mayer-Schönberger (2009), pp. 16–49, analyzing the importance of forgetting.

[67]Consider, for example, that in 1985, the Italian Supreme Court established that the right to a personal identity constituted an interest in ensuring against that identity's improper altering or prejudice. As the rapporteur explains, in a 2004 case, Italy's data protection authority ordered the de-indexing of prejudicial search links—years before the *Google Spain* case. *See* D'Antonio, Pollicino, Italy Report, p. 2.

[68]According to Post (2018), p. 1060.

[69]Ibid.

sense. The cases on which Post relies are cases in which the court's reasoning for not allowing publication was based on the fact that the information in question resulted from offensive intrusions.[70]

Be that as it may, in the European context, it makes no difference to the application of the right to be forgotten whether a state expressly inscribes a right to privacy in its constitution or not. Under the European Convention on Human Rights, all European countries recognize a right to private life, which then determines the extent of their recognition of the right to be forgotten. While upholding this principle, recent decisions from the ECtHR have generally shown deference to national courts on the proper balancing of competing background rights in challenges implicating the right to be forgotten.[71] As Post notes, this is because the "newsworthiness" standard at the fulcrum of the right to be forgotten's application is itself a balancing test between the descriptive and normative meanings inherent in the term news*worthy*.[72] What end of this sliding scale courts generally hue to in cases concerning the right to be forgotten is often predicated on fundamental cultural understandings and the existence and relative weight of certain background rights (namely, the rights to privacy, dignity, expression, and information.) Such an approach is analogous to the application of other legal standards in the light of concrete circumstances, negligence being perhaps the most prominent one.

Furthermore, it makes no difference in our evaluation whether a state's laws formally recognize the right to be forgotten for the purpose of acknowledging the right's actual existence. Swiss statutes, for instance, have traditionally not mentioned the right to be forgotten, but courts have never hesitated to affirm that right as a manifestation of the rights of the personality—these rights being themselves the reflection in private dealings of constitutional entitlements to the right of private life.[73] The same can be said in the European Union, where the CJEU recognized the right to be forgotten in 2014, that is before the adoption of the GDPR and, arguably, independently from the Data Protection Directive of 1995.

More important for the recognition of the right to be forgotten is whether other entitlements, such as free speech, have the tendency to trump the protection of privacy. An analysis through the lens of the freedom of speech, and the economic value it entails for the media, ultimately gives a better picture of the practical applications of the right to be forgotten. Aside from its potential for the proper functioning of democracy, one can see free speech as a powerful engine for the privatization of profits. The expression "marketplace of ideas" to designate the virtues of this right is quite revealing in that respect. Free speech must therefore be

---

[70]*See Shulman v. Grp. W. Prods., Inc.*, 955 P.2d 469, 485 (Cal. 1998); *Bollea v. Gawker*, 913 F. Supp. 2d 1325 (M.D. Fla. 2012).

[71]*See, e.g.*, ML and WW v. Germany, App. Nos. 60798/10 and 65599/10 (Eur. Ct. HR. 2018); Satakunnan Markinaporssi Oy and Satamedia Oy v. Finland, App. no. *931/13*, (Eur. Ct. HR. Jun. 27, 2017); Furst-Pfeifer v. Austria, App. nos. 33677/10 and 52340/10 (Eur. Ct. HR May 17, 2017); Axel Springer AG v. Germany, App. No. 39954/08 (Eur. Ct. H.R. Feb. 7, 2012).

[72]Post (2018), pp. 1058–1059.

[73]*See* Art. 28 CC.

put in parallel with free enterprise and capitalism, as well as the interests of media owners that may well trump those of the general population.[74] In those countries with a different attachment to capitalism, the tolerance for imposing limits to free speech is greater. In this respect, Professor James Q. Whitman's distinction between those countries favoring liberty, as in entrepreneurial U.S., over dignity, as in more statist Europe, is quite insightful.[75]

In parallel with the civil law-common law divide, a distinction can be made between those countries that favor free speech over the protection of private life and those countries that balance privacy against free speech. Under some vestigial influence of the British Empire, the laws in entrepreneurial former colonies would, no doubt, confirm this reality.

3.2 Dignity, Balancing Free Speech and the Right to Privacy, and the Recognition of the Right to Be Forgotten

As all other continental European countries, Switzerland constitutionalizes its right to privacy.[76] The Swiss Civil Code of 1912 further provides a general clause providing for the protection of the personality in Article 28. Swiss courts had no trouble recognizing that this general norm implies a specific right to be forgotten, even before Swiss courts began to interpret private law in the light of constitutional entitlements.[77]

Based on the same principles as laws protecting individuals' personality, the European Data Protection Directive of 1995 merely reinforced this approach.[78] At the same time, it may also be that the rights to deletion, anonymization, or opposition given under this directive helped the national European courts embrace the right to be forgotten beyond the scope of this directive.[79]

In 1992, Belgium, for example, passed a statute to ensure the protection of private personal data. In this statute, updated in 1998 to implement the 1995 directive, the rights to deletion, anonymization, and opposition were already partially recognized.[80] Perhaps as a result, Belgian courts embraced a right to be forgotten more

---

[74]*See* Stein (2010). *See also* Cohen (2017b), p. 56, who analyzes the unquestioning deference to the political power of money in free speech jurisprudence.

[75]Whitman (2004), pp. 1151, 1171–1188, with interesting insights into the specificities of German and French capitalism.

[76]On constitutionalization, see Brüggemeier et al. (2010), p. 31. See Art. 13 of the Swiss federal constitution, and its express reference to the "respect de la vie privée."

[77]*See, e.g.*, Werro (2009), pp. 290–291.

[78]This appears to be true also outside the EU; s*ee* Berk Kapanci, Sermin Paksoy, Turkey Report, pp. 2, 6.

[79]Directive 95/46/EC Art. 12, 14.

[80]Wildemeersch, Belgium Report, p. 2. In 1997, Belgium's Privacy Commission formally recommended the anonymization of personal details in judicial decisions.

than two decades before the *Google Spain* case.[81] In 2000, a Belgian court of first instance held against a local television station that replayed images of a theft and mentioned the then-accused thief's name ten years after the theft had occurred.[82] The court found little value in highlighting the thief's involvement ten years later, and considered instead that the television station had caused him moral prejudice in calling for damages.[83]

In Germany, the "general right to privacy...[is] enshrined in art 2 § 1 read together with art 1 § 1 [of the] Basic Law (*Grundgesetz*)."[84] Despite its derivation from the right to privacy, the right to be forgotten represents a distinct notion.[85] While a "general right to privacy...[is] enshrined in art 2 § 1 read together with art 1 §1 Basic Law (*Grundgesetz*)" ..."no specific right to be forgotten [exists]."[86] Like in other countries within the European Union, however, the German Federal Data Protection Act (*Bundesdatenschutzgesetz*)—implementing the European Data Protection Directive of 1995[87]—provided a framework to solicit deletion by search engines where they "transmit content from third parties that infringes personality rights."[88] In substance, the right to deletion in Germany amounts to a manifestation of the right to be forgotten, even if not labeled as such.

A similar approach appears to exist under Irish law. In identifying a source of law for the right to be forgotten, the Irish rapporteur explains: "We might say that a right to be forgotten already exists in Irish law if we understand it as an alternative label for the right to erasure under s 6(1) DP Act, which implements Article 12(b) DPD."[89]

Similarly, Italy guarantees a right to personality in Article 2 of its constitution.[90] Italian courts cite the right to privacy and the right to personality as their basis for the right to be forgotten.[91] Italy formally introduced the right to be forgotten into its laws over a decade before the *Google Spain* case when it passed legislation to implement the European Data Protection Directive of 1995. In 2003, Italy adopted a Privacy

---

[81] *See* Wildemeersch, Belgium Report, p. 11. Nevertheless, the rapporteur notes that the Belgian reaction to the ECJ's formal declaration of the right to be forgotten was one of surprise.

[82] Wildemeersch, Belgium Report, p. 11.

[83] Wildemeersch, Belgium Report, p. 11.

[84] Kühling, Germany Report, p. 2.

[85] Note that although we recognize a practical difference between the right to be forgotten and the right to privacy, we do not wish to assert that the right to be forgotten represents a self-substantiating right independent of the right to privacy. We maintain that the right to be forgotten is a derivative of the right to privacy. For a parallel discussion arguing against the recognition of data protection as its own right, see Poscher (2017), p. 129.

[86] Kühling, Germany Report, pp. 1–2.

[87] Directive 95/46/EC.

[88] Kühling, Germany Report, p. 1.

[89] O'Callaghan, Ireland Report, p. 8.

[90] D'Antonio, Pollicino, Italy Report, p. 1.

[91] D'Antonio, Pollicino, Italy Report, p. 1.

Code, under which individuals were granted the rights to erase, update, and contextualize their data.[92]

Likewise, Turkey explicitly codified the right to privacy in Article 20 of its constitution.[93] In 2016, the Turkish Constitutional Court formally recognized the right to be forgotten as derivative of personality rights, while acknowledging that the right to be forgotten was nowhere explicitly codified in national law.[94]

Similarly, Argentina, which constitutionalizes the right to privacy, has created the novel *writ of habeas data*, which allows parties to request judicial intervention specifically aimed at the removal of certain information on the internet.[95]

3.3 Liberty and Free Speech As Limits to the Recognition of the Right to Be Forgotten

One cannot say that in countries that give precedence to free speech, there is no recognition of privacy. However, in such countries, recognition remains limited to scattered alcoves of actionable legal intervention. Privacy there typically does not enjoy an explicit constitutional recognition, and it does not amount to an overall or overarching entitlement. It is also in these countries that one can see some reluctance in accepting or acknowledging the existence of a right to be forgotten. In the United States, the European Court of Justice's recognition of the right to be forgotten in *Google Spain* was met with surprise, if not derision, and is overall dismissed as inapplicable.[96]

The United Kingdom traditionally similarly favored free speech, and, at the same time, refused the implications of the protection of private life. As the rapporteur explains, the United Kingdom legislature and judiciary were "rather reluctant to recognise a right to be forgotten, preferring to advance freedom of expression."[97] Nevertheless, after the European Court of Justice's decision in the *Google Spain* case, United Kingdom "courts seem more receptive to privacy rights concerns."[98] Individuals have also acquired some degree of control over their information under the Data Privacy Protection Act of 1998 (successor to the Data Privacy Protection

[92] D'Antonio, Pollicino, Italy Report, p. 2.

[93] Kapanci B, Paksoy S, Turkey Report, p. 2.

[94] Kapanci B, Paksoy S, Turkey Report, p. 6.

[95] *See* Alfonsín ML, Argentina Report, p. 1, 4.

[96] *See, e.g.*, Bhardwaj (Feb. 28, 2018, 12:06 PM), http://www.businessinsider.com/google-right-to-be-forgotten-law-in-america-2018-2; *see generally* Gajda (2018), p. 93; Post (2018), p. 67.

[97] *See* Jacques, United Kingdom Report, p. 12.

[98] *See* Jacques, United Kingdom Report, p. 11.

Act of 1984, repealed in 2000).[99] However, that law only applies to "processing that causes unwarranted and substantial damage or distress."[100]

As a mixed jurisdiction, Canada provides a less unilateral approach than the one traditionally adopted in these common law jurisdictions. As the Canadian rapporteur explains, the right to be forgotten's earliest appearances date to 1889. At that time, the Superior Court of Québec ruled against a newspaper for publishing certain "*accusations depuis longtemps oubliées*."[101] More recently, in *Ouellet c. Pigeon*, a man sued a newspaper for publishing an article describing a crime committed by his late wife ten years prior.[102] Reflecting the ECtHR's approach, the Court of Québec ruled against the newspaper on the grounds that the article was better characterized as sensationalism than strict reporting of fact.[103] At the same time, in a 2013 decision, the Supreme Court of Canada unanimously struck down a provincial data protection law to the extent that it prohibited videotaping or photographing individuals in public spaces without their consent.[104]

In a more one-sided way, Singapore's Parliament has rejected the notion that the right to privacy is part of domestic law or its national Constitution.[105] Further, and arguably as a consequence of this rejection, there is no case law suggesting the existence of the right to be forgotten in Singapore law.[106] Yet, while Singapore's Personal Data Protection Act (2012) allows removal only of confidential identifying information, it does not apply to the data collected by the public sector.[107] Further, the Singapore rapporteur further expresses doubt that the present law could be used against search engine operators that have a presence in Singapore.[108]

As one can expect, public policy is a driving factor behind Singapore's approach to the right to privacy. As the rapporteur explains, The government is favourably disposed towards a growing role for technology in daily life. . . . In a similar vein, the government has always been keen to cultivate a pro-business environment and the antagonistic attitude towards in *Google v González* is strong in the jurisdiction.[109] In fact, Singapore's Personal Data Privacy Commission encourages businesses to use consumer data for purposes of business expansion.[110] As noted above, Singapore's

---

[99] Jacques, United Kingdom Report, p. 1.

[100] Jacques, United Kingdom Report, p. 1.

[101] Eltis, Trudel, Canada Report, p. 3.

[102] Eltis, Trudel, Canada Report, p. 4.

[103] Eltis, Trudel, Canada Report, p 4. Note that Oullet c. Pigeon was heard in 1997 — prior to the *von Hannover* case. But compare with the Court's reasoning in Axel Springer AG v. Germany, App. No. 39954/08 (Eur. Ct. H.R. Feb. 7, 2012).

[104] *See* Eltis, Trudel, Canada Report, p. 6.

[105] De Visser, Singapore Draft Report, p. 6. (on file with the author of the general report).

[106] De Visser, Singapore Draft Report, p. 2.

[107] De Visser, Singapore Draft Report, p. 4.

[108] De Visser, Singapore Draft Report, pp. 5–6.

[109] De Visser, Singapore Draft Report, p. 9.

[110] De Visser, Singapore Draft Report, p. 9.

preference, like that expressed by U.S. courts, for strong entrepreneurial capitalism causes a less hospitable environment for the recognition of the right to be forgotten.

4 The Core Justifications of the Right to Be Forgotten: Self-realization, Dignity, Personal Freedom, and Control over Information About Oneself

As previously mentioned, the right to be forgotten can no longer be seen merely as a right to delete information or to preclude its diffusion, as it was originally the case. In a rapidly advancing technological world, one must understand the right in a more multifaceted way. At its core, the right to be forgotten expands and defines itself as an entitlement for individuals to better control their personal data. Just as for the original pre-Internet right to be forgotten, this entitlement finds its justification in the recognition of the right to personal freedom, dignity, and self-realization. With privacy as its justification, the right to be forgotten manifests itself further as a right to control personal information and to take back or erase information even after having communicated it. As Whitman frames it, the right to privacy appears as a right to informational self-determination—the right to control the sorts of information disclosed about oneself.[111] This approach certainly does not mean that individuals can retain control of information that is in the public interest.[112] In order to shed some light on the analysis, this section will open with some general remarks (4.1). It will then explicate the polygonal nature of the right to be forgotten (4.2), and look at ways to expand future conceptions of the right to be forgotten (4.3).

4.1 General Remarks

From the foregoing survey, we see that a link exists between the right to be forgotten and a jurisdiction's balancing of the freedom of expression against the protection of private life. Where freedom of expression prevails over the protection of private life, the right to be forgotten tends to enjoy extremely limited or no protection. This finding parallels those of Whitman, whose work links the preference given to liberty over dignity with that given to free enterprise over a certain belief in the legitimacy of state control.[113] There can be no doubt that free speech is the constitutional

---

[111] Whitman (2004), p. 1161 (referencing German literature).

[112] As EU law shows, a right to be forgotten cannot trump the public interest in receiving information that contains historic value. On this question, amongst others, see Vivian Reding, as cited in Post, in footnote 314; for a critique of the way the CJEU handled the question, see Post (2018), p. 1051.

[113] Whitman (2004), pp. 1186, 1210.

currency of the United States' "marketplace of ideas." Thus, it will come as no surprise that freedom of speech carries a special weight in discounting the strength of the right to privacy and the right to be forgotten in that country. Yet, one can only hope that the development of technology and the omnipresent undisclosed surveillance of consumers and citizens will lead to some changes.[114] While one might be tempted to throw in the towel and "forget about the right to be forgotten," as an American colleague once suggested, one might also envision some mobilization following a continuing emerging consciousness around a cascading Orwellian threat.

As noted above, neither the right to free speech, nor the right to privacy, nor the right to be forgotten are advents of the modern world. The intrinsic connection and balance between these entitlements has been tested for centuries. Indeed, there were cases as early as the 1800s requesting the erasure of publicly available information.[115] The advent of the Internet, allowing for instant widespread communication, alongside increased digital storage capacity (i.e. information clouds) has, however, increased the frequency of conversations, images, and representations that an individual would want to disassociate him or herself from. Traditionally, the four walls of one's home served as the sole demarcation of the bounds of acceptable outsiders and governmental probing. Obviously, the digital world calls the sufficiency of this physical border into question. With the Internet, private life has become threatened as it never was before. One's home no longer delineates the boundaries necessary for full protection from invasions of privacy. At the same time, we easily fathom the great number of people that underestimate the utility of their computers and personal devices as prime conduits of unwelcomed personal data appropriation and spying.

As technology advances, questions regarding ownership and use of personal data abound. In this light, the advent of technology may be thought of as having a catalytic role in causing states to formally reconsider the object of property and to recognize, as part of one's personality, the right to own and keep control over one's personal information. Therefore, it is possible that societies will grow to find that sharing personal information should not be seen as an act of alienation or renouncement, and that individuals who give up such information keep the right to take it back. In this context, one would have to revisit the notion that social media companies are at liberty to use and sell information entrusted to them. As a means of protecting modern private life, the adoption of the GDPR in 2018 in the European Union can be seen as an attempt to recognize that individuals' control over personal information is paramount.[116]

In this vein, and from its core substance, the right to be forgotten can be seen as the individual's ability to *control* the public availability of information that relates to

---

[114] *See* Cohen (2017), pp. 230–231.

[115] *See* Eltis, Trudel, Canada Report, p 3; *see generally* Gajda (2018).

[116] Art. 17 GDPR (" Right to Erasure").

him or herself.[117] Furthermore, the effort to recapture control over one's personal data cannot be conceived of as limited to one particular course of action (e.g. the right to erasure.) As the Belgian rapporteur puts it, the right to be forgotten must be understood as a multifaceted entitlement: "*le droit à l'oubli se décline au pluriel; c'est un droit 'multi-facettes'. Les différents droits en cause poursuivent néanmoins le même objectif: permettre aux personnes physiques de (re)prendre et conserver le contrôle sur des informations privées et des données personnelles.*"[118] At the same time, this control-based right does not rest on roots distinct from those supporting the traditional right. As mentioned above, this report does not adhere to the distinctions made by Post between control-based and dignitarian privacy.[119] Data protection legislation's regulation of fair practices with respect to information and its removal of a harm requirement for standing does not sever the essential antecedent tether between a consumer's right to control and erase personal information and his or her right to exist as a free and dignified person. While data privacy regulation's stated aims may appear distinctive, it relies on the same fundamental justification: the protection of human dignity and self-determination. Consequently, and unlike Post, I do not think that the CJEU conflated the traditional right to be forgotten with a "bureaucratic" RTBF that applies to the protection of data within the meaning of Art. 8 of the Charter of Fundamental Rights and data protection legislation.[120] *Google Spain* focused both on the processing of personal data collected by the search engine and on the constraints that Google must respect when engaging in public communication, and to that effect linked them.

4.2 The Multiple Facets of the Right to Be Forgotten

If dignity, self-realization, freedom, and control over one's personal information determine the right to be forgotten's central meaning, it will come as no surprise that the right to be forgotten manifests itself in more than one way. Various reports show that conclusion across regions and continents. For example, Argentina has created the *writ of habeas data*, which, as noted above, allows individuals to petition courts to intervene in data processors' handling of their information.[121] Similar judicial

[117] *Cf.* Wildemeersch, Belgium Report, p. 2 ("L'article 8 de la Charte des droits fondamentaux de l'Union européenne a même fait du droit à la protection des données à caractère personnel un droit autonome, toute personne ayant, selon l'article 8, paragraphe 2 de la Charte, « le droit d'accéder aux données collectées la concernant et d'en obtenir la rectification »"); D'Antonio, Pollicino, Italy Report, p. 6 (right to contextualize); for a convincing comparison between the European and the American approaches with respect to control over oneself, see Whitman (2004), p. 1161, n. 44, 1169, n. 76, 1182, nn. 127-29.

[118] Wildemeersch, Belgium Report, p. 1.

[119] *See* Post (2018), p. 985.

[120] Post (2018), p. 985.

[121] *See* Alfonsín ML, Argentina Report, pp. 4–5 (discussion of habeas data).

recourse exists in Turkey.[122] Laws in Brazil[123] and Taiwan[124] allow for similar requests. In Italy and Denmark, individuals have the right to contextualize or update their publicly-available personal information.[125] Denmark, Canada and Japan, allow individuals to require the "de-indexation" of their information from search engine results, at least when the information is illegal.[126] Regardless of legality, individuals may, as is pointed out in the Finish report, "prohibit processing of [their] personal data 'for purposes of direct advertising, distance selling, other direct marketing, market research, opinion polls, public registers or genealogical research.'"[127] That statement is true across the board in the European Union.

The same standard is now reinforced through the European GDPR. This regulation explicitly establishes both a "right to erasure,"[128] referred to as the "right to be forgotten" (in parantheses in the title of the law), and a "right to rectification."[129] According to the GDPR, the right to erasure entails the ability of a data subject "to obtain from the [data] controller the erasure of personal data concerning him or her without undue delay,"[130] as conditioned upon a number of factors, such as the data's irrelevance, the data subject's withdrawn consent regarding the use of data, and the unlawful processing of the data.[131]

On the other hand, the "right to rectification" entails a person's ability "to obtain from the [data] controller without undue delay the rectification of inaccurate personal data concerning him or her."[132] This "right to rectification," however, is not preconditioned. It is a guaranteed right under the law. Despite the fact that the GDPR labels them distinctly, both of these rights fall within the penumbra of the right to be forgotten, and, in some respects, surpass it.

---

[122] Kapanci B, Paksoy S, Turkey Report, p. 1 ("Personal data which are processed in accordance with this law or relevant other laws shall be deleted, destroyed or anonymized either ex officio or upon request" with the right to be forgotten receiving constitutional backing recognized by the Constitutional Court in 2016).

[123] Gonçalves R, Brazil Report, p. 3 (Law No. 12965 of April 23, 2014).

[124] Chiou, Taiwan Report, p 1 (conventional right to request deletion of personal data stated in Personal Information Protection Act at article 11).

[125] *See* D'Antonio, Pollicino, Italy Report, p. 6 (right to contextualize); Motzfeldt, Naesborg-Andersen, Denmark Report at 3 (data must be up to date).

[126] *See* Court of Justice of the European Union, Judgment of 13 May 2014, *Google Spain and Google*, C-131/12, EU:C:2014:317; Eltis, Trudel, Canada Report, p. 5 ("Par contre, lorsqu'il est démontré que le propos contrevient à une loi ou viole un droit fondamental, les tribunaux canadiens n'ont aucune hésitation à ordonner le déréférencement."); Motzfeldt, Naesborg-Andersen, Denmark Report, p. 2 (Processing of Personal Data Act of 2000 allowing right to demand in a reduction in searchability); Yamaguchi, Japan Report at 12 (Nov. 6, 2009 Tokyo Dist. Ct. case requiring Google to delist 122 URLs because they infringed on the right to personality of the petitioner).

[127] Alén-Savikko, Finland Report, p. 5.

[128] Council Regulation 2016/679, art. 17, 2016 O.J. (L 119) 1 (EU) [hereinafter *GDPR*].

[129] GDPR, art. 16.

[130] GDPR, art. 17(1).

[131] *See* GDPR, art. 17(1)(a).

[132] GDPR, art. 16.

4.3 Expanding the Scope of the Original Right to Be Forgotten

From the foregoing analysis, we can conceive of the right to be forgotten as manifold.[133] Founded on the traditional dignity justifications listed above, the right's scope has expanded to offer individuals a combination of different entitlements allowing them to control or regain control over information that they offered to others and is now publicly available (i.e. de-indexation, erasure, or contextualization, etc.)

Because the right to be forgotten relates to an individual's ability to control the presence of publicly available information about him or herself, a number of socio-political considerations come into play, and the degree to which the right will manifest itself in national laws depends on them.

5 The Implications of the Right to Control Information About Oneself

If one accepts that, as an expression of one's dignity and privacy, a person's control of information touching upon his or her personality is the core feature of the right to be forgotten, where countervailing interests do not require disclosure, it is useful to imagine how this right could gain some recognition where it had none and increased recognition where it already existed. The next section will explore how individuals might be empowered to exercise control (5.1), and what obstacles they face (5.2). It will also examine the role of search engines (5.3), and briefly tackle the question of adjudication (5.4). To that end, this section will conclude with the question of search engine liability (5.5).

5.1 Individuals' Ability to Control Their Personal Data

Individual control of personal information seems to be finding its way to statutory protection, in some places more than others. At the international level, a certain recognition has seen the light of day. The relevant source of law on the subject is the International Covenant on Civil and Political Rights ("ICCPR"). The ICCPR currently counts 170 party states and six signatories. In 2013, the United Nations General Assembly passed a Resolution entitled "The right to privacy in the digital

[133] See Wildemeersch, Belgium Report, p. 1, stating that the right to be forgotten "regroupe en réalité plusieurs droits qui reposent sur des fondements législatifs différents."

age," where it "reaffirm[ed] the right to privacy" as recognized in international law under Article 17 of the ICCPR.[134]

Article 17(1) of the ICCPR guarantees that "[n]o one shall be subjected to arbitrary or unlawful interference with his privacy, family, home or correspondence...."[135] In that vein, the following activities may constitute an interference with the right to privacy: intercepting, collecting (especially bulk-collection, that is collection of big data without any suspicion against the person whose data are collected), storing, and the further distribution of communications data.[136]

Notably, though, what Article 17(1) outlaws is *unlawful* interference: the very same act of unlawful interference may be rendered lawful by subject individual consent.[137] Consent is the epitomical acknowledgement of control.[138] The role of individual control in matters of privacy—and by consequence, as related to the right to be forgotten—cannot be understated.[139] Nevertheless, the amount of control states afford data subjects varies dramatically state-to-state.[140] Certain states acknowledge that consent is not irrevocable, and certainly not where it has been given mechanically. Indeed, more often than not, consent does not rest on a real choice. One is no longer at liberty to have an email address or not, and the kind of mouth click one engages in when "consenting" certainly does not add up to historical conceptions of control and freedom.[141]

As already shown above, the 1995 Data Protection Directive in Europe gave some remedies to Internet users. Individuals have direct recourse against data controllers who use their information in a manner beyond that which they assented to, and may demand an explanation from the data administrator or demand that the

---

[134] G.A. Res. 68/167 at 2 (Dec. 18, 2013).

[135] International Covenant on Civil and Political Rights, art. 17(1), Dec. 16, 1966, 999 U.N.T.S. 171.

[136] *See* Peters (2017), pp. 145, 149.

[137] *See generally* Peters (2017), pp. 145, 149.

[138] For a recent critique of consent in the digital environment, *see* Richards and Hartzog (2019), p. 96.

[139] See also OECD Guidelines Governing the Protection of Privacy and Transborder Flows of Personal Data ¶ 10 ("Personal data should not be disclosed, made available or otherwise used for purposes other than those specified in accordance with Paragraph 9 except: a) with the consent of the data subject....").

[140] A deeper analysis on the extent to which the ICCPR and similar international instruments have affected conceptions of the right to be forgotten falls beyond the scope of this report, which is styled as a survey of the current status of the right to be forgotten.

[141] Whitman (2004), pp. 1193–1194 makes an interesting point when he remarks that "consumers need more than cheap goods and services, just as they need more than easy credit. They need dignity. If your consumer profile is floating around somewhere in cyberspace, you are not in control of your image.... This sort of thinking has far less resonance in America than it does in Germany and France." But, of course, this is because we have so much less of the continental sense that "a just world [] is a world in which everybody's respectability is carefully protected."

administrator block, correct, contextualize, or delete the data.[142] It will come as no surprise that the GDPR adds some focus to the role of data processers in the handling of personal information. As the German rapporteur explains: "[t]he processing of data from third parties is only legal within the meaning of Art. 6 [of the] GDPR if there is a legitimate interest in the processing of the data when the respective interests and affected fundamental rights are balanced."[143]

This is also true outside the European Union. Taiwan provides a right to request deletion of personal data where a data controller utilizes information for no valid purpose.[144] Again, in Argentina, the *writ of habeas data* allows data subjects to collect data about themselves from private and public databases, and request, where certain requirements are met, the suppression, rectification, sealing, or updating of data.[145]

5.2 Obstacles to Individuals' Control

However, questions remain on the effectiveness of individuals' control under these laws. Directly preceding implementation of the GDPR, Facebook relocated 1.5 billion user accounts from EU servers to non-EU servers. The move shows the limits of the GDPR's protection. Commentators' criticism and concern are unlikely to change that reality.[146] At the same time, whatever Facebook's motivations may have been, it is undeniable that Europeans have persevered in enforcing their data privacy rights. Between the European Court of Justice's ruling in the *Google Spain* case in 2014 and 2017, Google received requests to de-index over 2.3 million URLs from European users.[147] As individuals grow increasingly cautious about their online data presence, the question as to how far search engines must go in de-indexing valid requests under the Court's ruling has surged.

[142] See Hurdík, Czech Republic Report, p 4; the Alén-Savikko, Finland Report, p. 5, explains that individuals may prohibit processing of their personal data "for purposes of direct advertising, distance selling, other direct marketing, market research, opinion polls, public registers or genealogical research," See also Kühling, Germany Report at 13; Jacques, United Kingdom Report, pp. 7–8.

[143] Kühling, Germany Report, p. 13.

[144] Yet note that it is not clear under Taiwanese law that withdrawal of consent guarantees a right to erasure of personal data—the law merely requires that the data processor cease processing the data. *See* Chiou, Taiwan Report, p. 1 (discussing Article 11 of the Personal Information Protection Act).

[145] Alfonsín ML, Argentina Report, pp. 1, 4.

[146] *See* Hern (Apr. 19, 2018, 7:03 AM), https://www.theguardian.com/technology/2018/apr/19/facebook-moves-15bn-users-out-of-reach-of-new-european-privacy-law?CMP=share_btn_link.

[147] Bertram et al. (2018), p. 17, available at: https://www.elie.net/static/files/three-years-of-the-right-to-be-forgotten/three-years-of-the-right-to-be-forgotten-paper.pdf.

Because search engines like Google operate through country-code top level domains ("ccTLDs") in Europe,[148] the question has become: is an individual in France, validly entitled to the de-indexing of his or her personal data from search engine results, only entitled to have that data de-indexed under the search engine's French ccTLD (i.e. google.fr), or does that right to de-indexation extend beyond the French ccTLD?[149] France's highest administrative law tribunal, the Conseil d'État, lodged this very question before the European Court of Justice in August of 2017,[150] and the ECJ issued its judgement on September 24, 2019.[151] However, the experience of the Canadian Supreme Court shows that concepts of sovereignty and extraterritoriality are not easily applied in the digital realm. In an order noting that the Internet "*n'a pas de frontières—son habitat naturel est mondial*,"[152] the Canadian Supreme Court ordered Google to globally delist certain search results related to a Canadian company's trade secrets. In the United States, Google turned to an American federal district court for injunctive relief against the Canadian judgment's effect outside Canadian borders and Google's Canadian ccTLD (Google.ca), citing the presumption in favor of freedom of expression under American law.[153]

The federal district court granted Google's request, undercutting the Canadian Supreme Court's purportedly global effect.[154] The effective overruling of Canada's highest court by a court of first instance in the United States points to a gaping hole in the law of cross-border digital data privacy, which, until remediated, will continue to

---

[148]For example: google.fr, google.es, google.it, etc.

[149]In other words, to include websites such as google.es, google.it, or even the U.S. google.com.

[150]The case was lodged before the European Court of Justice after the French Data Protection Authority ordered Google to comply with universal delisting of certain URLs. See Request for a preliminary ruling from the Conseil d'État (France) lodged on 21 August 2017—*Google Inc. v Commission nationale de l'informatique et des libertés (CNIL)*, Case C-507/17. Question referred number 1: "Must the 'right to de-referencing', as established by the Court of Justice of the European Union in its judgment of 13 May 2014 on the basis of the provisions of Articles 12(b) and 14(a) of Directive [95/46/EC] of 24 October 1995, be interpreted as meaning that a search engine operator is required, when granting a request for de-referencing, to deploy the de-referencing to all of the domain names used by its search engine so that the links at issue no longer appear, irrespective of the place from where the search initiated on the basis of the requester's name is conducted, and even if it is conducted from a place outside the territorial scope of Directive [95/46/EC] of 24 October 1995?"

[151]*See* Judgment of the Court (Grand Chamber) of 24 September 2019 *Google LLC, successor in law to Google Inc. v Commission nationale de l'informatique et des libertés (CNIL)* Case C-507/17. At this late stage, the author regrettably cannot address the outcome of this case. The decision was rendered after the manuscript was submitted to the publisher, and the author regrets the outcome of the case, but has no space to fully comment on it further. For a similar view, see Marc Rotenberg, Google's Position Makes no Sense: Opposing view, at https://www.usatoday.com/story/opinion/2015/01/22/. http://curia.europa.eu/juris/document/document.jsf;jsessionid=D3B6FA325F40E190 00A709ED4DF087BB?text=&docid=218105&pageIndex=0&doclang=EN&mode=lst&dir=&occ=first&part=1&cid=3535949.

[152]Eltis, Trudel, Canada Report, p. 18.

[153]Eltis, Trudel, Canada Report, pp. 18–19.

[154]Eltis, Trudel, Canada Report, p. 19.

frustrate transnational advances in the right to be forgotten. The question of courts' ability to require extraterritorial de-indexing has also been entertained in Taiwan, whose tribunals have shied away from extraterritorial rulings.[155]

5.3 The Role of Search Engines/Data Processors

The question of tribunals' authority to order extraterritorial de-indexation is only one of many that call for answers. Additional concerns regarding the roles search engines and data processors should play with respect to individuals' control over their own information abound. One of them is the extent to which private actors such as Google should be in charge of deciding whether an individual has a right to be forgotten or not. Another question is that of the sufficiency of these private actors' protocols for evaluating requests under the right to be forgotten, and the extent to which these protocols are standardized, objective, and consistently applied.[156] These questions must also be seen in the light of the financial burdens that solving them will impose on the search engine operators and society.

The answer to at least some of these questions indicates some reluctance to grant Google decision-maker status with respect to individuals' right to be forgotten, as manifested through petitions for de-indexing search results.[157] For example, in *Rodriguez v. Google*, an Argentinian court held that determinations bearing on individuals' right to privacy are a state function that should not be left to private parties.[158] Academic commentators have also articulated discomfort with Google's role post-*Google Spain*. Post indicates that *Google Spain* implemented a "structure of enforcement [of the right to be forgotten that] is deeply flawed because it leaves important decisions about freedom of expression in the hands of an unaccountable private company with strong financial incentives to err on the side of censorship."[159] Indeed, in the *Google Spain* case, the Advocate-General cautioned the Court against leaving a balancing act between personal privacy and the public's access to

[155]*See* Chiou, Taiwan Report, pp. 2–3.

[156]Note, although Google does admittedly publish transparency reports with respect to its de-indexation requests, it is the Authors' opinion and that of many of the rapporteurs that more work to increase transparency may be done. *See, e.g.*, Alén-Savikko, Finland Report, p. 16; O'Callaghan, Ireland Report, p. 16; Kapanci B, Paksoy S, Turkey Report, p. 9 (calling for transparency at the level of individual applications).

[157]Notice and take down is defined as a process operated by online hosts in response to court orders or allegations that content is illegal. Content is removed by the host following notice. Notice and take down is widely operated in relation to copyright infringement, as well as for libel and other illegal content. Under U.S. and European law, this process finds its rules in the Digital Millennium Copyright Act 1998 and the Electronic Commerce Directive 2000.

[158]Marcelo Lopez Alfonsín, Argentina Report, pp. 3–4.

[159]Post (2018), p. 1067. The Canadian rapporteur also expresses some doubt as to allowing private third parties this role, *see* Eltis, Trudel, Canada Report, p. 11.

information in the hands of search engine providers, noting that "internet search engine service providers should not be saddled with such an obligation. . . ."[160]

However, Post's general view of the RTBF, based on *Google Spain*, does not consider other ECJ cases that address similar topics- notably the case of Salvatore Manni.[161] In *Manni*, the court explicitly considered a person's individual right in tandem with a public interest to access to data regarding business organizations.[162] The ECJ concluded that only in exceptional circumstances, when the data subject proves the existence of overriding and legitimate reasons to withhold disclosure, third parties might not be granted access to the data subject's personal information found in the company register.[163] *Google Spain* falls under the umbrella of "exceptional circumstances" because Google represents an internet service provider and the information was no longer reflective of the reality, and the "company was dissolved for a sufficiently long period of time."[164] However, in *Manni*, the public did have an interest in accessing the plaintiff's business history, and the platform on which they could seek said access's was that of a public record as opposed to an extraneous search engine. This difference between *Google Spain* and *Manni* shows that the ECJ, contrary to Post's assessment, does consider a public interest and a freedom of information through the lens of a balancing test as opposed to a mechanical process.

5.4 Adjudication of Control

As formulated, the critique above (5.3) falls short of full persuasion. As a matter of law, individuals who wish to challenge a determination made by Google concerning the appearance of their personal data in search results are not deprived of judicial recourse. Google's internal mechanisms "do not hinder the possibility for individuals to rely on national courts or the [corresponding data protection authority] in order to obtain a delisting order."[165] This very point was recently illustrated in the United Kingdom. In *NT1 and NT2 v. Google LLC*,[166] Google denied a businessman's petition to de-list search results related to his previous criminal convictions. The man sued Google and argued that the links were no longer relevant. The court agreed with the businessman and ordered Google to de-list them. The

[160]Case C-131/12, Google Spain SL v. Agencia Española de Protección de Datos, 2014 EUR-Lex Celex No. 62012CJ0131 (May 13, 2014), Opinion of Advocate General Jääskinen, Celex No. 612CC0131 at ¶¶ 133–34 (June 25, 2013).

[161]Case C-398/15 Camera di Commercio, Industria, Artigianato e Agricoltura di Lecce v Salvatore Manni ¶ 24 (March 9, 2017).

[162]For a comparison of the two cases, see Büyüksagis (2019), pp. 28–33.

[163]*Manni*. ¶ 64.

[164]*Id.*

[165]D'Antonio, Pollicino, Italy Report, p. 11.

[166][2018] EWHC 799 (QB).

same also recently occurred in France when Google refused to de-list links pertaining to a former chief financial officer's civil penalties in relation to insider trading.[167] Likewise, Google rapidly complied with an Italian court's order to de-list a number of links to pornographic videos of an Italian woman who committed suicide in connection with those videos' unintended mass dissemination.[168]

The same is true when it comes to giving individuals the possibility to challenge Google's decision to delist. Those individuals could always take court action to ask the reinstatement of the delisted information, by having the court state the illegality of the removal of the information. Alternatively, and in order to alleviate the costs and the burden on the person potentially affected by a delisting, one could conceive of the creation of an independent state authority in charge of handling such requests.

5.5 The Extent of Search Engines' Liability in Disputes over Control

In considering the extent of search engines' liability for disseminating defamatory material, some consider that it is disingenuous to characterize search results as the products of search engines. Under this conception, search results are the byproduct of a number of algorithmic mechanisms processed by bots equipped with artificial intelligence. That fact may, in part, be the reason why some jurisdictions place search engines within a safe harbor from legal liability.

The work of the national rapporteurs indicates that some jurisdictions will, accordingly, limit individuals' ability to pursue libel actions against search engines in the de-indexation context, through a scienter requirement.[169] In fact, prior to the *Google Spain* case, the Italian rapporteur notes that the case law of the Italian Supreme Court "highlighted the neutrality of search engine activities,"[170] and

[167]Browning and Sebag (Apr. 13, 2018, 9:49 AM), https://www.bloomberg.com/news/articles/2018-04-13/google-told-to-remove-links-to-businessman-s-criminal-conviction-jfy0dqv4.

[168]Warren (May 16, 2018), https://www.theatlantic.com/technology/archive/2018/05/tiziana-cantone-suicide-right-to-be-forgotten/559289/.

[169]*See, e.g.*, Alfonsín ML, Argentina Report, p. 2 ("Internet intermediaries become liable only upon obtaining 'effective knowledge' of the illegal content involving the notification by a court or other competent authority. . . ."); Gonçalves R, Brazil Report, p 8 (discussing Google's liability in the *Xuxa* case); Eltis, Trudel, Canada Report, p. 7 ("Selon les six juges majoritaires de la Cour, une personne ne peut en diffamer une autre simplement en publiant un hyperlien menant au site Web ou à un document d'un tiers qui contient des propos diffamatoires. . . ."); D'Antonio, Pollicino, Italy Report, p 3 (noting differing approaches to intermediaries' liability prior to the *Google Spain* decision); Chiou, Taiwan Report, p 4 ("Court decisions that uphold ISP's obligation to remove contents. . .are limited situations where the ISP knows that the contents or search results may infringe upon reputations of others or fails to know that as a result of gross negligence.").

[170]D'Antonio, Pollicino, Italy Report, p. 3.

perceived them as "mere intermediar[ies] which [made] accessible third-party websites without playing any 'active' role...."[171]

A strong counterargument—in favor of search engines' accountability—comes from the Japanese Supreme Court, which held that the "provision of search results" has an independent aspect of "expressive conduct by the search service provider itself" because its computer programs were made in a way to achieve the results in accordance with "the search service provider's policy."[172] The Italian and German rapporteurs suggest that the European Court of Justice would likely agree with this approach, as the ECJ impliedly attributed liability onto search engines in the *Google Spain* case, despite their "passive" role.[173]

That said, technological advancement complicates liability attribution. As the Japanese rapporteur advises: "[because] algorithms are increasingly programmed in ways that enable them to learn, decide, and constantly update by themselves, we ought to consider closely who exactly should be held liable for harms caused by algorithmic decision-making, among complexly-related multiple entities who are involved in the design and practical implementation of such algorithms."[174]

Nevertheless, it appears that liability regimes should consider the degree of data intermediaries' actual involvement in injuries committed upon private persons.[175] Furthermore, cases of actual mismanagement by data intermediaries should be distinguished from injuries they have "passively" facilitated. For example, Facebook recently stood at the center of several enormous data privacy scandals. Notably, the Cambridge Analytica scandal involved the misappropriation of 50 million Facebook users' online data to Cambridge Analytica, a United Kingdom-based political consultancy, which then sold this information to the Trump campaign for purposes of ad-targeting during the 2016 presidential campaign in the United States.[176] Another controversy involves Facebook's logging of text message and call histories for some of its users.[177] Questions surrounding Facebook's use—or misuse—of its users' data has prompted the United States Federal Trade Commission to open an investigation into the company's practices.[178] Facebook's purported mishandling of personal data is clearly and readily distinguishable from any liability that may be

---

[171] D'Antonio, Pollicino, Italy Report, p. 3.

[172] Yamaguchi, Japan Report, p. 3.

[173] German Report, p. 2 (noting that the issue is unsettled at the national level); D'Antonio, Pollicino, Italy Report, p 3 (discussing conflicting case law on the subject prior to the *Google Spain* decision).

[174] Yamaguchi, Japan Report, p. 20.

[175] In that respect, liability for damages should be limited.

[176] Rosenberg et al. (March 17, 2018), https://www.nytimes.com/2018/03/17/us/politics/cambridge-analytica-trump-campaign.html.

[177] McMillan (Mar. 26, 2018, 7:31 PM), https://www.wsj.com/articles/facebook-logs-text-call-histories-for-some-android-users-1522072657.

[178] Salinas (Mar. 26, 2018), https://www.cnbc.com/2018/03/26/ftc-confirms-facebook-data-breach-investigation.html.

attributed to Google for providing search results that link to a blog disparaging an individual.

6 Conclusion

In his recent testimony before the United States Congress, the CEO of Facebook, Mark Zuckerberg, emphasized the importance of users' control over their personal data: "every single time you go to share something on Facebook, whether it's a photo in Facebook, or a message, every single time, there's a control right there about who you're going to be sharing it with ... and you can change that and control that in line."[179] So *what is it* about controversies such as the Facebook/Cambridge Analytica incident that instills discomfort? Is it that individuals' personal data is available to third parties, or is it rather that, in a way contrary to their own personal freedom and respect for their autonomy, individuals lose control of how their information is utilized once in the hands of these third parties?[180]

In either case, the tension between free speech and personal privacy that animates the right to be forgotten seems likely to grow as a dominant feature of international public discourse. This preoccupation will only rise with the anticipated increases in access and interconnectivity. Although it manifests itself in different ways across national and regional laws, the right to be forgotten and its various facets are more ubiquitous than many would think—at least if we accept that it does not cover one single reality but rather a variety of entitlements, as we have tried to show. It also appears that, ultimately, recent societal concern for individuals' control over their own information may be a catalyst behind the recognition or reinforcement of some of these entitlements.

As the world continues into the modern age, questions surrounding the balance between personal freedoms, the public interest, and the role of commercial exploitation of personal data will fuel dialogue, keeping the right to be forgotten and its various manifestations at the center of attention.

Acknowledgements I wish to thank Ephraim David Abreu (JD Georgetown Law '19) for his major contribution to an early version of the report without which the current one would not exist. I also wish to thank Sibilla Grenon (JD Georgetown Law '21) for her thoughtful work on later drafts.

[179]Watson (Apr. 11, 2018 12:00 AM), https://www.theguardian.com/technology/2018/apr/11/mark-zuckerbergs-testimony-to-congress-the-key-moments.

[180]Facebook retains a great amount of data about its users that a standard consumer would not expect. That data includes users' phonebook contacts, a list of users' removed friends, and even the number of advertisers with users' personal information. *See* Chen (Apr. 11, 2018), https://www.nytimes.com/2018/04/11/technology/personaltech/i-downloaded-the-information-that-facebook-has-on-me-yikes.html?mtrref=undefined.

The Right to Be Forgotten: Questionnaire

Question 1

- Comment votre droit protège-t-il le droit à l'oubli ? Le droit à l'oubli est-il consacré de manière spécifique dans une loi ou découle-t-il de dispositions générales ?

How is the right to be forgotten protected under your law? Does your law specifically grant a right to be forgotten or does this right derive from a more general framework?

Question 2

- Quelles sont les limites au droit à l'oubli selon votre droit ?

What are the limits to the right to be forgotten under your law?

Question 3

- Quels sont, dans votre droit, les moyens de droit pour mettre en œuvre son droit à l'oubli ?

What are, in your law, the legal remedies available to enforce the right to be forgotten?

Question 4

- Dans le prolongement de la question précédente, est-ce que votre droit permet à une personne qui s'estime lésée par une information sur internet d'obtenir une réparation de son dommage ou de son tort moral ? Si oui, est-ce que la mise en œuvre d'une telle action en responsabilité est réalisable en pratique ?

As a follow-up to the previous question, does your law allow the plaintiff to receive material or immaterial damages? If yes, is such remedy realistic in practice?

Question 5

- De manière générale, comment évaluez-vous la mise en œuvre du droit à l'oubli dans votre droit ? Est-elle efficace ? Le droit à l'oubli est-il souvent utilisé en pratique ? Existe-t-il des obstacles particuliers à sa mise en œuvre ?

In general, how do you assess the implementation of the right to be forgotten in your law? Is it effective? Is it used in practice? Are there particular obstacles in the implementation of this right?

Question 6

- Comment les tribunaux et les auteurs de doctrine ont-ils accueilli la décision *Google c. González* de la CJUE dans votre État ?

How did courts and commentators in your country welcome the ECJ ruling on Google v González?

Question 7

- Pour les ressortissants d'un État qui ne fait pas partie de l'Union européenne, est-ce que les tribunaux de votre État ont suivi la décision de la CJUE ? Pensez-vous qu'ils vont le faire ?

For those who are from a country that is not part of the European Union, did your courts follow the ECJ ruling on the right to be forgotten? Is it likely Do that they will follow it?

Question 8

- Est-ce que votre droit accordait déjà un droit à l'oubli sur internet similaire à celui consacré par la CJUE ?

Did your law already grant a similar right to be forgotten than the one stated in the ECJ ruling?

Question 9

- Pour mettre en œuvre la décision de la CJUE, Google a mis en place un formulaire permettant à toute personne intéressée de déposer une requête pour déréférencer une information qui la concerne. Sur la base de cette demande, Google doit faire une pesée des intérêts entre l'intérêt privé de la personne à déréférencer son information et l'intérêt public à ce que l'information soit publique. Google ne rend toutefois pas publique la manière dont il traite les requêtes de déréférencement. En particulier, Google n'informe pas le public du nombre de demandes qu'il reçoit, du type de demande, du cercle des personnes concernées, du nombre d'acceptation et de refus et des raisons des refus. Pensez-vous que Google doive améliorer la transparence dans la mise en œuvre du droit à l'oubli ?

To implement the ECJ ruling, Google has created a form in which anyone interested can submit a request to have information about him-or herself be delisted. Based on this request, Google will weigh between the private interest of the petitioner and the public interest to be informed. Google does not disclose the ways in which it deals with requests. In particular, Google does fully not disclose, the category of requests that are excluded or accepted, the proportion of requests and successful de-listings and, among others, the reason for the denial of delisting. Do you think that Google should be more transparent about the ways it uses to implement the right to be forgotten?

Question 10

- Est-ce que les citoyens de votre État font usage du formulaire de Google pour mettre en œuvre le droit à l'oubli sur internet ?

Is the procedure prepared by Google used in your country?

Question 11

- Des réformes sont-elles prévues au niveau législatif pour renforcer ou modifier la protection du droit à l'oubli dans votre droit ?

Is there any upcoming legal reform in your country whose purpose is to reinforce or modify the right to be forgotten?

Question 12

- Quelle devrait être à votre avis la prochaine étape dans la protection du droit à l'oubli ? Pensez-vous que les États devraient protéger davantage la personnalité des utilisateurs sur internet ? Pensez-vous que l'Union européenne devrait modifier ou adapter ses normes qui protègent le droit à l'oubli ?

In your opinion, what should be the next step in the protection of the right to be forgotten? Do you think that one must go further and strengthen the right to be forgotten? Do you think that the European Union should modify or adapt its legislation on the right to be forgotten?

References

Case Law

Axel Springer AG v. Germany, App. No. 39954/08 (Eur. Ct. H.R. Feb. 7, 2012)
Case C-131/12, Google Spain SL v. Agencia Espanola de Proteccion de Datos, 2014 E.C.R. 317
Case C-398/15 Camera di Commercio, Industria, Artigianato e Agricoltura di Lecce v Salvatore Manni, 2017 E.C.J
Furst-Pfeifer v. Austria, Application nos. 33677/10 and 52340/10 (Eur. Ct. HR May 17, 2017)
Griswold v. Connecticut, 381 U.S. 479, 484–85 (1965)
Mapp v. Ohio, 367 U.S. 643, 656 (1961)
ML and WW v. Germany, Nos. 60798/10 and 65599/10 (Eur. Ct. HR. 2018)
Satakunnan Markinaporssi Oy and Satamedia Oy v. Finland, Application no. 931/13, (Eur. Ct. HR. Jun. 27, 2017)
Smith v. Daily Mail Publishing Company, 443 U.S. 308 (1977)
von Hannover v. Germany (59320/00), [2004] E.M.L.R. 21

Literature

Alston P (2005) Non-state actors and human rights. Oxford University Press, Oxford, p 2
Bennett SC (2012) The "Right to be Forgotten": reconciling EU and US perspectives. Berkeley J Int Law 30(1):161
Bertram T et al (2018) Three years of the right to be forgotten. Google, Inc, Menlo Park, p 17, Available at: https://www.elie.net/static/files/three-years-of-the-right-to-be-forgotten/three-years-of-the-right-to-be-forgotten-paper.pdf

Bhardwaj P (Feb 28, 2018, 12:06 PM) Millions of Europeans are asking Google to be "forgotten" — here's why Americans don't have that option, Business Insider, http://www.businessinsider.com/google-right-to-be-forgotten-law-in-america-2018-2
Browning J, Sebag G (Apr 13, 2018, 9:49 AM) Google has to hit delete after right to be forgotten turns to crime, Bloomberg, https://www.bloomberg.com/news/articles/2018-04-13/google-told-to-remove-links-to-businessman-s-criminal-conviction-jfy0dqv4
Brüggemeier G, Ciacchi AC, O'Callaghan P (eds) (2010) Personality rights in Europe. Cambridge University Press, Cambridge, p 31
Büyüksagis E (2019) Towards a transatlantic concept of data privacy. Fordham Law Rev
Chen BX (Apr 11, 2018) I Downloaded the information that Facebook has on me. Yikes., N.Y. Times, https://www.nytimes.com/2018/04/11/technology/personaltech/i-downloaded-the-information-that-facebook-has-on-me-yikes.html?mtrref=undefined
Clapham A (2006) Human rights obligations of non-state actors. Oxford University Press, Oxford
Cohen J (2017a) The biopolitical public domain: the legal construction of the surveillance economy. Philos Technol 31:230–231
Cohen JE (2017b) The Zombie first amendment. William & Mary Law Rev 56:119
Gajda A (2018) Privacy, press, and the right to be forgotten in the United States. Wash Law Rev 93:201
Hern A (Apr 19, 2018, 7:03 AM) Facebook moves 1.5bn users out of reach of new European privacy law, The Guardian https://www.theguardian.com/technology/2018/apr/19/facebook-moves-15bn-users-out-of-reach-of-new-european-privacy-law?CMP=share_btn_link
Heyman SJ (2008) Free speech and the natural rights tradition. In free speech and human dignity. Yale University Press, New Haven, pp 7–22
Legrand P (2017) Jameses at play: a tractation on the comparison of laws. Am J Comp Law 65(1):1
Mayer-Schönberger V (2009) Delete: the virtue of forgetting in a digital age. Princeton University Press, Princeton, pp 16–49
McLean I (2004) Thomas Jefferson, John Adams, and the Déclaration des Droits de L'Homme et du Citoyen. In: Fatton R, Ramazani RK (eds) The future of Liberal Democracy. Palgrave Macmillan, New York
McMillan R (Mar 26, 2018, 7:31 PM) Facebook logs text, Call histories for some android users, Wall St. J., https://www.wsj.com/articles/facebook-logs-text-call-histories-for-some-android-users-1522072657
Müller JP (2018) Verwirklichung der Grundrechte nach Art. 35 BV. Berne
National Conf. of State Legislatures, Privacy Protections in State Constitutions (May 5, 2017) http://www.ncsl.org/research/telecommunications-and-information-technology/privacy-protections-in-state-constitutions.aspx
Page JA (2010) In: Brüggemeier G, Ciacchi AC, O'Callaghan P (eds) Personality rights in Europe. Cambridge University Press, Cambridge, p 38
Peters A (2017) Privacy, Rechtsstaatlichkeit, and the legal limits on extraterritorial surveillance. In: Miller RA (ed) Privacy and power a transatlantic dialogue in the shadow of the NSA-Affair. Cambridge University Press, Cambridge, p 145, 149
Poscher R (2017) The right to data protection a no-right thesis. In: Miller RA (ed) Privacy and power a transatlantic dialogue in the shadow of the NSA-affair. Cambridge University Press, Cambridge, p 129
Post RC (2018) Data privacy and dignitary privacy: Google Spain, the right to be forgotten, and the construction of the public sphere. Duke Law J 67:1059–1061
Richards N, Hartzog W (2019) The pathologies of digital consent. Wash Law Rev:96
Rosenberg M et al (Mar 17, 2018) How Trump consultants exploited the Facebook data of millions, N.Y. Times, https://www.nytimes.com/2018/03/17/us/politics/cambridge-analytica-trump-campaign.html
Rotenberg M, Google's position makes no sense: opposing view, at https://www.usatoday.com/story/opinion/2015/01/22/

Salinas S (Mar 26, 2018) Facebook stock slides after FTC launches probe of data scandal, CNBC, https://www.cnbc.com/2018/03/26/ftc-confirms-facebook-data-breach-investigation.html
Stein L (2010) Speech right in America. University of Illinois, Chicago
Warren R (May 16, 2018) A Mother wants the Internet to Forget Italy's most viral sex tape, The Atlantic, https://www.theatlantic.com/technology/archive/2018/05/tiziana-cantone-suicide-right-to-be-forgotten/559289/
Warren SD, Brandeis LD (1890) The right to privacy. Harv Law Rev 4:193–220
Watson C (Apr 11, 2018 12:00 AM) The key moments from mark Zuckerberg's testimony to Congress, The Guardian, https://www.theguardian.com/technology/2018/apr/11/mark-zuckerbergs-testimony-to-congress-the-key-moments
Werro F (2009) The right to inform v. the right to be forgotten: a transatlantic clash. In: Ciacchi AC et al (eds) Haftungsrecht im dritten Millenium - Liability in the Third Millennium. Bremen. Nomos Publishing, Baden-Baden, p 285, 291, 299
Whitman JQ (2004) The two western cultures of privacy: dignity versus liberty. Yale Law J. 113:1180

Conventions/Statutes

Convention for the Protection of Human Rights and Fundamental Freedoms, art. 8, Nov. 4, 1950, 213 U.N.T.S. 222
International Covenant on Civil and Political Rights, art. 17(1), Dec. 16, 1966, 999 U.N.T.S. 171
OECD Guidelines Governing the Protection of Privacy and Transborder Flows of Personal Data ¶ 10
Organization of American States, Am. Conv. on Human Rights, Nov. 22, 1969, O.A.S.T.S. No. 36, 1144 U.N.T.S. 123

Part I
Europe

Le droit à être oublié en droit belge

Jonathan Wildemeersch

Abstract Unheard of until a few years ago, the "right to forget" is becoming a common expression in the legal world and beyond. It is important, however, to agree on what exactly this generic term covers. It groups together several rights based on different legal grounds.

Two sub-categories can be distinguished. The first covers all the legal instruments that allow the citizen to obtain or recover control of his personal data. This is what I term "droit à l'oubli", literally "right to forget". The legal basis for this comes from the legislation on the protection of personal data (such as, for example, the GDPR). The second sub-category is an expression of the right to private life. It comprises the tools that allow a person to refuse to be an object of information, either systematically or on a specific occasion. It is "the right to be forgotten". It is this second sub-category that I will mainly deal with.

The "right to be forgotten" limits the freedom of the press. It is therefore a question of balancing the interests protected by this fundamental right against the right to private life. The action that is likely to be brought with success after an infringement of the "right to be forgotten" is based on the common law of liability, namely Article 1382 of the Belgian Civil Code. Two specific conditions have to be met. The disputed facts have been lawfully published for the first time. This information must, then, be

---

Docteur en sciences juridiques. Professeur à l'Université de Liège (ULiège), chargé de cours invité à l'Université de Louvain (UCLouvain) et référendaire à la Cour de justice de l'Union européenne. La rédaction de cet article a été achevée le 20 décembre 2018. Les propos tenus n'engagent que l'auteur et non les institutions auxquelles il appartient. @ : jwildemeersch@uliege.be.

J. Wildemeersch (✉)
University of Liège (ULiège), Liège, Belgium

University of Louvain, Louvain-la-Neuve, Belgium

Court of Justice of the European Union, Luxembourg City, Luxembourg
e-mail: jwildemeersch@uliege.be

F. Werro (ed.), *The Right To Be Forgotten*, Ius Comparatum – Global Studies in Comparative Law 40, https://doi.org/10.1007/978-3-030-33512-0_2

disseminated a second time, in a way that may be different from the original disclosure. If these preliminary conditions are met, the judge has several parameters at his disposal to assess the balance of the interests involved.

1 Introduction

Peu connu il y a encore quelques années, le droit à l'oubli est en passe de devenir une expression courante dans le monde juridique et au-delà. Il convient toutefois de s'entendre sur ce que recouvre exactement cette appellation générique. Elle regroupe en réalité plusieurs droits qui reposent sur des fondements législatifs différents. En d'autres termes, le droit à l'oubli se décline au pluriel Tulkens and Sohier (2015) n°19 ; c'est un droit « multi-facettes » Defreyne (2013), p. 77. Les différents droits en cause poursuivent néanmoins le même objectif : permettre aux personnes physiques de (re)prendre et conserver le contrôle sur des informations privées et des données personnelles[1]. Le « droit à l'oubli » est ainsi susceptible de viser des réalités aussi différentes que le droit au déréférencement, le droit à l'effacement mais aussi le droit à l'anonymisation ou encore le droit d'opposition, le droit d'accès, etc.[2]

Deux sous-catégories peuvent être distinguées. La première vise « l'ensemble des instruments juridiques – éventuellement renforcés par des outils technologiques – qui permettent au citoyen d'obtenir ou de retrouver la maîtrise de ses données personnelles, dès lors qu'elles ont été dispersées, que ce soit volontairement ou involontairement » Jongen and Strowel (2017) n° 588. Cette sous-catégorie est parfois appelée « droit à l'oubli numérique »[3]. L'expression « droit à l'oubli de données personnelles » Montero and Van Enis (2016)[4] nous paraît plus précise et devoir être privilégiée dès lors que les outils numériques peuvent également être au centre de la deuxième sous-catégorie du droit à l'oubli. Le fondement du « droit à l'oubli de données personnelles » réside dans la législation relative à la protection des données personnelles. L'article 8 de la Charte des droits fondamentaux de

[1]En ce sens, Jongen and Strowel (2017) n° 587 ; Cruysmans (2014) n° 2 et n° 3. Pour une version actualisée de cet article, voir Cruysmans (2016b), pp. 403–418.

[2]À propos de ces différents droits, voir Dechenaud (2015).

[3]Selon ces auteurs, les lois ayant transposé la directive 95/46 ont dessiné, « de façon indirecte et implicite, les contours de ce qui s'apparente à un 'droit à l'oubli numérique' » (n° 17). À côté de celui-ci, ils identifient un premier droit, plus ancien et limité aux faits ayant fait l'objet d'une condamnation judiciaire : le « droit à l'oubli du passé judiciaire ». Ils envisagent, enfin un troisième droit, plus récent et plus large, de nature à viser, outre la presse, les éditeurs de sites d'archives en ligne et les gestionnaires de moteur de recherche (voir notamment n° 19). Preuve de l'imprécision liée à l'expression « droit à l'oubli numérique », c'est cette dernière catégorie qui est annoncée dans leur introduction sous le vocable de… « droit à l'oubli numérique » (voir dernière phrase du n° 13). Comme ils le reconnaissent eux-mêmes dans leur conclusion, « la dénomination 'droit à l'oubli numérique' est une commodité de langage qui ne devrait pas occulter les *multiples modalités d'exercice* dudit droit » (n° 43, souligné par les auteurs).

[4]Cruysmans (2016a), 618 à 620, n° 3.

l'Union européenne (la Charte) a même fait du droit à la protection des données à caractère personnel un droit autonome, toute personne ayant, selon l'article 8, paragraphe 2 de la Charte, « le droit d'accéder aux données collectées la concernant et d'en obtenir la rectification » [5]. Le « droit à l'oubli de données personnelles » est aujourd'hui au cœur du règlement général de l'Union européenne sur la protection des données (RGPD)[6]. Les différents droits couverts par cette appellation générale (comme, par exemple, le droit à l'effacement ou à l'anonymisation des données ou le droit au retrait ou d'opposition) étaient déjà néanmoins partiellement consacrés en droit belge par la loi du 8 décembre 1992 relative à la protection de la vie privée à l'égard des traitements de données à caractère personnel qui, à la suite de sa modification en 1998, assurait la transposition de la directive 95/46/CE, du Parlement européen et du Conseil, du 24 octobre 1995 relative à la protection des personnes physiques à l'égard du traitement de données à caractère personnel et à la libre circulation de ces données[7]. Cette première sous-catégorie « s'entend [donc] dans une perspective particulière – la finalité du traitement des données – et se traduit, d'une part, par une obligation incombant au responsable du traitement d'effacer les données lorsque la finalité du traitement est atteinte ou ne justifie plus la conservation de celles-ci, d'autre part, par le droit de toute personne à l'effacement des données la concernant en cas d'erreur, de péremption ou d'exploitation (utilisation, conservation, communication) non conforme à la loi ou encore en cas d'opposition à certains traitements » Montero and Van Enis (2016) n° 17.

La seconde sous-catégorie du « droit à l'oubli » est une expression du droit à la vie privée en ce qu'elle comprend les instruments qui « perm[etten]t à un individu de refuser – moyennant l'application d'une série de critères – d'être un objet d'information, que ce soit de manière systématique ou à l'occasion d'un événement particulier » Jongen and Strowel (2017) n° 588. Il s'agit davantage d'un « droit à l'oubli judiciaire » Cruysmans (2014) n° 2 ou, plus largement, d'un « droit à être

---

[5]Sur l'article 8 de la Charte comme fondement du « droit à l'oubli de données personnelles », voir Aramazani (2011), p. 40; Defreyne (2013), p. 86; Cruysmans (2016b), p. 406.

[6]Règlement (UE) n° 2016/679 du Parlement et du Conseil, du 27 avril 2016, relatif à la protection des personnes physiques à l'égard du traitement des données à caractère personnel et à la libre circulation de ces données, et abrogeant la directive 95/46/CE (*JOUE* 2016, L 119, p. 1).

[7]*Mon. b.*, 3 février 1999. C'est ainsi, notamment, que l'article 5, paragraphe 1er, 5°, de la loi prévoit que les données à caractère personnel sont « conservées sous une forme permettant l'identification des personnes concernées pendant une durée n'excédant pas celle nécessaire à la réalisation des finalités pour lesquelles elles sont obtenues ou pour lesquelles elles sont traitées ultérieurement ». L'article 12 peut également être vu comme une consécration du droit à l'oubli en ce qu'il prévoit que « toute personne a en outre le droit de s'opposer, pour des raisons sérieuses et légitimes tenant à une situation particulière, à ce que des données la concernant fassent l'objet d'un traitement » (en ce sens, Cassart and Henrotte 2014, p. 1191). Notons que la directive 45/96 a été abrogée par le RGPD.

oublié » – terminologie préférée par certains auteurs[8]. D'abord utilisé dans le cadre de la presse traditionnelle, il connaît de nouveaux développements à l'aire des archives numériques au travers du droit à l'anonymisation – dans le cadre d'une nouvelle diffusion, par le biais d'archives numériques, d'un ancien article paru plusieurs années auparavant dans la presse écrite par exemple[9] – ou du droit au déréférencement. Ce « droit à être oublié » est appréhendé comme une composante du droit au respect de la vie privée garanti par les articles 8 de la Convention européenne de sauvegarde des droits de l'homme et des libertés fondamentales, 17 du Pacte international relatif aux droits civils et politiques et 22 de la Constitution belge[10], auquel il convient d'ajouter l'article 7 de la Charte.

C'est de cette deuxième sous-catégorie que nous traiterons essentiellement.

2 La mise en œuvre du droit à être oublié en droit belge

2.1 Le contexte général

Le droit à l'oubli (ou « droit à être oublié ») constitue une restriction à la liberté de la presse. Il s'agit donc d'effectuer une balance entre les intérêts protégés par des droits fondamentaux qui sont chacun garantis au niveau constitutionnel et international : d'une part, la liberté d'expression [article 25 de la Constitution[11], article 11 de la

[8]En ce sens, Jongen and Strowel (2017) n° 588 et suivants. J. Le Clainche parle d'un « droit *subjectif* à être oublié » Le Clainche (2012), p. 44. Dans un jugement pourtant relativement récent (mais dans le cadre de la diffusion d'un reportage télévisé), le Tribunal de première instance de Bruxelles parle de l'exercice d'un « droit de repentir » qu'il définit d'une façon étonnamment proche du « droit à être oublié ». Qu'on en juge : « Le droit au repentir doit être compris comme permettant à une personne qui a été sous les feux de l'actualité d'en sortir après un certain temps, en raison du temps écoulé depuis les faits concernés » (Civ. Bruxelles, 25 mai 2011, *A & M*, 2011/4-5, p. 569). En l'espèce, le juge a refusé de reconnaître ce droit dans le chef des demandeurs après avoir constaté qu'ils avaient accepté d'apparaître dans un reportage qui ne concernait pas un sujet d'actualité mais un sujet de société.

[9]Le « droit à l'oubli judiciaire » n'est pas limité à la presse traditionnelle : il embrasse tous les types de presse, y compris la presse et les archives numériques (en ce sens, Cruysmans 2014, n° 3). Pour une illustration, voir, dans la même affaire, les jugements d'instance et d'appel : Civ. Neufchâteau, 25 janvier 2013, *J.L.M.B.*, 2013, p. 1182 et Liège, 25 septembre 2014, *J.L.M.B.*, 2014, p. 1952, note Cruysmans ; *R.G.D.C.*, 2016, p. 294, note Montero & Van Enis ; *NjW*, 2015, p. 26, note Van Eecke & Le Boudec, ainsi que Civ. Liège, 3 novembre 2014, *J.L.M.B.*, 2014, p. 1961, note Cruysmans.

[10]Cass., 29 avril 2016, *J.T.*, 2016, p. 609. Selon le premier alinéa de l'article 22 de la Constitution belge, « [c]hacun a droit au respect de sa vie privée et familiale, sauf dans les cas et conditions fixés par la loi ».

[11]Selon l'article 25 de la Constitution :

« La presse est libre ; la censure ne pourra jamais être établie ; il ne peut être exigé de cautionnement des écrivains, éditeurs ou imprimeurs.

Lorsque l'auteur est connu et domicilié en Belgique, l'éditeur, l'imprimeur ou le distributeur ne peut être poursuivi ».

Charte des droits fondamentaux de l'Union européenne – le second paragraphe de cette disposition prévoyant expressément que la liberté des médias (et leur pluralisme) doivent être respectés – et article 10 de la Convention européenne des droits de l'homme] et, d'autre part, le droit à la vie privée (article 22 de la Constitution, article 7 de la Charte et article 8 de la Convention européenne des droits de l'homme). Ces deux droits « constituent les fondements de toute société démocratique et ne sont ni absolus ni hiérarchisés, étant d'égale valeur »[12]. Pour justifier l'ingérence dans le droit à la liberté d'expression, les conditions de légalité, de légitimité et de proportionnalité imposées par l'article 10, paragraphe 2, de la Convention européenne des droits de l'homme doivent donc être réunies[13].

Le droit à l'oubli ne doit pas être confondu avec le « délit de presse ». Une telle qualification entraînerait, en Belgique, la compétence d'un jury d'assises en vertu de l'article 150 de la Constitution belge. Il ne s'agit pas d'une mise en cause du contenu de la publication. Comme rappelé par le tribunal de première instance de Liège, « le délit de presse est une infraction de droit commun qui se caractérise par son mode d'exécution, c'est-à-dire par la voie de la presse. Ainsi, pour qu'il y ait délit de presse, il faut notamment que puisse être reproché un comportement incriminé par la loi pénale tel que par exemple la calomnie, la diffamation, la provocation publique à commettre des crimes »[14]. En d'autres termes, pour qu'il y ait délit de presse, il est nécessaire que la manifestation de la pensée par la voie de la presse revête un caractère délictueux[15].

2.2 *L'action en responsabilité comme voie procédurale privilégiée*

Si le « droit à être oublié » découle du droit à la protection de la vie privée, c'est néanmoins l'article 1382 du Code civil qui constitue le véritable fondement de l'action susceptible d'être introduite avec succès à la suite d'une atteinte au « droit à être oublié »[16]. En effet, « prétendre efficacement à un oubli, c'est concrètement

[12]Civ. Liège, 3 novembre 2014, *J.L.M.B.*, 2014, p. 1961.

[13]En ce sens, Cass., 29 avril 2016, *J.T.*, 2016, p. 609. En revanche, le Tribunal de première instance de Namur avait jugé, dans un jugement plus ancien, que « [l]e respect de ce droit [à l'oubli], en ce compris par les journalistes se prévalant de l'exercice de la liberté de la presse, doit être considéré comme le principe », les possibilités d'y déroger apparaissant, par conséquent, comme des exceptions d'interprétation restrictive (Civ. Namur, 27 septembre 1999, *A & M*, 2000/4, p. 471).

[14]Civ. Liège, 3 novembre 2014, *J.L.M.B.*, 2014, p. 1961. Le tribunal renvoie à Bosly et al. (2014), p. 1084. Ces développements du tribunal seront repris par la Cour d'appel de Liège dans le cadre de la procédure d'appel (Liège, 4 février 2016, *A & M*, 2016/5-6, p. 462 ; *J.T.*, 2016, p. 710). Ils étaient d'ailleurs issus de l'un de ses arrêts antérieurs (Liège, 25 septembre 2014, *J.L.M.B.*, 2014, p. 1952).

[15]Cass., 29 avril 2016, *J.T.*, 2016, p. 609.

[16]Selon cet article, « [t]out fait quelconque de l'homme, qui cause à autrui un dommage, oblige celui par la faute duquel il est arrivé, à le réparer ».

demander réparation d'un dommage causé par l'accessibilité permanente d'informations qui auraient dû tomber dans l'oubli » Cruysmans (2016b), p. 407.

L'article 1382 du Code civil, qui constitue le droit commun de la responsabilité en droit belge, a été reconnu comme étant « une loi suffisamment accessible, claire, précise et prévisible au sens de l'article 10, paragraphe 2, de la Convention européenne des droits de l'homme pour justifier d'éventuelles restriction à la liberté d'expression »[17]. Il a donc été jugé applicable aux organes de presse « qui ne peuvent ignorer que leur responsabilité est susceptible d'être engagée si l'exercice de la liberté de la presse cause un préjudice découlant de l'atteinte à 'des droits d'autrui' (terminologie utilisée par l'article 10, paragraphe 2, de la Convention européenne des droits de l'homme) parmi lesquels figure le droit à la vie privée »[18].

En revanche, une demande d'anonymisation fondée sur le fondement de la loi du 8 décembre 1992 relative à la protection de la vie privée à l'égard des traitements de données à caractère personnel a été rejetée par le président du Tribunal de première instance de Bruxelles. Selon celui-ci, si l'action était recevable, la mise en ligne d'archives journalistiques répondait néanmoins à la définition d'un traitement effectué aux seules fins de journalisme pour lequel la loi admet un régime dérogatoire[19].

2.2.1 Les trois conditions de l'action en responsabilité

Concrètement, le requérant qui fonde son action sur l'article 1382 du Code civil doit démontrer une faute en lien causal avec le préjudice dont il demande la réparation[20]. Dans le cadre d'une action visant à réparer une atteinte au droit à l'oubli, la faute ne réside pas dans le fait de mettre ou de laisser accessible sur un site internet l'information préalablement publiée, mais bien dans le refus opposé, sans motif raisonnable, à la demande de retrait ou d'anonymisation[21]. Pour déterminer le caractère fautif d'un tel refus, le comportement de l'éditeur est comparé au modèle abstrait de « l'éditeur normalement prudent et diligent »[22].

---

[17]Civ. Neufchâteau, 25 janvier 2013, *J.L.M.B.*, 2013, p. 1182 ; *A & M*, 2013/6, p. 478. Voir également Civ. Bruxelles, 25 mars 2014, *A & M*, 2014/5, p. 419.

[18]Liège, 25 septembre 2014, *J.L.M.B.*, 2014, p. 1952. Voir, également, Civ. Liège, 3 novembre 2014, *J.L.M.B.*, 2014, p. 1961 ; Liège, 4 février 2016, *A & M.*, 2016/5-6, p. 462 ; *J.T.*, 2016, p. 710.

[19]Civ. Bruxelles (réf.), 9 octobre 2012, *A & M*, 2013/3-4, p. 267 avec note d'observations Cruymans. Le jugement sera confirmé en appel (le litige ayant toutefois ayant été limité, en appel, à l'application de l'article 15 de la loi du 8 juin 1992), voir Bruxelles, 21 mars 2013, *A & M*, 2014/5, p. 416. Pour une critique de cette approche, voir Montero and Van Enis (2016) n° 27, voir également Defreyne (2013), pp. 93 et 94.

[20]Liège, 4 février 2016, *A & M*, 2016/5-6, p. 462 ; *J.T.*, 2016, p. 710.

[21]En ce sens, Civ. Neufchâteau, 25 janvier 2013, *A & M*, 2013/6, p. 478 ; Civ. Liège, 3 novembre 2014, *J.L.M.B.*, 2014, p. 196 (confirmé en appel par Liège, 4 février 2016, *A & M.*, 2016/5-6, p. 462 ; *J.T.*, 2016, p. 710).

[22]En ce sens, Cruysmans (2016b), p. 410.

En pratique, l'examen de la faute se réalise au travers d'une pondération entre le droit à la liberté d'expression et le droit à la vie privée. De cette analyse dépend l'existence ou non de la faute[23]. Conformément à l'article 10, paragraphe 2, de la Convention européenne des droits de l'homme, pour obtenir l'indemnisation de la violation de son droit à être oublié – l'indemnisation pouvant être une anonymisation des données qui permettent d'identifier le requérant dans un article de presse –, le citoyen doit prouver que l'ingérence est prévue par une loi, qu'elle poursuit un but légitime et que la mesure est nécessaire dans une société démocratique.

1. *Être prévue par la loi :* la Cour de cassation a approuvé la jurisprudence qui voyait dans l'article 1382 du Code civil, combiné avec les dispositions protégeant le droit à la vie privée, une consécration légale du droit à l'oubli[24].
2. *La poursuite d'un objectif légitime :* il ne fait guère de doute que les mesures permettant de mettre en œuvre le droit à l'oubli (comme le déréférencement ou l'anonymisation) poursuivent un but légitime en protégeant le droit à la vie privée de la personne concernée[25], le droit à l'oubli pouvant lui-même constituer l'objectif légitime[26].
3. *Le caractère nécessaire de la mesure dans une société démocratique :* c'est dans le cadre de l'examen de cette troisième condition que les juridictions doivent opérer un contrôle de proportionnalité à propos duquel il est possible d'identifier une méthodologie propre au « droit à être oublié ». Cette méthodologie requiert la réunion de deux conditions cumulatives, la juridiction ayant, ensuite recourt à plusieurs paramètres d'appréciation non cumulatifs[27].

2.2.2 Les conditions spécifiques au « droit à être oublié »

Tout d'abord, deux conditions cumulatives doivent être réunies. Qui dit « droit à être oublié » dit « information préalable ». En d'autres termes, pour invoquer le droit à être oublié, les faits litigieux doivent avoir été divulgués licitement une première fois, par quelque moyen que ce soit. L'information doit, ensuite, avoir été divulguée

[23]En ce sens, Cruysmans (2016a) 618 à 620, n° 5.

[24]Cruysmans (2016a) 618 à 620, n° 5.

[25]En ce sens, Montero and Van Enis (2016) n° 29.

[26]En ce sens, Cruysmans (2016c) www.justice-en-ligne.be/article908.html.

[27]Certains auteurs ne font pas de distinction entre les conditions liminaires et les paramètres d'appréciation, ces derniers étant alors appréhendés comme des conditions cumulatives (en ce sens, voir Tulkens and Sohier 2015, n° 19). Une telle interprétation nous semble pourtant contraire à l'utilisation qui en est faite par les juridictions. En effet, contrairement aux deux premières conditions qui sont nécessaires à l'existence d'un « droit à être oublié », les autres paramètres n'interviennent que dans un second temps. Ils sont les critères qui permettent d'opérer la balance d'intérêts entre les deux droits fondamentaux en présence. Or, cet exercice ne devra pas être effectué si les deux premières conditions ne sont pas remplies. En d'autres termes, contrairement aux paramètres d'appréciation, les deux premières conditions – une publication et une redivulgation – sont des conditions *sine qua non* du « droit à être oublié ».

une deuxième fois, selon un procédé qui peut être différent de la première divulgation. En effet, comme l'a expliqué la Cour d'appel de Liège, le « droit à être oublié » ne peut être limité à une nouvelle publication *stricto sensu* de faits antérieurement diffusé dans un article de presse. En effet, « [à] côté de la traditionnelle facette du droit à l'oubli, liée à la redivulgation par la presse d'un passé judiciaire d'une personne, [il] existe une seconde facette liée à l'effacement des données numériques et, en particulier, des données disponibles sur internet ». C'est ainsi que, s'inspirant expressément du raisonnement utilisé par la Cour de justice de l'Union européenne (CJUE) dans l'affaire *Google Spain et Google*[28], la Cour d'appel de Liège a jugé que la condition relative à la nouvelle divulgation était également rencontrée lorsqu'un éditeur « permet une mise en une de l'article litigieux *via* le moteur de recherche de son site consultable gratuitement, mise en une qui est par ailleurs multipliée considérablement par le développement des logiciels d'exploration des moteurs de recherche du type Google »[29]. Si elles ont pu être critiquées par certains auteurs[30], la condition de « redivulgation » et l'interprétation particulière que la Cour d'appel de Liège en a donnée dans la sphère numérique ont été confirmées par la Cour de cassation. En effet, aux termes de son arrêt, la plus haute juridiction judiciaire du pays a considéré que la Cour d'appel de Liège avait « décid[é] ainsi légalement que l'archivage en ligne de l'article litigieux constitu[ait] une nouvelle divulgation du passé judiciaire du défendeur pouvant porter atteinte à son droit à l'oubli »[31]. Par conséquent, cette façon d'appréhender la condition de la nouvelle divulgation permet de pouvoir revendiquer un « droit à être oublié » pour des écrits qui auraient, dès le départ, été diffusés dans la presse écrite *et* sur le site internet du journal. Une telle définition de la « redivulgation » étend donc sensiblement le sens courant de cette notion[32].

Ensuite, la juridiction dispose de plusieurs paramètres d'appréciation, non cumulatifs, pour opérer la balance d'intérêts entre les droits en présence. En effet, une fois les deux conditions cumulatives démontrées, il appartient au juge de déterminer lequel des deux droits fondamentaux en jeu doit prévaloir dans le cas d'espèce soumis à son jugement. Six critères d'appréciation, non cumulatifs et non exhaustifs, peuvent être dégagés de la jurisprudence. Ils s'inspirent de ceux utilisés pour résoudre les litiges classiques entre le droit à la liberté d'expression et le droit au respect de sa vie privée[33] et sont en partie similaire à ceux énoncés par la CJUE dans l'arrêt *Google Spain et Google*[34].

---

[28]CJUE, arrêt du 13 mai 2014, *Google Spain et Google*, C-131/12, ECLI:EU:C:2014:317.

[29]Liège, 25 septembre 2014, *J.L.M.B.*, 2014, p. 1952, spéc. p. 1957. Voir, également, Liège, 4 février 2016, *A & M*, 2016/5-6, p. 462 ; *J.T.*, 2016, p. 710.

[30]Voir Montero and Van Enis (2016) n° 40 et n° 41.

[31]Cass., 29 avril 2016, *J.T.*, 2016, p. 609.

[32]En ce sens, Cruysmans (2014) note 26.

[33]En ce sens, Jongen and Strowel (2017) n° 604.

[34]CJUE, arrêt du 13 mai 2014, *Google Spain et Google*, C-131/12, ECLI:EU:C:2014:317, points 97 à 99.

Primo, jouera en faveur du droit à l'oubli, *le laps de temps écoulé entre les deux divulgations* ou, dans le cas de l'archivage numérique, entre la première divulgation et la demande d'anonymisation formulée[35].

Secundo, si l'écoulement du temps aura tendance à conforter le droit à être oublié, *l'intérêt historique des informations rappelées* peut, en revanche, venir contrebalancer ce premier critère. En effet, il se peut qu'au fil du temps les faits rappelés aient acquis une dimension historique. Dans ce cas, ils peuvent toujours être repris librement par la presse[36]. Ce critère permet donc d'écarter le droit à l'oubli lorsqu'il aboutirait à effacer complètement l'histoire Cruysmans (2014) n° 9.

Tertio, l'intérêt contemporain de l'information rappelée peut prévaloir sur le droit à l'oubli.

Quarto, le caractère public de la personne visée par la publication est un autre facteur pouvant influencer la balance d'intérêts entre la liberté d'expression et le droit à la vie privée dont découle le droit à l'oubli. Une personnalité publique ne disposerait pas du « droit à être oublié » pour des faits qui relèvent de son activité publique[37]. C'est ainsi que la Cour d'appel de Liège a, en revanche, considéré que « [v]ingt ans après les faits, l'identité d'une personne *qui n'est pas une personne publique* n'apporte aucune valeur ajoutée d'intérêt général à l'article litigieux »[38].

Quinto, le type d'informations rappelées constitue un paramètre d'appréciation supplémentaire. La portée de ce critère reste encore imprécise. Dans sa conception traditionnelle, le « droit à l'oubli judiciaire » viserait le droit à l'oubli d'une personne *condamnée* judiciairement[39]. Il serait donc restrictif dans son champ d'application mais ne viserait que les demandes ayant pour but d'empêcher ou sanctionner une nouvelle mise en lumière de faits anciens. En revanche, l'exigence de condamnation ne serait pas requise dans le contexte des outils numériques, le « droit à être oublié » visant uniquement à obtenir la suppression d'informations maintenues en permanence sur internet[40]. Dans cette vision extensive du « droit à être oublié », tout

---

[35]En ce sens, Liège, 25 septembre 2014, *J.L.M.B.*, 2014, p. 1952.

[36]En ce sens, Jongen and Strowel (2017) n° 598.

[37]En ce sens, Cruysmans (2014) n° 9. Voir, également, Civ. Bruxelles, 25 mars 2014, *A & M*, 2014/5, p. 419.

[38]Liège, 25 septembre 2014, *J.L.M.B.*, 2014, p. 1952 ; c'est nous qui soulignons.

[39]En ce sens, Civ. Bruxelles, 25 mars 2014, *A & M*, 2014/5, p. 419. Voir, également, Defreyne (2013), p. 81.

[40]En ce sens, Civ. Liège, 3 novembre 2014, *J.L.M.B.*, 2014, p. 1961. Cette distinction du Tribunal de première instance de Liège se fonde sur Defreyne (2013). Le propos de cet auteur nous semble pourtant se limiter à la distinction entre les deux grandes sous-catégories du droit à l'oubli (« droit à l'oubli de données personnelles » et « droit à être oublié »), le critère du passé *judiciaire* ne trouvant pas à s'appliquer dans la première hypothèse. Or, le tribunal importe cette distinction dans une affaire relative à la republication d'un passé judiciaire pour différencier les hypothèses de redivulgation *stricto sensu* des hypothèses modernes de republication par un moyen informatique. Il semble que la Cour de cassation ait néanmoins reconnu la légalité de cette définition dans son arrêt du 29 avril 2016 (en ce sens, Cruysmans 2016b, n° 6). En effet, selon la Cour de cassation, l'arrêt de la Cour d'appel de Liège (confirmant le jugement du Tribunal de première instance de Liège du 3 novembre 2014) a décidé légalement que l'archivage en ligne de l'article litigieux

élément du passé qui aurait fait l'objet d'une médiatisation pourrait bénéficier d'une anonymisation[41]. Dans une interprétation intermédiaire du critère relatif aux informations rappelées, les faits relatés doivent être « d'ordre judiciaire »[42] mais cela signifie simplement qu'il faut avoir été « concerné » par une affaire judiciaire relatée dans les médias pour pouvoir bénéficier du droit à l'oubli. Dans cette interprétation intermédiaire, les *victimes* peuvent, par exemple, bénéficier du droit à être oublié[43]. Ne sont alors pas seuls visés les articles relatant des condamnations judiciaires, mais « tout écrit qui explique peu ou prou un élément d'une procédure judiciaire, qu'elle aboutisse ou non à une condamnation » Cruysmans (2014) n° 9[44]. Cette interprétation mérite d'être privilégiée. En effet, une information relative à une enquête criminelle suffit souvent à insinuer dans l'esprit des lecteurs un doute ou une

---

constituait une nouvelle divulgation du passé judiciaire du défendeur pouvant porter atteinte à son droit à l'oubli en jugeant que « le litige concerne 'une [...] facette du droit à l'oubli qui vise 'la possibilité pour une personne de demander l'effacement des données qui la concerne, et plus spécialement des données mises en ligne, après une période donnée', 'l'enjeu n'[étant] plus d'empêcher ou de sanctionner la mise en lumière de faits anciens mais d'obtenir la suppression d'informations disponibles sur internet' ». Nous partageons cependant l'avis d'E. Cruymans selon qui « cette précision révèle peut-être une confusion dans les types de droit à l'oubli : en effet, cette définition plus large pourrait en fait viser davantage le 'droit à l'oubli de données personnelles' que le 'droit à être oublié' » Cruysmans (2016a) note 25. Pour reprendre l'expression qu'il utilise, avec cette définition large, la Cour de cassation « ouvre le champ des possibles ». Cela étant dit, E. Defreyne n'est pas non plus exempte d'ambiguïté dans son approche puisqu'elle estime également, dans la suite de son article, que « [s]i une personne subit un préjudice en raison du rappel de son passé judiciaire dans un nouvel article, elle peut invoquer le droit à l'oubli 'traditionnel' en postulant l'octroi de dommages et intérêts. Par contre, si une personne subit un préjudice en raison d'un article écrit dans le passé mais toujours disponible en ligne, elle pourra invoquer, de manière plus adéquate, le droit à l'oubli numérique, en ce qu'il concerne précisément la problématique de la conservation des données sur internet » Defreyne (2013), p. 88. Ce point de vue est confirmé ultérieurement, l'auteure affirmant que la problématique des archives de presse numérique relève de la facette du droit à l'oubli que nous avons qualifiée « droit à l'oubli de données personnelles » Defreyne (2013), p. 95 (voir également sa conclusion). Néanmoins, E. Defreyne estime que, dans le cadre d'une action fondée sur l'article 1382 du Code civil, le juge peut apprécier la faute de l'éditeur « en s'inspirant des critères dessinés par la jurisprudence pour le droit à l'oubli du passé judiciaire, moyennant certaines adaptations », lesquelles ne viseraient cependant pas la condition qui veut que « les faits soient de nature judiciaire » Defreyne (2013), p. 95... L'auteure approuvant *in fine* la décision qui « consacre pour la première fois le droit à l'oubli en matière d'archives journalistiques, en appliquant les critères définis pour la version 'traditionnelle' du droit à l'oubli au droit à l'oubli numérique » Defreyne (2013), pp. 95 et 96 (voir également sa conclusion).

[41]C'est ainsi qu'E. Cruysmans interprète le jugement du tribunal de première instance de Liège du 3 novembre 2014 (*J.L.M.B.*, 2014, p. 1961). Prenant appui sur l'arrêt de la Cour de justice de l'Union européenne *Google Spain et Google*, E. Montero et Q. Van Enis sont également favorables à ce que le droit à l'oubli dans le contexte numérique ne soit pas limité au droit à l'oubli « strictement judiciaire » Montero and Van Enis (2016) n° 40.

[42]En ce sens, Liège, 25 septembre 2014, *J.L.M.B.*, 2014, p. 1952 ; Liège, 4 février 2016, *A & M.*, 2016/5-6, p. 462 ; *J.T.*, 2016, p. 710.

[43]En ce sens, Jongen and Strowel (2017) n° 601 ; Cruysmans (2014) n° 9.

[44]Voir également, Jongen and Strowel (2017) n° 601.

présomption de culpabilité susceptible d'entacher de façon irrémédiable l'honneur ou la réputation de la personne concernée. Le droit à être oublié nous paraît donc d'autant plus justifié qu'il n'y a pas eu de condamnation.

Sexto, l'intérêt à la resocialisation de la personne (condamnée ou non) peut également jouer un rôle dans l'appréciation des intérêts en présence.

2.3 La réparation de la violation du « droit à être oublié »

Dès lors que la mise en œuvre du droit à l'oubli se fait au travers du droit de la responsabilité, la réparation du dommage résultant de la violation du « droit à être oublié » est tout à fait réalisable. Les réparations accordées sont de différents types, notamment :

- L'anonymisation de l'article de presse litigieux figurant dans les archives numériques d'un journal[45]. En revanche, le droit de rectification numérique n'est pas admis comme une solution adéquate s'agissant de la problématique d'un article relatant une information devenue préjudiciable par l'écoulement du temps[46].
- L'obligation de solliciter de certains moteurs de recherche la désindexation du nom du demandeur[47];
- Un euro (à titre provisionnel ou définitif) au titre de la réparation du dommage moral.

Le dédommagement d'un dommage matériel n'est pas exclu, notamment lorsque le refus d'anonymisation d'un article de presse peut constituer un frein au développement de l'activité professionnelle du requérant[48]. Notons encore qu'avant le développement d'internet, des décisions de justice avaient déjà accordé ce type de dédommagements pour des publications d'articles de journaux ou de reportages télévisés rappelant des faits anciens[49].

---

[45]Civ. Liège, 3 novembre 2014, *J.L.M.B.*, 2014, p. 1961, confirmé en appel (Liège, 4 février 2016, *A & M*, 2016/5-6, p. 462 ; *J.T.*, 2016, p. 710).

[46]Liège, 4 février 2016, *A & M.*, 2016/5-6, p. 462 ; *J.T.*, 2016, p. 710.

[47]Civ. Liège, 3 novembre 2014, *J.L.M.B.*, 2014, p. 1961. Cet aspect du jugement a toutefois été réformé en appel, la Cour d'appel de Liège considérant qu'il n'appartenait pas aux éditeurs de formuler pareille demande, celle-ci devant émaner du requérant, « suivant les procédures de notification proposées en ligne par les différents moteurs de recherche, s'il souhaite, complémentairement à l'anonymisation, la suppression de toute trace de référencement » (Liège, 4 février 2016, *A & M*, 2016/5-6, p. 462 ; *J.T.*, 2016, p. 710).

[48]Civ. Liège, 3 novembre 2014, *J.L.M.B.*, 2014, p. 1961, confirmé en appel (Liège, 4 février 2016, *A & M*, 2016/5-6, p. 462 ; *J.T.*, 2016, p. 710).

[49]Pour une illustration d'une condamnation à un euro symbolique en réparation du dommage moral subi à la suite de la violation du droit à l'oubli, voir Civ. Namur, 27 septembre 1999, *A & M*, 2000/4, p. 471.

Toutefois, sans statistiques précises, il est impossible d'évaluer la mise en œuvre du droit à l'oubli. La jurisprudence publiée tend à démontrer que ce type de demandes est rare, même si l'indemnisation du préjudice subi (laquelle peut revêtir plusieurs formes) est possible. Par ailleurs, la possibilité de s'adresser directement à Google au moyen d'un formulaire en ligne est sans doute perçue comme une alternative moins chère, plus rapide et plus efficace qu'une longue et coûteuse procédure judiciaire.

2.4 L'utilisation du formulaire Google comme alternative à la réparation judiciaire

Depuis la mise en ligne du formulaire qui permet de demander à Google la suppression d'un contenu indexé dans la recherche Google au titre de la législation européenne en matière de protection des données, Google a mis à la disposition des citoyens un large éventail de statistiques et d'informations relativement précises comme le nombre de demandes introduites, le nombre d'adresses URL concernées, le nombre d'URL supprimées ou encore les catégories de demandeurs, des sites hébergeant du contenu faisant l'objet d'une demande de suppression dans les résultats de recherche ou de contenu faisant l'objet d'une demande de suppression dans les résultats de recherche. Les statistiques relatives à chacun de ces critères peuvent être affinées pays par pays.

D'après Google, les demandes introduites via leur formulaire sont évaluées « manuellement » et la décision est communiquée à la personne concernée par e-mail. Si l'URL n'a pas été supprimée, Google s'engage à leur « expliqu[er] brièvement pourquoi »[50]. Ces données, accessibles à tous et régulièrement mises à jour, nous semble être suffisamment précises et transparentes pour assurer l'information générale du public, le demandeur ayant de son côté une décision individuelle qu'il peut contester[51].

[50]Voir https://transparencyreport.google.com/eu-privacy/overview (sous le titre : « Examen des demandes »).

[51]Selon les données de Google pour la Belgique, du 25 mai 2014 au 20 décembre 2018, 19 977 demandes de suppression des résultats de recherche avaient été introduites auprès de Google, le tout concernant 78 070 adresses URL. Malgré un léger tassement, le nombre de demandes apparaît, par ailleurs, suivre une progression constante. Cela tend donc à démontrer une connaissance accrue du mécanisme. Entre le 28 mai 2014 et le 20 décembre 2018, 31 439 adresses URL auraient été supprimées pour la Belgique, soit 47,5 % (contre 44,1 % pour l'ensemble des pays).

3 L'incidence limitée de l'arrêt Google c. González

Bien que le droit à l'oubli sur internet ne soit pas consacré en tant que tel dans le droit belge (sous réserve de la loi du 8 décembre 1992 relative à la protection de la vie privée à l'égard des traitements de données à caractère personnel)[52], les juridictions belges n'ont pas attendu l'arrêt de la CJUE dans l'affaire *Google Spain et Google*, pour consacrer l'existence d'un tel droit[53].

La doctrine avait également soutenu une adaptation du droit à l'oubli à l'environnement numérique[54]. La Commission de la protection de la vie privée avait aussi rendu, dès 1997, un avis sur la diffusion des décisions juridictionnelles par le recours aux technologies de l'information et de la communication. Elle y recommandait une anonymisation des décisions de justice tout en précisant que les solutions proposées devaient tenir compte de l'équilibre entre le droit du public de savoir et le droit des personnes à la protection des données les concernant, cet équilibre pouvant « varier en fonction de différents critères objectifs liés à la nature du litige, à la juridiction en cause et aux catégories de personnes concernées »[55].

Par ailleurs, il n'est pas anodin de constater que des décisions de justice avaient déjà consacré, avant l'avènement d'internet mais dans des conditions largement similaires à ce contexte (notamment dans le cadre d'une rediffusion d'un reportage télévisé), l'existence du droit à l'anonymisation, en tant que composante du droit à l'oubli. C'est ainsi que le Tribunal de première instance de Namur avait considéré que si la rediffusion d'images relatant un événement régional dix ans après les faits n'était pas fautive dans le chef d'une chaîne de télévision locale, il ne percevait pas l'intérêt qu'il y avait à reproduire intégralement le commentaire fait à l'époque par le journaliste qui reprenait à plusieurs reprises le nom de l'auteur des faits condamnés. Le Tribunal ajoutait encore « [q]u'à supposer même que la défenderesse ait estimé devoir reproduire le commentaire d'époque, il lui était particulièrement facile techniquement, lors de la rediffusion de la séquence ancienne, de ne pas reproduire

---

[52] À ce propos, voir Defreyne (2013); Aramazani (2011).

[53] Voir, par exemple, Civ. Neufchâteau, 25 janvier 2013, *A & M*, 2013/6, p. 478. Pour un rejet du droit à l'oubli numérique dans le cadre d'une demande fondée sur la loi du 8 décembre 1992 relative à la protection de la vie privée à l'égard des traitements de données à caractère personnel (en raison de l'application de l'exception en matière de journalisme aux archives journalistiques prévue à l'article 3, paragraphe 3, de ladite loi), Civ. Bruxelles (réf.), 9 octobre 2012, *A & M*, 2013/3-4, p. 267 (confirmé en appel – le litige ayant été limité à l'application de l'article 15 de la loi du 8 juin 1992 – par Bruxelles, 21 mars 2013, *A & M*, 2014/5, p. 416) ; pour un refus en raison de l'absence de condamnation judiciaire et de divulgation antérieure, voir Civ. Bruxelles, 25 mars 2014, *A & M*, 2014/5, p. 419.

[54] Voir, notamment, Aramazani (2011); Le Clainche (2012); Defreyne (2013).

[55] Avis n°42/97, du 23 décembre 1997, sur la diffusion des décisions juridictionnelles par le recours aux technologies de l'information et de la communication (disponible sur le site internet de la Commission de la protection de la vie privée : https://www.privacycommission.be/sites/privacycommission/files/documents/avis_42_1997_0.pdf).

à l'antenne le nom du voleur »[56]. En agissant comme elle l'a fait, la chaîne de télévision a manqué de prudence et a causé au requérant un préjudice (moral) dont il peut obtenir réparation[57].

C'est donc assez naturellement que les juridictions belges ont intégré dans leur jurisprudence l'arrêt *Google Spain et Google.* C'est ainsi que les principes dégagés par la CJUE à l'occasion de ce litige (opposant un citoyen espagnol à l'exploitant d'un moteur de recherches) ont été transposés – et donc élargis – à l'archivage en ligne des articles par les éditeurs de presse.

En effet, comme nous l'avons vu précédemment, après avoir rappelé les considérations de la Cour d'appel de Liège selon lesquelles la mise en ligne d'un article a permis « une mise en 'une' de cet article via le moteur de recherche de son site consultable gratuitement, mise 'en une' par ailleurs multipliée considérablement par le développement des logiciels d'exploitation des moteurs de recherche du type Google », la Cour de cassation a estimé que l'arrêt attaqué avait décidé légalement que « l'archivage en ligne d'un article constitue une nouvelle divulgation du passé judiciaire pouvant porter atteinte au droit à l'oubli de la personne visée par l'article »[58].

4 L'incidence du RGPD en droit belge

À notre connaissance, aucune réforme nationale n'est envisagée pour renforcer ou modifier la protection du droit à l'oubli. L'application du RGPD ne devrait pas fondamentalement modifier l'approche suivie par les juridictions belges[59]. En effet, si l'article 17, paragraphe 1, du RGPD garantit désormais expressément à toute personne « le droit d'obtenir du responsable du traitement l'effacement, dans les meilleurs délais, de données à caractère personnel la concernant » notamment lorsque ces données « ne sont plus nécessaires au regard des finalités pour lesquelles elles ont été collectées ou traitées d'une autre manière », le troisième paragraphe de l'article 17 précise que ce droit au déréférencement ne s'applique pas lorsque le traitement des données en cause est justifié par l'exercice du droit à la liberté d'expression et d'information.

---

[56]Civ. Namur, 27 septembre 1999, *A & M*, 2000/4, p. 471.

[57]Voir, également, à propos de l'interdiction de diffusion d'une émission en raison du fait que celle-ci ne présentait pas d'intérêt contemporain quatre ans après les faits relatés, Civ. Bruxelles, 30 juin 1997, *J.T.*, 1997, p. 710. Pour d'autres références, voir Cruysmans (2016b) note 18.

[58]Cass., 29 avril 2016, *J.T.*, 2016, p. 609. L'arrêt de la Cour d'appel de Liège en cause est un arrêt rendu à peine quatre mois après l'arrêt de la CJUE (Liège, 25 septembre 2014, *J.L.M.B.*, 2014, p. 1952). Dans le même sens, voir aussi Liège, 4 février 2016, *A & M*, 2016/5-6, p. 462.

[59]Conformément à l'article 99, paragraphe 1, du RGPD, le règlement est entré en vigueur le vingtième jour qui a suivi sa publication au Journal officiel de l'Union européenne (c'est-à-dire le 25 mai 2016). Toutefois, aux termes de l'article 99, paragraphe 2, ce n'est que le 25 mai 2018 qu'a débuté son *application*.

Par conséquent, la recherche de l'équilibre entre le droit à la vie privée et le droit d'accès à l'information du public au moyen des critères énoncés par la CJUE dans l'arrêt *Google Spain et Google*[60] – similaires à ceux utilisés par les cours et tribunaux belges – devrait se poursuivre[61]. En outre, le refus opposé à la demande de déréférencement ou le délai mis par le responsable du traitement pour faire droit à la demande de déréférencement pourrait toujours être constitutif d'une faute au sens de l'article 1382 du Code civil belge et entraîner l'indemnisation d'un dommage moral et/ou matériel. En effet, l'article 82, paragraphe 1, RGPD prévoit désormais que « [t]oute personne ayant subi un dommage matériel ou moral du fait d'une violation du présent règlement a le droit d'obtenir du responsable du traitement ou du sous-traitant réparation du préjudice subi ». Cette possibilité est confortée par le considérant 146 du RGPD aux termes duquel « [l]e responsable du traitement ou le sous-traitant devrait réparer tout dommage qu'une personne peut subir du fait d'un traitement effectué en violation du présent règlement [et cela] sans préjudice de toute action en dommages-intérêts fondée sur une infraction à d'autres règles du droit de l'Union ou du droit d'un État membre ». Or, si la directive 95/46 disposait d'un article similaire relatif à la responsabilité du responsable du traitement[62], le droit à l'effacement n'était pas envisagé comme tel par la directive. En revanche, le règlement prévoit désormais que toute personne concernée peut, sans frais, faire effacer dans les meilleurs délais les données à caractère personnel qui se rapportent à elle « lorsque la conservation de ces données constitue une violation du présent règlement, du droit de l'Union ou du droit d'un État membre auquel le responsable du traitement est soumis »[63].

5 Conclusion

Le choix du « règlement » plutôt que de la « directive » pour régir la protection des données à caractère personnel au niveau de l'Union européenne est un pas important dans une meilleure protection des utilisateurs et du droit à l'oubli[64]. En effet, il supprime en grande partie la possibilité de divergence entre les États membres puisque le règlement « est obligatoire dans tous ses éléments et [qu']il est directement applicable dans tout État membre »[65]. L'efficacité du RGPD se mesurera, cependant, à son effectivité, laquelle dépendra de l'étendue «

[60]CJUE, arrêt du 13 mai 2014, *Google Spain et Google*, C-131/12, ECLI:EU:C:2014:317, points 97 à 99.

[61]C. de Clercq et F. Dechamps relient également l'exception de l'article 17, paragraphe 3, RGPD et les développements de l'arrêt *Google Spain et Google* (de Clercq and Dechamps 2017, p. 673).

[62]Voir article 23 de la directive 95/46.

[63]Considérant 65 du RGPD. En ce sens, de Terwangne et al. (2017), p. 313.

[64]En ce sens, de Terwangne et al. (2017), p. 302.

[65]Article 288, alinéa 2, TFUE.

géographique » des droits qu'il consacre[66] et de la reconnaissance en dehors de l'Union européenne des décisions de justice qui l'appliqueront. Ces questions ne dépendent donc pas tant d'une adaptation de la législation de l'Union que de son interprétation par la CJUE et, surtout, d'une collaboration au niveau internationale (notamment au niveau du Conseil de l'Europe).

Celle-ci apparaît indispensable car il est certain que la « demande de protection » va connaître un développement nouveau avec l'application du RGPD mais aussi (et surtout ?) la prise de conscience croissante des dangers de l'utilisation des données personnelles dans le chef des utilisateurs des réseaux sociaux[67]. Il est donc probable que le droit à l'oubli traditionnel – c'est-à-dire celui visant à faire oublier un *comportement* passé et, en principe, condamné judiciairement – se déplace vers une demande accrue du « droit à l'oubli de données personnelles ».

Dans ces circonstances, le risque d'aboutir à une application systématique du droit des données à caractère personnel en vue d'obtenir le déréférencement ou l'effacement de pages disponibles sur internet alors même que les informations qu'elles contiennent ne portent pas atteinte au respect de la vie privée ou ne sont pas diffamatoires n'est pas exclu[68]. L'équilibre entre les droits en présence passera donc encore et nécessairement par le respect de deux exigences, l'une préventive et l'autre sanctionnatrice. En amont, il est plus que jamais indispensable d'améliorer l'apprentissage de l'usage d'internet et des technologies de l'information. En aval, il convient de préserver la possibilité de

[66]Voir, à cet égard, la demande de décision préjudicielle du Conseil d'État français à la CJUE, du 21 août 2017, relative à la portée du droit au déréférencement (sur la base de la directive 95/46) (aff. C-507/17). Par ses deux premières questions, le Conseil d'État demande, en substance, si le droit au déréférencement tel qu'il a été consacré par la CJUE dans l'arrêt *Google Spain et Google* implique un déréférencement sur tous les noms de domaine exploités par le moteur de recherche concerné, de telle sorte que les liens litigieux n'apparaissent plus *quel que soit le lieu à partir duquel la recherche lancée sur le nom du demandeur est effectuée*, y compris hors du champ d'application territorial de la directive ou si le déréférencement s'applique uniquement aux liens des résultats affichés à la suite d'une recherche effectuée sur le nom de domaine correspondant à l'État où la demande est réputée avoir été effectuée ou, plus généralement, sur les noms de domaine du moteur de recherche qui correspondent aux extensions nationales de ce moteur pour l'ensemble des États membres de l'Union européenne. Dans son arrêt, rendu le 24 septembre 2019, la Cour a refusé d'étendre au-delà du territoire de l'Union l'obligation de déréférencement. Elle retient néanmoins une interprétation qui peut être qualifiée de large de cette obligation puisqu'elle l'étend à l'ensemble des États membres. En effet, selon la Cour de justice, "lorsque l'exploitant d'un moteur de recherche fait droit à une demande de déréférencement (...), il est tenu d'opérer ce déréférencement non pas sur l'ensemble des versions de son moteur, mais sur les versions de celui-ci correspondant à l'ensemble des États membres, et ce, si nécessaire, en combinaison avec des mesures qui, tout en satisfaisant aux exigences légales, permettent effectivement d'empêcher ou, à tout le moins, de sérieusement décourager les internautes effectuant une recherche sur la base du nom de la personne concernée à partir de l'un des États membres d'avoir, par la liste de résultats affichée à la suite de cette recherche, accès aux liens qui font l'objet de cette demande" (CJUE, arrêt du 24 septembre 2019, Google (Portée territoriale du déréférencement), C-507/17, ECLI:EU:C:2019:772, point 73 et dispositif).

[67]Nous songeons notamment aux révélations relatives à l'utilisation des données des utilisateurs de Facebook par Cambridge Analytica en mars 2018 (https://www.nytimes.com/2018/03/17/us/politics/cambridge-analytica-trump-campaign.html?smid=tw-share, consulté pour la dernière fois le 4 avril 2018).

[68]En ce sens, Le Clainche (2012), p. 56.

contrôler la liberté d'expression, grâce au contrôle de proportionnalité, sous la seule responsabilité d'un juge indépendant et impartial.

Principales revues

A & M	Auteurs & Médias
J.D.E.	Journal de droit européen
J.L.M.B.	Revue de jurisprudence de Liège, Mons et Bruxelles
J.T.	Journal des tribunaux
NjW	Nieuw Juridisch Weekblad
R.B.D.C.	Revue belge de droit constitutionnel
R.D.T.I.	Revue du droit des technologies de l'information
R.E.D.C.	Revue européenne du droit de la consommation
R.G.D.C.	Revue générale de droit civil belge

References

Aramazani A (2011) Le droit à l'oubli et internet. R.D.T.I. 43:34–49
Bosly H, Vandermeersch D, Beernaert M-A (2014) Droit de la procédure pénale, 7é éd. La Charte, Bruges
Cassart A, Henrotte J-Fr (2014) Arrêt Google Spain : la révélation d'un droit à l'effacement plutôt que la création d'un droit à l'oubli. J.L.M.B. 1183–1191
Cruysmans E (2013) Le traitement de données à caractère personnel effectué à des fins de journalisme : la consécration de la liberté d'expression. A & M 3–4:267
Cruysmans E (2014) Liberté d'expression, archives numériques et protection de la vie privée : la conciliation de trois réalités divergentes grâce au droit à l'oubli. J.L.M.B. 1972–1980
Cruysmans E (2016a) Le droit à l'oubli devant la Cour de cassation. J.T. 618–620
Cruysmans E (2016b) Oubliez-moi ! Droit à l'oubli, déréférencement, anonymisation et archives numériques. In: Hoc A, Wattier S, Willems G (eds) Human rights as a basis for reevaluating and reconstructing the law. Bruylant, Bruxelles, pp 403–418
Cruysmans E (2016c) Quand la Cour de cassation de Belgique se met à oublier… In: Justice en ligne. www.justice-en-ligne.be/article908.html Consulté pour la dernière fois le 20 décembre 2018
de Clercq C, Dechamps F (2017) Internet à l'épreuve du droit ou le droit à l'épreuve d'internet. J.T. 669–681
de Terwangne C, Rosier K, Losdyck B (2017) Le règlement européen relatif à la protection des données à caractère personnel : quelles nouveautés ? J.D.E. 302–316
Dechenaud D (ed) (2015) Le droit à l'oubli numérique. Données nominatives – Approche comparée. Larcier, Bruxelles
Defreyne E (2013) Le droit à l'oubli et les archives journalistiques. R.D.T.I. 51:75–98
Jongen Fr, Strowel A (avec la coll. de Cruysmans E) (2017) Droit des médias et de la communication. Larcier, Bruxelles
Le Clainche J (2012) L'adaptation du 'droit à l'oubli' au contexte numérique. R.E.D.C. 1:39–60
Montero E, Van Enis Q (2016) Les métamorphoses du droit à l'oubli sur le net. R.G.D.C. 5:243–255
Tulkens Fr, Sohier J (2015) Les cours et tribunaux. Chronique de jurisprudence constitutionnelle 2013–2014. R.B.D.C. 2:289–323
Van Eecke P, Le Boudec (2015) Recht op vergetelheid ook van toepassing op krantenarchieven. NjW 26

The Right to Be Forgotten in the Czech Republic

Jan Hurdík

Abstract The author of this chapter focused his attention on the new wave in the frame of privacy protection in the EU law: from co called "right to be forgotten" (case Google vs. Gonzáles) to General Data Protection Regulation (DCFR). Author explained especially the impact of the recent European development of the personal data protection on the Czech law including the newly prepared Czech personal Data Protection Act, the changes of doctrinal thinking and the challenges to the juridical practice in the Czech Republic.

1 The Right to Be Forgotten Under Czech Law: General Framework

The Czech law is nowadays built on an essential legislative reform the grounds of which are concentrated in the recodificated Civil Code (Act No 89/2012 Coll.). The main part of recodification, incl. the Civil Code, has been in force since January 1st, 2014. One part of the Civil Code (Sections 81–117) concerns the protection of human personality. Due to the traditional conception of the Civil Code, the protection of personal data is established in a general framework and in the "traditional" parts of the protection of human personality (protection of life, health, corporal and mental integrity of human beings, face, privacy etc.). A special protection of personal and sensitive data is not expressly incorporated into the text of the Civil Code. Thus, personal data can be protected only indirectly, through the broader concept of privacy protection (Sections 86–90, Civil Code). The framework of protection of human privacy is formulated only in Section 90, Civil Code: "*Legal grounds for interference with the privacy of another or for the use of his/her image, documents of personal nature or audio or video recordings may not be used unreasonably and in conflict with the legitimate interests of individuals.*" The term "privacy" is not defined in the text of

J. Hurdík (✉)
Faculty of Law, Masaryk University, Brno, Czech Republic
e-mail: Jan.Hurdik@law.muni.cz

F. Werro (ed.), *The Right To Be Forgotten*, Ius Comparatum – Global Studies in Comparative Law 40, https://doi.org/10.1007/978-3-030-33512-0_3

the Civil Code. Even if privacy seems to be rather an unclear concept in the Czech law and "nobody can articulate what it means,"[1] the Czech law apparently inclines to a broader concept of privacy, including (but not being limited to) the right to protection of confidentiality of communication.[2] It generally means that this right also includes privacy on the Internet[3] but for now without a greater support of the Czech courts in their practice. At present, there are no significant legal initiatives the aim of which would be protection of privacy on the Internet. The discussion about the freedom of information on the Internet and its limits has only been developing in the Czech Republic in the past 10 years.[4]

More specific rules about responsibility for the content of the data on the Internet were introduced into the Czech law by the Act No 480/2004 Coll., on Some Services of Information Society, regulating the activities relating to the use of certain personal data on the Internet. This Act (implementing the Directive of the European Parliament and the Council 2000/31/EC of 8 June 2000 and the Directive of the European Parliament and the Council 2002/58/EC of 12 July 2002) only expressly establishes the responsibility of the provider of services to include the information given by the user (Section 5).

The Section 3 of the Act establishes the responsibility of the provider of services, which are based either on the data transfer or on the searching of the data on the Internet. This kind of responsibility arises only in cases when the provider: (a) initiates the data transfer; (b) chooses the user of the transfered information, or (c) chooses or changes the content of transfered information.

Under Section 5.1 of the Act, the operator of a website is responsible for processing the personal data which appear on the website, even if they are published from other sources. If the operator does not react or if he has rejected the applicant's request, the applicant is entitled to take a legal action against the website operator.

A special legal regulation of protection of personal and sensitive data (but without specifically focusing on the data on the Internet) is included in Act No 101/2000 Coll., on Protection of Personal Data. This act is in accordance with the Directive of the European Parliament and Council 95/46/EC of 24 February 1995 and regulates the general framework of collecting and processing of personal and sensitive data. The applicability of this act for the purposes of "the right to be forgotten" on the Internet is limited: This act excludes its own applicability on a haphazard (accidental) collection of personal data unless these data are subsequently processed (Sections 3 al. 4). The duty of the administrator or the processor to delete the personal data arises at the moment when the aim of the data collecting or processing

[1]The citation has been used as the hyperbole. *See* Solove (2006); see also Melzer F, Tégl P, a kol. (2013) Občanský zákoník - velký komentář, svazek I. § 1–117. Praha, Leges, p. 550.

[2]See for example the Decision of the Constitutional Court of Czech Republic from 22 March 2011, Pl. ÚS 24/10.

[3]*See also* Melzer F, Tégl P, a kol. (2013) Občanský zákoník - velký komentář, svazek I. § 1–117. Praha, Leges, p. 555.

[4]See for example Polčák (2012), p. 95.

disappears, or at the request of the person subject to the data collection (Section 20). (See also the text above relating to the website operator's responsibility for its content.)

2 Czech Law and the Limits to Right to Be Forgotten

The general limits to protection of human personality are formulated in the Civil Code as follows:

(a) the consent of the person whose privacy has been interfered with. See Section 86, Civil Code: *"Without an individual's consent, it is prohibited in particular to intrude on his/her private premises, to watch or to take audio or video recordings of his/her private life, to use such or other recordings made by a third person about the private life of an individual, or distribute such recordings about his/her private life. Private documents of personal nature are protected to the same extent."* and
(b) the so-called legal reasons (licences) to affect the privacy of persons. The situations in question are as follows: (a) if an image, or an audio or video recording, is made or used to exercise or protect other rights or legally protected interests of others; (b) where an image, or a document of a personal nature, or an audio or video recording, are made or used for official purposes based on the law, or (c) where someone performs a public act in matters of public policy. (d) where something is used for scientific or artistic purposes and for the press, radio, television or similar coverage (see Sections 88–89 Civil Code).

The are also limits to the extent of the legal reasons for an interference with privacy, based on the defense of right abuse, see Section 90 Civil Code: *"Lawful reasons for an interference with the privacy of another or for the use of his/her image, documents of a personal nature or audio or video recordings shall not be used unreasonably and in conflict with the legitimate interests of individuals"*.

In the Czech law, there are also specially formulated limits to the right to be forgotten. They can be found in the text of the Act No 101/2000 Coll., on Protection of Personal Data (see above). Section 20 al. 2 mentions special acts establishing exceptions for the archival purposes and for the purposes of judicial or administrative proceedings.

In accordance with the EU judicial practice, the limits of the extent of the right to be forgotten have to be applied in relation to the official registers where the public interest prevails over the protection of personality.

To sum up: The right to be forgotten is not considered absolute (but only is this sense). It may only be applied under the following conditions: (a) the personal data are no longer needed for the purpose for which they were collected or processed, (b) if the consent has been withdrawn, (c) some other legal reason arises for erasing the personal data, as, for example, in the case of a legal regulation on an archive, the duty to save the data for the exercising one's rights in civil, criminal or administrative proceedings, etc.

3 The Czech Legal Remedies Available to Enforce the Right to Be Forgotten

The Czech law recognizes a general way to protect "the right to be forgotten" in the form of a remedy which is part of private law. This essential instrument of legal protection is to be used through a civil action. The aggrieved natural person can choose, as the aims of the proceedings while having to formulate it in the action, between *restitutio in integrum* and/or *actio negatoria* (see Section 82 al. 1 Czech Civil Code: *(1) "An individual whose personality rights have been affected has the right to claim that the unlawful interference be refrained from or its consequence remedied".).* After the death of the aggrieved (natural) person, the protection of his/her personality rights may be sought by any of his/her close persons (see Section 82 al. 2, Civil Code). ("A *close person*" means, under Section 22 al. 1, Civil Code, *"a relative in the direct line, sibling or spouse, or a partner under another statute governing the registered partnership (hereinafter a "partner"); other persons in the familial or similar relationship shall, with regard to each other, be considered to be close persons if the harm suffered by one of them is perceived as his/her own harm by the others. Persons related by affinity and persons permanently living together are also considered to be close persons.*"

The specific legal steps, aimed at the protection of the rights of the person subject to data collection, are included in Section 21, the Act 101/2000 Coll., on Personal Data Protection. Any person subject to data collection, who finds out or supposes that the data administrator or processor is dealing with his/her data in contravention to the protection his/her personal or private life, or contrary to the law, is entitled:

(a) to demand an explanation from the data administrator or the processor;
(b) to demand a rectification of the situation from the data administrator or the processor; he/she is entitled especially to require blocking or correcting the data, filling the missing data, or deleting the data.

If the data administrator or the processor does not meet the demands, the person subject to collection is entitled to turn directly to the national authority on personal data protection—The Bureau for the Personal Data Protection (its legal regulation is contained in Section 28 ff. Act No 101/2000 Coll., on Personal Data Protection. If the applicant is rejected by the Bureau for the Personal Data Protection, he/she is entitled to request an inquiry made by the President of the Bureau (§ 152 Act 500/2004 *Coll*). If the inquiry is unsuccessful, the applicant can request a judicial inquiry made by the Administrative Court, and subsequently to file an appeal complaint with the Supreme Administrative Court, and after that a constitutional complaint with the Constitutional Court.

The second option how the affected person may assert his/her right to be forgotten is to make use of the general personality protection, also mentioned above. The affected person can, pursuant to Section 82 al. 1 Czech Civil Code (2012), demand from the website operator/keeper to remove his/her personal data from the website. The liability of the search engine operator has not been established yet in the application of this regulation by the Czech courts (see also below). The current Czech legal practice and the principal juridical opinion draw on the presumption that it is the operator of the website who is responsible for the content of the website (see also Section 5 al. 1 Legal Act No 480/2004 Coll., on Some Services of Information Society). From that follows that legal actions are brought against the website operators.

The development of the responsibility of the website operators for the content of the websites may also be seen in the earlier Czech case law. The decision of the Czech Supreme Court No 23 C 70/2003 declared the primary responsibility of the website operators for the content of their websites.[5]

4 The Possibility to Receive Material or Immaterial Damages: Legal Regulation and Reality in Practice

In case when an injury arises as a consequence of interference with the protection of personal data, the injured person is entitled to a remedy for both material and immaterial injuries. The legal regulation of this instrument has the legal support among the special acts only in Sections 25–26 Act No 101/2000 Coll. (see above). These rules enable to claim compensation for an injury arisen, e.g. due to an infringement of the right to be forgotten, but these rules refer to a general regulation, i.e. to the rules of the Civil Code. Because of the position of the right to be forgotten as one of the fundamental rights, the compensation for an injury arisen due to its infringement has a special and more favorable position among the various constructions of responsibility for injury. The right to be forgotten is considered an absolute right. The concept of the "absolute" nature of the right means that this right may be asserted against anyone, or that this right belongs only to its holder (compare the absolute nature of this right, which means that this right is not without limits—see in the question 2). Consequently, *"(A) tortfeasor who is at fault for breaching a statutory duty, thereby interfering with an absolute right of the victim, shall provide compensation to the victim for the harm caused."* (Section 2909, Civil Code).

In the case of interference with the right to be forgotten, the injured person is entitled to demand compensation for both material damage and immaterial harm (see Sections 2956 – 2957, Civil Code: *"Where a tortfeasor incurs a duty to compensate*

[5]See the website of the Czech Supreme Court: http://www.nsoud.cz/: *"Responsibility for the content of Internet websites is borne primarily by the person whose expression of will is at issue. This person used to be described in the professional literature as the content provider (also the operator of the websites or the keeper of the websites). The content provider is a specific person who hasn created or have to create the content (for example the websites or another data). The provider located this content (the website or another data) on the storage space of the disk reserved for this purpose by the provider of free space. The legal person has its own expression of will, too, and bears also the responsibility for the content of its websites. The entity providing hyperlinks to its own websites or links to other documents accesible on the Internet is called the hyperlink supplier."*

an individual for harm to his natural right protected by the provisions of the first part of this Act, he shall compensate the individual for the damage as well as for non-pecuniary harm thus caused; the compensation for the non-pecuniary harm shall also include mental suffering. The manner and amount of adequate satisfaction must be determined so as to also compensate for the circumstances deserving special consideration. These circumstances shall mean causing intentional harm, including, without limitation, harm caused by trickery, threat, abuse of the victim's dependence on the tortfeasor, multiplying the effects of the interference by making it publicly known or as a result of discrimination against the victim with regard to the victim's sex, health condition, ethnicity, creed, or other similar serious reasons. The victim's concerns about losing their life or about serious damage to their health are also taken into account if such concerns are caused by a threat or otherwise."

The application of the rules enabling to demand also non-pecuniary compensation for harm depended in the past on the courts' decisions (of the Czech Supreme Court or the Constitutional Court)[6] whether the right to be forgotten was among "*the natural rights protected by the provisions of the first part of Civil Code*". Actually, this question is dealt with in Art. 1, Regulation of the European Parliament and Council 2016/679 of 27 April 2016, which declares the protection of natural persons when processing personal data to be an essential right.

The manner and extent of the compensation are regulated in Sections 2951–2952, Civil Code, which make distinction, in pecuniary harm, between "restitutio in integrum" and monetary compensation, and in non-pecuniary harm, between non-monetary or monetary satisfaction.[7]

5 The Effectivity and Practical Usefulness of the Implementation of the Right to Be Forgotten in Czech Law

As also mentioned above, the implementation of the right to be forgotten in the Czech law existed in the past, too, before the decision in *Google v. Gonzáles* (based generaly on the Civil Code and especially on the special legal regulation of

[6]The development of the application of non-pecuniary harm besides the pecuniary harm is aparent, among othečrs, in the decision of the Constitutional Court No Pl. ÚS 16/04 which declared the need of the application of not only pecuniary but also non-pecuniary harm in the case of personality protection. For this, see also the decision of the Supreme Court No 30 Cdo 1315/2008, which sumarized the conditions establishing the duty to cover the non-pecuniary harm.

[7]See the text of Sections 2951–2952, Civil Code: "*Damage is compensated for by the restoration to the original state. If this is not reasonably possible, or if so requested by the victim, damage is payable in money. Non-pecuniary harm is compensated for by an appropriate satisfaction. The satisfaction must be provided in money, unless a real and sufficiently effective satisfaction for the harm incurred can be provided otherwise. The actual damage and what the victim has lost (lost profit) is paid for. If the damage results in a debt, the victim has the right to be released from the debt or provided with compensation by the tortfeasor....*"

protection of personal and sensitive data which is included in Act No 101/2000 Coll., on Protection of Personal Data. The extent of this kind of protection was, nevertheless, limited: its application was not extended in practice to the responsibility of search engine operators.

But this protection has been significantly influenced and enlarged, in the past 2–3 years, by the decision in *Google v. Gonzáles,* and also by the need to implement the Regulation of the European Parliament and Council 2016/679 of 27 April 2016.

The novelty of the decision in *Google v. Gonzáles* is for the Czech law mainly in the responsibility of the internet search engine operators: *"The Court stated that an internet search engine operator is responsible for processing the personal data which appear on the web pages published by other sources. Operating a search engine is a different activity to that of publishing content on a website, and search results can undermine a person's right to privacy. Thus, the operator acts as a processor of personal data and must comply with legislation that protects individuals in this regard (Directive 95/46/EC).*

The Court ruled that the search engine operator could, in some circumstances, be obliged to remove links to certain web pages from the list of results that appear when a search is conducted for a particular name."

The decision in *Google v. Gonzáles* is not the only which affects the right to be forgotten. This right has also been affected a new European Regulation: Regulation of the European Parliament and Council 2016/679 of 27 April 2016—General Data Protection Regulation (GDPR).

This Regulation came in force on 25th May 2018 in all member states of the EU and changed fundamentally the procedure of collecting and using personal data.

In the Czech Republic, the Regulation brought about a heavy burden for the State and local administration as well as for business corporations.

According to the Union of Industry and Transport of the Czech Republic (hereinafter referred to as UITCR), the majority (cca 60%) of all Czech enterprises could get into serious difficulties, which is due to the short deadline to recognize and make preparations for the implementation of the new European Regulation in practice. It concerns not only the mapping and integration of data protection into business operations, but also of the right "to be forgotten." A significant number of commercial entities did not know, just by the date of inforce coming of GDPR about the forthcoming rules. The deadline for to putting the new rules into practice (1.5 year) seems to be too short in the light of this information. In the opinion of the Union of Industry and Transport of the Czech Republic (UITCR), the new rules are a significant burden for the industry and trade of the Czech Republic, in terms of both organization and finances.

Also, the level of the knowledge of some aspects (details) of the GDPR, just before it came into force, was very low. For example, only each 10th business/ enterpreneur in the Czech Republic knows about the possible amount of penalties. The amount of all the penalties is estimated at 530 milions of Czech crowns (CZK). The majority of the businesses fear mainly the danger that their employees could cause a loss of the protected personal data. Some commentators estimate that such a heavy burden of penalties could liquidate a large number of businesses.

The range of big challenges for the Czech businesses, as far as the implementation of GDPR is concerned, seems to be as follows:

(1) Lack of control mechanisms in the case of breach of personal data protection
(2) Uncertainty as for identifying the person responsible for a problem arisen
(3) Lack of sufficient protection of the IT and personal data
(4) A limited operation of the existing IT systems
(5) Limited sources to improve the existing practices
(6) Lack of the existing formal processes enabling the identification of the position and ownership of the data
(7) Lack of financial means

The newly established position of the data protection authority (DPA) is also considered to be a heavy burden because of additional costs bound with the necessity of its embodiment into the structure of the companies or other establishments.[8]

According to the information of the Union of Industry and Transport of the Czech Republic (UITCR) the new European Regulation is bringing both administrative and financial burden to the Czech businesses. Despite of the businesses having less time for the implementation of the changes, 60% of them do not have the elementary information about the content of the Regulation.

The new European rules on the personal data protection also include and strengthen the right to be forgotten in practical activities of businesses. Not only enterprises but also other persons sharing their personal data with them have to delete the data of the affected persons.

It is not only the banks or insurance companies that are forced make changes: they also concern technological and manufacturing companies, if these are able identify the customers, monitor their behaviour or monitor the behaviour of their employees, for example during the manufacturing processes.

The new concept of the right to be forgotten affect also small businesses, mainly those producing various applications. What is seen as the main problem in the current shortage of labor, mainly in the branch of information technologies, is, according to the Czech business, the lack of professionals who could join the existing staff in companies who are in charge the data protection.

Despite the difficulties mentioned above, the approach of the Czech government and the state administration to the implementation of the changes concerning data protection, including the right to be forgotten, is appreciated by the UITCR as an example of a good practice. Their steps when implementing the Regulation are carefully consulted with all participants including entrepreneurs under the auspices of the digital coordinator. The UITCR takes part in these activities organizing seminars on the forthcoming changes.[9,10]

[8]Source: Ipsos (the leading agency for the market and opinion research in the Czech Republic; www.ipsos.cz.

[9]Source: Profimedia.cz.

[10]The EU commissioner from the Czech Republic, Věra Jourová, significantly contributed to the approval of GDPR by the EU member countries.

6 Right to Be Forgotten in the Czech Juridical Practice and Commentaries

As it is apparent from the published articles, not numerous ones, and short commentaries, the first wave of them was focused mostly on the interpretation of the content and the practical effects of the *Google v. Gonzáles* on personal data protection in the Czech Republic. There has been a lack of professional articles or commentaries or books on this subject-matter so far. The reasons seems to be, among others, the still ongoing all-embracing reform of the entire Czech legal order having the following main goals: (a) to complete the europeanization of the Czech legal order started in 2004; and (b) to overcome the remnants of the pre-1989 Communist era. In their current situation, it is almost impossible for the Czech lawyers to cope fully with the incoming challenges.

The application of the decision in Google v. Gonzáles by the courts is a matter for the future: the period elapsed from May 25th 2014, was too short for some applicants to reach a court decision. In addition, the author of this text has not found any relevant statistical data about the claims filed.

In the past, the Czech courts dealt with the data protection on the Internet only rarely. Their decisions were based on the legal regulation mentioned in the previous questions (see the explanation to the question 1). Also, these cases only concerned the "traditional" means of information such as the printed mass media.[11]

There are actually no decisions of the Czech Supreme Court, the Czech Supreme Administrative Court and/or of the Czech Constitutional Court, concerning the right to be forgotten in the narower sense of the term (after the decision in *Google v. Gonzáles).*[12]

7 The Actual Solution of Google and the Czech Approach

In the Google activities there seems to be a significant development towards improvements, and—according to the decision in *Google v. Gonzáles*—to find a better balance between the two groups of fundamental rights. Thus, the Google addresses in particular the questions of finding a path to the best solution.

[11]See the decisions of the Supreme Court No 28 Cdo 9312/2002 and Cdo 2162/2002 which declared the responsibility of both persons: the owner of the mass-media and the provider of information.

[12]See the sources: http://www.nsoud.cz/judikaturans_new/; http://www.nssoud.cz/main0Col.aspx?cls=JudikaturaSimpleSearch&pageSource=0&menu=188; http://nalus.usoud.cz/Search/Search.aspx).

Nevertheless, some commentators draw attention to the possible dangerous impact of any legal regulation narrowing the scope for resolving the conflict between general legal principles or fundamental rights.

In the Czech Republic, however, more attention is paid to the new European legal regulation (GDPR) or the new Czech legal regulation (the draft of the new personal data protection act) than to the impact of the decision in *Google v. Gonzáles* on the Czech society.

In the Czech Republic, 3 years after the decision in *Google v. Gonzáles,* more than 25,000 requests have been submitted, concerning about 10,000 web sites. The applicants were successful in 51% of the total number of the requests (the data comes from the Google on-line information). This number of successful requests is above average: the European statistics indicate that in the whole of Europe only 43% of requests have been successful.

If the operator of a search engine rejects an applicant's request, the applicant is entitled to address the Bureau for the Personal Data Protection (its legal regulation is contained in Sections 28 ff., Act No 101/2000 Coll., on the Personal Data Protection), which is the Czech national authority on the personal data protection. If the applicant's request has been dismissed by the Bureau for the Personal Data Protection, the applicant is entitled to request the President of the Bureau for an inquiry (§ 152 Act 500/2004 *Coll*).

The Bureau has recently announced it monthly examines several tens of questions and requests; only 2 requests turned out to be really relevant: (a) In the first case the applicant was complaining that the Google in its electronic form also required a copy of the applicant's identity card in order to identify him. This request was successful: The requirement of the Google is newly modified so that the applicants' personal data is better protected. (b) In the second case, the applicant demanded the deleting of the link with an article in the Czech tabloid Blesk (both in the electronic and printed forms) magazine Blesk which stated that the applicant had committed a crime many years previously, and that subsequently he had been convicted. The Google dismissed the applicant's request so he turned to the Bureau for the Personal Data Protection. The Bureau asked the Google about its standards used in similar cases. According to the response, these standards include factors such as seriousness of crime, the offender's age at the time when he/she commited the crime, whether the crime was commited long time ago, etc. The Bureau found this answer satisfying and rejected the applicant's request.

The Czech Republic is among the few countries in the world where the Google does not have an overwhelming majority in the field of search engines. It is due to the search engine called Seznam. The operator of the Seznam has received, during the period after the decision in *Google v. Gonzáles*, no more than a few tens of individual requests for the deletion, granting cca 70% of them. In the practice of the search engine Seznam the sensible information is left on the website, too, but it cannot not to be searched by name (personal data). The Seznam also agreed with the opinion expressed by the Google that the European law obliges only the EU member countries and "*. . . there is no reason for applying these rules outside the European Union*", as declared by the spokesperson of the Seznam, Irena Zatloukalová.[13]

It seems that the whole process of "erasing the memory" now continues without problems. Nevertheless, a lot of unanswered questions remain, giving rise to a discussion about both the general approach to this right and the specific form of the impact of this right on the real life in the Czech Republic and also in other countries.

8 Upcoming Legal Reform in Czech Law to Reinforce the Right to Be Forgotten and Next Steps

On the one hand, the ECJ decision itself (*Google vs. Gonzáles*) did not provoke a need for a legal reform in the Czech Republic.

On the other hand, it is certain that the reform of the Czech legal order, mainly concerning the harmonization of the entire Czech legal order with the changes of the European law in the field of personal data protection, is considered necessary.

The government of the Czech Republic found the legal situation in the country incompatible with the EU law, esp. with the Regulation of the European Parliament and Council (EU) 2016/679 of April, 27th 2016 (GDPR) and the Directive of the European Parliament and Council (EU) of April, 27th 2016, and prepared a draft of a completely new act on processing personal data. The draft includes the following: the general provisions (Chapter I); the special provisions on personal data protection, based on the directly applicable legal act of the EU (GDPR); personal data protection during their processing in preventing, ínvestiging and prosecuting crimes, and in securing the execution of punishments and protective treatment measures, an in securing the safety of the Czech Republic, the public order and internal safety (Chapter III); and personal data protection in securing the defense of the Czech Republic.

The draft has been an object of a hot debate because of many additional duties and costs incurred by many natural and legal persons, municipalities, etc. Nevertheless, the passing of the draft by the Parliament of Czech Republic is inevitable due to the duty to harmonize the Czech legal order with the EU law. Unfortunately, the legislative procedure has been still not completed.

The future development of the right to be forgotten (at least in the Czech Republic, in my view) can be foreseen as follows: in part as a prolongation of the development in the past 2–3 years, and in part as a result of the public debate of this topic among lawyers and politicians.

[13]See: http://byznys.lidovky.cz/kdo-ma-na-googlu-pravo-byt-zapomenut-cesi-uz-podali-25-tisic-zadosti-1dc-/firmy-trhy.aspx?c=A170724_105901_firmy-trhy_onv, listed 02. 09. 2017.

Even though the discussion on this topic is still ongoing and there are many critical voices, the development in the Czech Republic proceeds to the acceptance of the right to be forgotten, many particular and additional questions have arisen and have to be solved.

In accordance with the development in other European countries, the debate is focused mainly on the possible problems with the practical application of the decision in *Google v. Gonzáles*. There is a discussion about the possible enforcement of the new rules by the European Union not only against the "European" domains like Google.cz but also against foreign domains like the American Google.com. There is also another question: with what accuracy the decision in Google v. Gonzáles is capable of being interpreted and applied, especially in hard cases, if the court must gently balance the individual right to privacy on the one hand, and on the other hand, the interest of the public in accessing information, or the right to be informed. In the Czech Republic, this is the case of "hot" data such as the information about the membership of various persons in the former Communist Party of Czechoslovakia or the activities of various persons connected with the former (Communist) secret service. Historically conditioned data, in particular, are related to the question of how long it will take before their potential is exhausted.

In the Czech Republic, the position of the Americans is also take into account as they perceive the right to be forgotten as part of the efforts of some EU politicians to subordinate the American technology companies under the extreme European regulation to prevent their development (see also the statement of European Commissioner for the Digital Economy and Society G. Oettinger. The Czech commentators mean mainly that this endeavour cannot be successful. It seems to be the dangerous tendency to the reciprocal reaction of some countries (as USA, China etc.).

One question arises from the conflict between the development of the technologies and their subsequent legal regulation and control. There is a remarkable succession and reversibility, as the technological development gives rise to the legal regulation in the sense of action-response: also, the new legal regulation often puts pressure on the subsequent technological developments. For example: Are the current technology systems capable of identifying without doubt which data are of a personal nature? The same word can mean both the name of a person and the designation of a certain thing.

One may think that the trouble with the right to be forgotten is the conflict between the nature of the information society technologies on the one hand and the extensive development of fundamental rights on other hand. Even if the European legislature and judiciary react to the technical development and possible ways of using the Internet and regulate possible negative impacts on individuals' personal data, their abuse may still occur. Thus, even if the citizens are entitled to exercise the right to be forgotten against all the operators of the web search engines working on the territory of European Union, it is not possible to have an absolute or prevailing judicial control in that respect. The development of technologies will still be ahead of the development of the legal and judicial protection. What is the implication? Both the legislation and the judiciary in the field of the right to be forgotten will be forced to develop continuously and dynamically in the future.

References

Decision of the Constitutional Court of Czech Republic from 22 March 2011, Pl. ÚS 24/10
Melzer F, Tégl P, a kol. (2013) Občanský zákoník - velký komentář, svazek I. § 1–117. Praha, Leges, p 550, 555
Polčák R (2012) Internet a proměny práva. Auditorium, Praha, p 95
Solove DA (2006) Taxonomy of privacy. Univ Pa Law Rev 154(3):477

The Right to Be Forgotten in Denmark

Hanne Marie Motzfeldt and Ayo Næsborg-Andersen

Abstract The General Data Protection Regulation (hereinafter the GDPR) and the Danish Data Protection Act (hereinafter the DDPA) has been effective since 25 May 2018 (Regulation (EU) 2016/679 of the European Parliament and of the Council of 27 April 2016 on the protection of natural persons with regards to the processing of personal data and on the free movement of such data, and repealing Directive 95/46/EC (General Data Protection Regulation), see: http://eur-lex.europa.eu/legal-content/EN/TXT/HTML/?uri=CELEX:32016R0679&from=ENG. The Danish Data Projection Act is available in English at https://www.datatilsynet.dk/media/6894/danish-data-protection-act.pdf.). However, the interpretation of the regulation still raises questions in Denmark. Thus, in this article, predictions of future case law are mainly based on the work of the Danish Ministry of Justice on adapting Danish law to the GDPR. This work resulted, among others, in a white paper encompassing more than 1000 pages, published on 24 May 2017 (White paper no. 1565, GDPR – and the legal framework of Danish legislation, available at: http://justitsministeriet.dk/nyt-og-presse/pressemeddelelser/2017/nye-regler-styrker-beskyttelsen-af-persondata-i-europa.). In Denmark, case law traditionally stays close to the evaluations and considerations set forth in such white papers and other preparatory works. The predictions must therefore be considered to be founded on a realistic and sound basis.

It should be noted that, in accordance with the systematics of the GDPR, the right to be forgotten is in the following regarded as a right related to information which is *correct*, and otherwise handled legally. In connection to this distinction, reference is made to chapter II of the GDPR on the rights of the data subjects, section 3. This states that article 16 regulates the correction of inaccurate personal data, while article 17 establishes a right to erase correct personal data—and in the header's brackets, reference is made to the "right to be forgotten".

H. M. Motzfeldt (✉)
Faculty of Law, University of Copenhagen, Copenhagen, Denmark
e-mail: hanne.marie.motzfeldt@jur.ku.dk

A. Næsborg-Andersen
Department of Law, University of Southern Denmark, Odense, Denmark
e-mail: ayo@sam.sdu.dk

F. Werro (ed.), *The Right To Be Forgotten*, Ius Comparatum – Global Studies in Comparative Law 40, https://doi.org/10.1007/978-3-030-33512-0_4

1 How Is the Right to Be Forgotten Protected Under Your Law? Does Your Law Specifically Grant a Right to Be Forgotten or Does This Right Derive from a More General Framework?

In general, Danish legal scholars consider it misleading that the European Court of Justice's prejudicial ruling in case C-131/12 (Google Spain and Google) in the eye of the public is said to include "a right to be forgotten".[1] According to Denmark's most prominent researcher in the field of data protection, Peter Blume, this ruling only establishes "a right to be remembered with greater difficulty".[2]

> This is supported in particular by the European Court of Justice holding that search engines, according to the circumstances, can be obliged to erase the link to contents of older origin on websites etc. From a Danish point of view, the ruling does not indicate an obligation to erase or edit the underlying content, which the deleted link lead to. The personal data can still be found by using keywords other than the said person's name or a search directly on the web page, etc, where the information is stored. In Denmark, Google Spain's "right to be forgotten" is therefore considered as a right for the data subject to become less searchable, i.e. less exposed to the public.

The Danish Act on Processing of Personal Data (hereinafter the PPD) was effective from 2000 and until 25 May 2018, and was then replaced by the Data Protection Act (hereinafter the DDPA).[3] The PPD Act was an implementation of the Data Protection Directive.[4] Until 25 May 2018, Denmark did not provide express statutory provisions on the right of data subjects to be forgotten, even if this right is defined as a right to reduce the searchability of data relating to at person (the data subject). The PPD Act and the Data Protection Directive contained, however, a number of provisions which corresponded to the provisions applied by the European Court of Justice in Google Spain. These provisions were used to grant data subjects the right to demand a reduction in the searchability of personal data pertaining to him/her since the adoption of the PPD Act in 2000—without, however, describing it as "a right to be forgotten".

These provisions were article 35 and 37 of the Danish PPD Act and they were supplemented with article 5 of the Act. Article 35 of the Danish PPD Act was an implementation of litra (a), point 1, paragraph 1 of article 14 of the Data Protection Directive, cf. the Directive's preamble, no. 45. Article 37 of the Act was an

[1]Hereinafter referred to as Google Spain.

[2]Blume (2014), p. 74.

[3]Act No. 429 of 31 May 2000 as amended by section 7 of Act No. 280 of 25 April 2001, section 6 of Act No. 552 of 24 June 2005, section 2 of Act No. 519 of 6 June 2007, section 1 of Act No. 188 of 18 March 2009, section 2 of Act No. 503 of 12 June 2009, section 2 of Act No. 422 of 10 May 2011, section 1 of Act No. 1245 of 18 December 2012 and section 1 of Act No. 639 of 12 June 2013.

[4]Directive 95/46/EC of the European Parliament and of the Council of 24 October 1995 on the protection of individuals with regards to the processing of personal data and on the free movement of such data, available at: http://eur-lex.europa.eu/LexUriServ/LexUriServ.do?uri=CELEX:31995L0046:EN:HTML.

implementation of litra (b) of article 12 of the Data Protection Directive. Article 5 of the PPD Act contained the principles related to personal data processing, as reflected in article 6 of the Data Protection Directive (and after 25 May 2018 article 5 of the GDPR).

Article 35 of the Danish PPD Act provided the data subject with the right to object if personal data related to him or her were used for data processing. An objection was to be handed in to the data controller. If the objection was found to be justified, the data *processing* could no longer continue. Further, article 37 of the Act provided that, at the request of a data subject, the controller had to correct, erase or block personal data concerning him or her, if the information proved to be false or misleading or in any similar way was treated in violation of the law. Article 5 of the PPD Act contained the general principles relating to personal data processing.

The present (as well as former) legislation in Denmark states i.a.

- *that* the data being processed must be adequate, relevant and limited to what is necessary in relation to the purposes for which they are processed,
- *that* processing of data must be organised in such a way that the data is updated,
- *that* the necessary controls and checks are carried out to ensure that no false or misleading data is processed,
- *that* any data which appears incorrect or misleading are corrected or deleted as soon as possible, and finally
- *that* data are not stored in a way which allows the person to be identified for a longer period than is necessary for the purposes for which the data is processed.

Since article 5 of the General Data Protection Regulation and article 5 of the present Data Protection Act entered into force, appropriate security of the personal data and the principle of accountability has been added.

Back in year 2000, the Danish Data Protection Agency (hereinafter the DPA) clarified how the interaction between these provisions should be interpreted and applied in Denmark, -14 years before the European Court of Justice's preliminary ruling in the Google Spain case. The legal position remains the same after the GPDR and the DDPA. Thus, the former official guidance on data subjects' rights is still relevant (hereinafter the rights guide).[5] The rights guide stated i.a. that:

> It follows from subsection 1 of article 35 of the Act that the data subject may object at any time to the controller against the processing of data about the person concerned.
>
> In cases where the controller finds that an objection to the processing of information is justified, it follows from the provision in section 2 of article 35 of the Act that the processing may no longer take place as far as data about the data subject is concerned.
>
> ...
>
> An objection against processing of personal data will of course be justified if processing is unlawful, i.e. takes place against the regulations outlined in the PPD Act or other legislation.
>
> An objection may however also be considered eligible, even if processing is otherwise lawful. This will be the case if the data subject has presented compelling reasons in support of the request that processing should not take place due to the data subject's special

[5]Guide no. 126 of 10/07/2000 on data subjects' rights under the rules in chapters 8-10 of the Act on processing of personal data, see: https://www.retsinformation.dk/Forms/R0710.aspx?id=852.

individual situation. The controller must therefore carry out an assessment of the existence of such special circumstances surrounding the data subject and decide whether the processing of data concerning the said person should be restricted or completely cancelled. This could e.g. be the case for an employee of an authority or company who, due to harassment from a former spouse, does not want his name made public in a staff listing on a website on the internet.

An objection shall i.a. not be upheld if the data processing about the data subject is prescribed by law. Also, as a general rule, in the case of processing for statistical or scientific purposes, an objection should not be accepted.

...

According to subsection 1 of article 37 of the Act, the controller shall correct, erase or block information, which is false, or misleading, or otherwise treated in violation of law or provisions issued by law, if a registered person makes a request for this.

...

Whether data, which is found to be false or misleading or in any similar way treated in violation of the law, must be corrected, erased or blocked is determined by the controller based on the specific circumstances. In some cases, however, it may be stated in the legislation that a specific method of correction should be applied. It is assumed that such legislation precedes the provision in subsection 1 of article 37 of the Act, and there is thus no obligation to correct, erase or block data pursuant to this provision in cases where otherwise – for particular reasons – provided for in the legislation.

A case law example illustrating the interaction between the above-described provisions and the fundamental principles of data protection regulation is found in three consecutive cases from the Danish DPA regarding a museum's establishment of a publicly available database.[6] The database was scheduled to include ordinary information about members of the resistance (the organised opposition against the German occupation of Denmark) during World War II. The information was already available to the public in printed works and other sources. Under a number of preconditions, the DPA accepted the establishment of the database within the framework of article 5 and section 1 of article 6 of the PPD Act, corresponding to article 6, subsection 1 of article 7 of the Personal Data Directive. Among these preconditions was:

The DPA also requires that the Museum of Danish Resistance (Frihedsmuseet), prior to the opening of the database, inform the public and, in particular, those directly concerned, about the establishment of the database, the website and the possibility of objecting to the disclosure of personal data. This may be done on the museum's website and through the relevant associations' networks, magazines, etc.

[6]The DPA file number 2006-321-0457, 2007-321-0039 and 2008-321-0134, available at: https://www.datatilsynet.dk/tilsyn-og-afgoerelser/historiske-afgoerelser/2006/okt/offentliggoerelse-af-oplysninger-om-modstandsfolk-paa-internettet-i/, https://www.datatilsynet.dk/tilsyn-og-afgoerelser/historiske-afgoerelser/2007/sep/offentliggoerelse-af-oplysninger-om-modstandsfolk-paa-internettet-ii/ and https://www.datatilsynet.dk/tilsyn-og-afgoerelser/historiske-afgoerelser/2009/apr/offentliggoerelse-af-oplysninger-om-modstandsfolk-paa-internettet-iii/.

> The DPA therefore also requires the Museum of Danish Resistance to remove personal data from the database if this relates to living persons, if they so request, or to deceased, if their close relatives make the request.[7]

Crucially, in none of the three cases were the personal data intended at any time to be erased from the publicly available sources, in which they already appeared. On the contrary, the DPA assigned the right of objection to the data subjects in relation only to the *processing* within the publicly available database—which would increase the searchability of the personal data. The Danish resistance fighters thus had a right to "be remembered with greater difficulty" which is strikingly comparable to the result of the Court of Justice of the European Union in Google Spain. Here, Google Spain was not required to erase the original personal data, only lower the searchability. At the same time, both the Danish case law and Google Spain show that the right of objection to the processing of personal data is strengthened with (a) the age of information and (b) the increase in searchability through e.g. a publicly available database or indexing on the web.

This understanding of the right to be forgotten, where the data subject can object to the part of processing which involves high searchability of data concerning him or her is not, however, limited to processing of personal data of older origin, at least not in Denmark. In Denmark, the right to object is not limited to processing which will eventually become incompatible with data protection regulation when they are no longer necessary for the purposes for which they have been processed so far, see Google Spain (92). Other factors than searchability and time may justify an objection.

A Danish DPA case from 2004 illustrates how an objection can be just even shortly after a processing of personal data is initiated. In this case, an elderly retired physician had protested against a municipality's plan for future use of an area in the municipality in a letter.[8] The municipality had addressed the physician's letter and subsequently published, among other things, the physician's name and address on the municipality's website.

The physician objected under article 35 of the Danish PPD Act and requested that the data should be removed. In his job as a physician, he argued, he had previously been the victim of several burglaries from drug addicts, and additionally he did not want to be contacted in his spare time. First, the DPA stated that the processing of these correct, ordinary personal data, had legal basis in subsection 1 of article 6 of the PPD Act, which corresponded to article 7 of the Data Protection Directive.[9] There

[7]DPA file number 2007-321-0039, available at: https://www.datatilsynet.dk/tilsyn-og-afgoerelser/historiske-afgoerelser/2007/sep/offentliggoerelse-af-oplysninger-om-modstandsfolk-paa-internettet-ii/.

[8]DPA file number 2004-313-0247, available at: https://www.datatilsynet.dk/tilsyn-og-afgoerelser/historiske-afgoerelser/2005/jun/klage-over-offentliggoerelse-af-navn-og-adresse-paa-kommunes-hjemmeside/.

[9]In other words, the processing was legal, in that it was necessary for the purpose of performing a task in the public interest, necessary for the purpose of the public authorities in connection with

was therefore no basis for his objection on the basis that the data was incorrect, or the processing in itself was unlawful. Nevertheless, the DPA further stated:

> An objection may, however, also be considered justified, even if the processing is otherwise legal. This will be the case if the data subject has presented compelling reasons in support of the request that processing should not take place due to the data subject's special individual situation. The controller must, therefore, assess the circumstances surrounding the particular processing of personal data, and whether these circumstances should lead to the processing of data about the person concerned being restricted or completely stopped.

The DPA assessed whether the physician had presented such arguments in order for his objections to be met. Here, the DPA paid specific attention to the fact that the municipality could replace the physician's name and address with the term "residents within the municipality" or similar. Such a pseudonymisation would allow the public continued access to the protest concerning the use of land made by the physician, without compromising his interests. Therefore, the objection was justified. The DPA held that the municipality as a minimum should erase the physician's name from the website. In accordance with the Danish understanding of the right of objection, the underlying processing of the physician's data could of course continue, meaning the municipality's original documents were preserved.

In the private sector Danish law requires that the processing intensity—the searchability—may be reduced in some cases as well. This applies to both personal data of older origin and more recently collected data. An illustrative example is found in a case from the DPA from 2009, in which, however, the consideration of predictability and transparency of the data processing also played a part. The case concerned the erasure of a profile and a number of posts in a discussion forum at D. dk. In the specific forum, users had the right to delete their posts themselves, but only as long as their profiles were active. In this case, however, a user, K,—after having stated the intention of deleting his posts—was deactivated by the administrator of the forum. The user objected and expressed that he wanted the said posts deleted. The data controller refused to meet this objection. First, the DPA noted that an objection may be regarded as justified, even if the processing is otherwise legitimate, if weighty reasons relating to the particular situation of the data subject are present. The agency then said:

> "It is the opinion of the DPA that public profiles in a discussion forum etc in many cases are created without further consideration of whether a user wants to have a permanent publicly available user profile on the forum. It can therefore be seen as burdensome if a user cannot have his profile information deleted upon request.
>
> The DPA therefore finds that a request to have a profile deleted or anonymised must often be considered legitimate and should be met. This may be done by continuing to let the user appear under a general and neutral term such as "Guest" or "Previous User"".
>
> With regards to the question of the erasure of personal data from posts, the DPA considers the expectations the forum has provided users with in terms of such possibilities.

exercising authority and necessary for those who had access to data to pursue a legitimate interest, and the consideration of the data subject would not normally exceed that interest by such disclosure.

An expectation can e.g. arise if it at the time of the posting is possible for the user to delete or edit his or her posts. The expectation may also arise if such a possibility is stated in the Terms of Use.

If the user has such a justified expectation, it would normally constitute specific and legitimate circumstances which indicate that the user's request for erasure must be met, c.f. article 35 of the PPD Act. Depending on the circumstances, the user's request can be met by preserving the posts, but in such a way that it is not listed under the user profile.

In the specific case, the DPA assumes that it is possible for users at D.dk to delete their own posts ... On this basis, the DPA is of the opinion that the data controller is not entitled to maintain K's profile and posts in a form that is attributable to him as a person after he has made a request for erasure.

As far as K's profile ["OMITTED"] is concerned, the DPA assumes that the profile has been deleted and that the posts linked to the profile are listed as written by a deleted user. The DPA therefore assumes that the posts have been maintained in such a way that they cannot be directly attributed to K.

Thus, in the private sector, Danish regulation in some cases requires otherwise lawfully processed information to be erased, made less exposed or less searchable. In the above described case, the decisive element was the expectation brought about by the controller. The right to be forgotten—or remembered with greater difficulty—is, however, seen as a subset of the broader right of objection in Danish legislation. Data becoming "old" is, in other words, just one of several factors which may justify an objection to a lawful processing of correct data. Most often, the consequence of an objection being considered justified is that the high searchability must be lowered or the data be pseudonymised. Erasure of personal data in original documents and sources is, however, less relevant.

In view of the above, it is hardly surprising that Danish legal literature assumes that the provisions of article 35-37 of the PPD Act must be read in the light of recent practice of the European Court of Justice in i.e. Google Spain.[10] In accordance with the European Court of Justice's understanding of litra (b) of article 12 and litra (a) of section 1 of article 14 of the Data Protection Directive, article 37 of the Danish PDD Act is therefore interpreted as stating that a search engine provider, in order to comply with the provisions in question, may be required to remove links to websites from the list of results displayed after a search on a natural person's name—including the cases in which the name or data is not removed from that website and even when the publication on the website is legitimate.

The provisions presented by the DPA were and are still status quo in Denmark. Though the DPA in the future will refer to article 16 and 17 of the GDPR, the legal position remains unchanged. In the Danish white paper on the GDPR it is stated[11]:

4.7.3. The Data Protection Regulation

It is clear from the Commission's proposal for The Data Protection Regulation that article 17 elaborates and describes the right of erasure referred to in litra (b) of article 12 of the Data Protection Directive and sets out the conditions for the right to be forgotten, including the

[10] Waaben and Nielsen (2015), p. 538.

[11] White paper no. 1565, GDPR – and the legal framework of Danish legislation, p. 330 et seq.

duty of the controller who has published personal data to inform third parties of the data subject's request to delete all links, copies, or representations of the personal data in question.

4.7.3.1. Right to erasure — section 1 of article 17

Section 1 of article 17 of the Regulation regulates when a data subject has the right to have personal data erased by the controller without undue delay, and when the controller is simultaneously obliged to erase personal data without undue delay.

. . .

This provision applies to everyone, but in preamble no. 65, it is emphasised that the right is particularly relevant when the data subject gave consent as a child and was not fully aware of the risks associated with processing, and later wishes to remove such information, in particular on the Internet. It is also stated that the data subject should be able to exercise this right, regardless of whether he or she is no longer a child.

The data subject's right to erasure, and the controller's related obligation to erase, arises pursuant to section 1 of article 17, when one of the following conditions apply:

a) The personal data is no longer necessary to fulfil the purposes for which they were collected or otherwise processed.
b) The data subject withdraws the consent which is the basis for the processing, c.f. litra (a) of section 1 of article 6 or litra (a) of section 2 of article 9, and there is no other legal basis for the processing.
c) The data subject objects to the processing pursuant to section 1 of article 21 and there are no legitimate reasons for the processing which precede the opposition, or the data subject objects to the processing pursuant to section 2 of article 21.
d) Personal data have been processed illegally.
e) Personal data must be erased to comply with a legal obligation under EU law or the national law of the Member States to which the controller is subject.
f) Personal data have been collected in connection with the offer of information society services as referred to in section 1 of article 8.

With regards to litra (a), the wording of the provision corresponds to the wording of section 5 of article 5 of the Danish PPD Act. The provision is likely to be independently relevant in relation to litra (e) of section 1 of article 5 in cases where the data subject makes use of his right to erasure without undue delay in cases where the controller could otherwise await e.g. the expiry of a generally fixed deadline for erasure.

The provision in litra (b) must be assumed to correspond to article 38 of the PPD Act on withdrawal of consent. Regarding the obligation in subparagraph (c), reference is made to section 4.11. on article 21 and the right of objection. The provision in litra (d) should be read in conjunction with article 5 of the regulation on principles relating to personal data processing, and it must be assumed that the obligation to erase the information pursuant to litra (d) will apply to the controller, regardless of whether the data subject makes use of his right in accordance with the provision, and, where appropriate, the controller may be obliged to erase without delay pursuant to the provision of litra (d) of section 1 of article 5.The same is likely to assert itself in relation to erasure in the situations referred to in litra (e), however, the deadline in article 17 (without undue delay) may give the provision independent content. Litra (f) is likely to supplement litra (b) in cases where the custodian has given consent on behalf of a data subject, who at the time of registration was a child under the age of 16, and the data subject now, regardless of whether he or she is no longer a child, c.f. preamble no. 65, wishes to make use of the right to erasure.

. . .

4.7.3.3. Exceptions – section 3 of article 17

Section 3 of article 17 provides a number of exceptions to the right of erasure for data subjects and the controller's obligation to erase without undue delay, as well as expectations to the controller's obligation to notify the subject n and thus from the "right to be forgotten."

The exceptions in section 3 of article 17 are significant.

...

4.7.4. Considerations

In view of the extensive exceptions to the right of erasure and the "right to be forgotten" as well as the interpretation in the draft proposal of the regulation which states that article 17 elaborates and describes the right of erasure, it is likely that article 17 of the regulation is generally a continuation of current law.

As can be seen, the Danish Ministry of Justice does not assume that the GDPR imposes amendments to Danish law regarding the "right to be forgotten". In other words, the right to correction, limitation and erasure can still be seen in conjunction with the right of objection—and the right to be forgotten is thus merely a subset of the common right of objection.

To the white paper's interpretation of the GDPR, it may be added that in relation to Danish law – both current and past – article 18 of the GDPR is not without interest. Article 18 of the GDPR is interpreted as regulating the right to limitations imposed on the processing of both correct and incorrect data – thereby preserving the nuanced approach to the right to be forgotten" in Danish data protection regulation.

This approach provides a balanced and nuanced "right to be forgotten" since opposite considerations can be taken into account and the result adapted to the specific situation. Depending on the weight of data protection rights in the individual case, both the assessment of the objection's justification *and* the choice of legal consequence can be adapted to the specific situation.

The balancing of interests in data protection rights and freedom of information and expression is illustrative. Correction of information by altering or anonymising an original source must be considered the strongest means of ensuring data protection interests, but at the same time it is the deepest interference with the freedom of expression and information. This is because such a change involves editing the past. Very strong privacy interests must therefore be at stake before changes can be made to an original source. A scaled down operation will usually be preferable. Such scaling can e.g. consist of an update rather than a correction – i.e. add new, additional information to the original source. Further down the scale is a decrease in searchability, which may also be more or less interfering. De-indexing must currently be considered as the most interfering means of lowering searchability, since "de-indexing" means that the script from the website, where the content in question is located, will be changed. De-indexing will affect all search engines – and all search engines will cease to refer to the source. Less intrusive is the erasure of links, wherein a person can be searched for by using an identifier such as name or national identification number.

However, it is important to emphasise that the right of objection (the right to be forgotten) is not unconditional in Denmark, regardless of the means used to reduce searchability. There are wide exceptions in a number of areas which prevent the data subjects from using the right of objection, and the right contained therein to be forgotten in the form of reduced searchability.

2 What Are the Limits to the Right to Be Forgotten Under Your Law?

At first glance, Danish law consists of a confusing patchwork of legislation and unwritten regulation. National general regulation, regulated and developed both by legislation as well as judicial and the Danish ombudsmans practice, interfere with national specific rules, parallel rules and principles of EU law and international law. Therefore, the following describes only five areas where the right of objection is limited. These main limitations of the right of objection—and as a subset hereof the right to be forgotten—are all justified by the national weighing of the protection of personal data for opposing (conflicting) interests.

The chosen areas concern the freedom of expression and information (Sect. 2.1), the securing of possibilities for research and statistics (Sect. 2.2), documentation of public administration activities (Sect. 2.3) and the consideration of intelligence work (Sect. 2.4). In addition, special legislative provisions apply to the Law enforcement authorities' processing of personal data (Sect. 2.5).

2.1 Freedom of Expression and Freedom of Information

With regard to balancing the interests of freedom of expression and information and the protection of personal data, the national starting point is found in article 2, section 1 and 4-8 of the DDPA. Based on the preliminary announcements, it is expected that this legal position will, in general, also continue after 25 May 2018, where the GDPR takes effect in Danish law.[12]

With regards to balancing the interests of freedom of expression and information and the protection of personal data, the national starting point is found in section 3 and sections 6-10 of the current PPD Act. It is worth noting here, that the regulation has not fundamentally changed with the implementation of the GDPR.As the Danish regulation is somewhat complex, the following description of the importance of the provisions is categorised as follows for ease of reading:

- Mass media's processing of personal data in databases (Sect. 2.1.1)
- Processing of personal data for journalistic or literary purposes, not constituting processing in a database (Sect. 2.1.2)
- Processing, which is not made for journalistic purposes and not by a mass medium, but where considerations of freedom of expression and information results in denying the right of objection under the PPD Act to the data subject, and the DPA will therefore refuse to process the cases (Sect. 2.1.3)
- Processing, not carried out for journalistic purposes and not by a mass medium, in which the data subject can invoke the right of objection, but where the interests of

[12]White paper no. 1565, GDPR – and the legal framework of Danish legislation, p. 958 et seq.

freedom of expression and freedom of information affect the outcome of DPA assessments of whether an objection must be met (Sect. 2.1.4).

2.1.1 Mass Media's Processing of Personal Data in Databases

As far as mass media's processing of personal data in databases is concerned, the rules contained in subsections 4-6 of section 3 of the PPD Act determines whether the PPD Act may be invoked against these media. The provisions read as follows:

(4) This Act shall not apply to the processing of data covered by the Act on information databases operated by the mass media.
(5) This Act, as well as chapters II-VII and IX of the GDPR, shall not apply to information databases which exclusively include already published periodicals or sound and image programmes covered by paragraphs 1 or 2 of section 1 of the Act on Media Responsibility, or part hereof, provided that the data is stored in the database in the original version published. However, Article 28 and 32 of the GDPR shall apply.
(6) This Act, as well as chapters II-VII and IX of the GDPR, shall not apply to information databases which exclusively include already published texts, images or sound programmes covered by paragraph 3 of section 1 of the Act on Media Responsibility, or part hereof, provided that the data is stored in the database in the original version published. However, Article 28 and 32 of the GDPR shall apply.

As can be seen, it follows that the right of objection cannot be invoked when processing of personal data takes place in databases governed by the Danish Act on mass media databases.

The act relating to mass media databases comprises, pursuant to subsection 1 of section 2 of the Act, databases operated by mass media as defined in the Danish Media Liability Act. According to section 1 of the Danish Media Liability Act, mass media is interpreted as national periodic journals, including images and similar representations, which are printed or otherwise reproduced.[13] Included are also audio and video programmes broadcast by DR, TV2 A/S, regional TV2 companies and other entities authorised or registered to conduct radio or television activities. Finally, texts, images and audio programmes regularly distributed to the public are included, provided they can be characterised as news communication comparable to the dissemination activities listed above.[14]

The databases governed by the Act on mass media databases, are, first of all, the internal systems of the mass media when these are only applied for journalistic and editorial work (editorial information databases), cf. subsection 3 of section 2 of the Act relating to mass media databases. Secondly, the act regulates the publicly available information databases, cf. subsection 2 of section 2 of the Act. These latter databases are pre-notified systems using electronic data processing in connection with the dissemination of news and other information, and they are available to anyone under normal business conditions, cf. subsection 1 of section 6 of the Act.

[13]Waaben and Nielsen (2015), p. 121 et seq.

[14]Jakobsen and Schaumburg-Müller (2016), p. 263 et seq.

Thus, the vast majority of online media will fall within the definition of publicly accessible information databases, provided they have previously been notified as such to the Danish Press Council.

Accordingly, the data subjects cannot in the context cited here invoke the time limitation principle in section 5 of the PPD Act, or the right of objection under Articles 16-18 of the GDPR. There are, however, special provisions in the mass media databases Act which, to some extent, seek to limit the indefinite storage of personal information. It follows from subsection 3 of section 8 of the Act that data on purely private conditions can only be stored for 3 years following the event in these publicly available databases or, if the time of the event cannot be confirmed, from the time of storage. This starting point is waived if the interest in the public availability of the information is found to be of more importance than the interest of the individual in having the information removed, as a consequence of the interest of the freedom of information. In light of, in particular, the practice of the European Court of Human Rights, the importance of the freedom of information will significantly limit, if not outright prohibit, the erasure of information about politicians, opinion-formers and other celebrities within the 3-year limit.

Another form of database used by the media are the so-called unedited full-text databases. These are databases where only previously published content is processed and entered into the information database unchanged from the original publication. It is assumed in the literature that at least the archives of old articles, audio and video programmes etc on television channels' and newspapers' publicly available web pages are counted as unedited full-text databases.[15]

From the above-mentioned provisions of subsections 4–6 of section 3 of the PPD Act, it is clear that neither the time-limitation principle of the PPD Act nor the data subjects' right of objection apply in connection with the processing of these unedited full-text databases. Mass media's use of such databases is also not governed by the law on mass media databases, cf. subsections 2-3 of section 1 of the law on mass media databases.

The reason why unedited full text databases are not covered by neither the act of mass media databases nor the PPD Act is that the original publication is subject to the rules of the Danish Criminal Code and the Danish Media Liability Act. Therefore, the legislature has not seen any need for regulating further processing in the form of continued public disclosure.

A form of the right of objection can however be found in the rules on "sound press ethics". In 2013, a provision was introduced in the Advisory Rules for Sound Press Ethics, which forms the basis for the Danish Press Council's assessment of the general clause on "sound press ethics", cf. subsection 1 of section 34 of the Danish Media Liability Act. The provision, as set out in point B.8, reads as follows:

> Messages published in digital media will often be available long after they are published. Upon request to the media, the availability of such previously published, sensitive or private information may be blocked to the extent possible and reasonable.

[15]Jakobsen and Schaumburg-Müller (2015), pp. 176–186.

The Association of Danish Media and the Danish Union of Journalists' Guide to the Advisory Rules for Sound Press Ethics states that point B.8 of the guidelines for sound press ethics may apply when people want to have old information which causes them difficulties removed.[16] This could be in connection with a job search. Information about public figures may nevertheless be relevant long after publishing. According to the guide, it does not however automatically follow that e.g. convicted persons have the right to have articles about their long-term imprisonment for serious crime removed. In relation to the way in which the media may block the availability of the article etc in question, the guide states that media can decode, anonymise or entirely de-publish an article, i.e. remove the article from the website. In other words, here too, as in Danish law in general, a practice can be established which focuses on reducing the searchability rather than actual de-publishing.

Two Danish authors, Professor Søren Sandfeld Jakobsen and Professor Sten Schaumburg-Müller, reviewed the application of this rule in 2015. The authors concluded that the Danish Press Council had imposed restrictive practices—referring i.a. to the fact that, in principle, the council has stated that the new rule "should, as a rule, be used only in the case of information which is particularly stressful for the person or company mentioned, and in cases where a publication of the council's ruling on criticism cannot be considered sufficient to take account of the consideration for the person or company".[17] In other words, the considerations of freedom of expression and information are strongly placed in the committee's weighing against the right to be forgotten.

In summary, in Denmark the data subject cannot rely on the data protection right of objection against media's use of databases—and in reality, there is no right to be forgotten in these databases. Other controllers who further enhance the searchability of personal data in publicly accessible media databases may in contrast hereto be obliged to meet the objections raised, cf. the ruling of the EU Court of Justice in Google Spain and the description of Danish practice under question 1.

2.1.2 Processing of Personal Data for Journalistic or Literary Purposes Within or Outside of a Mass Media Enterprise

Subsection 7-8 of section 3 of the PPD Act show that the right of objection does not apply in the listed situations:

> (7) This Act, as well as chapters II-VII and IX of the GDPR, shall not apply to manual archives containing clippings from published printed articles, provided that the data is processed for journalistic purposes exclusively. However, Article 28 and 32 of the GDPR shall apply.

---

[16] Advisory Rules for Sound Press Ethics – on media ethics and how to file media complaints, available at: https://journalistforbundet.dk/sites/default/files/inline-files/Guide_Presseetiske-regler.pdf.

[17] Jakobsen and Schaumburg-Müller (2015), pp. 176–186.

(8) This Act, as well as chapters II-VII and IX of the GDPR, shall not apply to other kinds of processing of information, provided that the data is processed for journalistic purposes exclusively. However, Article 28 and 32 of the GDPR shall apply.

In other words, the right of objection under the PPD Act—and consequently the right to be forgotten in the form of a right of reduced searchability—does not apply to manual archives of clips from published, printed articles etc. In today's digital society, the provision stating that the right of objection does not apply to processing, which takes place exclusively for journalistic purposes, is, however, the widest exception.

There is no requirement for a special professional status, if only the controller can prove that the purpose of the activity is disclosure, cf. the judgment of the European Court of Justice in C-73/07 Satamedia Oy (hereinafter Satamedia). According to Danish law, it is sufficient that the activity is intended to disseminate views, information or ideas to the public. It is irrelevant whether the purpose is to achieve a financial gain.[18]

In summary, as long as there is no processing for "mixed" purposes, the provision in subsection 8 of section 3 of the PPD Act will include quite a large part of the daily personal data processing. Thus, the data protection right of objection can rarely be applied to e.g. the processing of personal data made by the increasing amount of bloggers.[19]

The right of objection under the personal data protection rules can still, however, be relied on against other controllers who, with their services, further increase the searchability of the personal data concerned, cf. under Sect. 2.1.1. Therefore, there may be an obligation e.g. to delete links referring to the processed personal data, cf. the description of Danish practice under question 1, and for older information, the ruling of the European Court of Justice in Google Spain.

2.1.3 Processing, Which Is Not Made for Journalistic Purposes and Not by a Mass Medium, but Where Considerations of Freedom of Expression and Information Results in Denying the Right of Objection Under the PPD Act to the Data Subject, and the DPA Will Therefore Refuse to Process the Cases

In Danish law, an assessment of whether there is a right of objection to the processing of ordinary personal data not made by mass media and not for purely journalistic or literary purposes but for freedom of expression and freedom of information, is made on the basis of subsection 1 of section 3 of the PPD Act. The provision reads as follows:

[18] The same will apply to online books, including specialist books, cf. for the latter, the DPA file number 2002-082-0075, available at: https://www.datatilsynet.dk/tilsyn-og-afgoerelser/historiske-afgoerelser/2004/mar/spoergsmaal-om-persondatalovens-anvendelse-paa-faglitteratur/.

[19] If the starting point is waived based on a concrete assessment, an assessment must be made in accordance with subsection 1 of section 3 of the PPD Act, cf. Jakobsen and Schaumburg-Müller (2016), p. 255.

(2) This Act shall not apply where this will be in violation of the freedom of information and expression, cf. Article 10 of the European Convention for the Protection of Human Rights and Fundamental Freedoms or Article 11 of the European Charter of Fundamental Rights.

The provision is incorporated in the Act to emphasize that the PPD Act does not restrict the freedom of information and expression resulting from Article 10 of the European Convention on Human Rights, or Article 11 of the EU Charter.[20] As can be seen, the data subject is completely prevented from invoking the right of objection under the PPD Act and the right to be forgotten, if the conditions contained in subsection 1 of section 3 of the PPD Act are met.

The provision is, as most provisions referring to another law, not particularly informative. The DPA has, however, established a fairly clear practice. A weighty consideration of freedom of information and expression in relation to expression of opinions and attitudes, leads to the agency refusing to process such cases with reference to the fact that it follows from subsection 3 of section 1 of the PPD Act that the law does not apply.[21] As a modification, the law applies if judgments, opinions, attitudes etc are presented as facts. In other words, the PPD Act will apply in situations where, using the popular expression of today, there is a risk of "fake news" concerning natural persons, see Sect. 2.1.4 below.

There are a number of examples of situations where the agency has refused to apply the PPD Act with reference to subsection 1, section 3 of the Act. One of these is a number of linked cases relating to a website. On the website, information about named public employees was published. It appeared, however, from the website, *that* the officials in question had not necessarily been charged or convicted of the counts in question and *that* registration could happen solely because an individual had felt unfairly treated by the person concerned.[22] The DPA refused to take measures against the website. The DPA referred to the website's reservations for the accuracy of the information, to the subjective nature of the assessments and value-charged statements, which were part of the expression of opinions on the site -and *that* this was clear to all users of the site. The owner of the website in question was subsequently convicted for defamation according to the Danish Penal Code, and was furthermore sentenced to remove part of the information. Through a subsequent complaint, the DPA was later requested to make a ruling on the website's compliance with the PPD Act. The agency again refused to make a ruling about the website on the basis that it upheld its previous assessment that the law, due to freedom of

[20]Bill no. 147 of 31 May. 2000 (the Folketing Hansard 1999/00): the special notes to section 2.

[21]In case of rejection, the DPA is not the competent authority, as the DPA supervises compliance with the PPD Act, but does not have the jurisdiction to decide on the Danish Penal Code's rules on defamation etc.

[22]DPA file number 2011-215-0874, available at: https://www.datatilsynet.dk/tilsyn-og-afgoerelser/historiske-afgoerelser/2012/jun/offentliggoerelse-af-personoplysninger-paa-hjemmeside-i/, and DPA file 2011-222-0094, available at: https://www.datatilsynet.dk/tilsyn-og-afgoerelser/historiske-afgoerelser/2012/jun/offentliggoerelse-af-personoplysninger-paa-hjemmeside-ii/.

expression and freedom of information issues, did not apply, cf. subsection 2 of section 2 of the PPD Act, now subsection 1 of section 3.[23]

In summary, the consideration of freedom of expression and freedom of information is a very crucial factor in Danish law and it will prevent data subjects from invoking their legal right to data protection, as long as the published information is part of expressions of opinion etc and do not appear as something more than such. As discussed above under Sects. 2.1.1 and 2.1.2, this, however, does not mean that other controllers, who further enhance the searchable nature of such personal data, are not obliged to erase links or otherwise have the right of objection in connection with similar services.

2.1.4 Processing Not Carried Out for Journalistic Purposes and Not by a Mass Medium in Which the Data Subject Can Invoke the Right of Objection, but Where the Interests of Freedom of Expression and Freedom of Information Affect the Outcome of DPA Assessments of Whether an Objection Must Be Met

When information, contrary to the information described above, appears as descriptions of facts and with no reservation for their accuracy, the Danish DPA tends to agree to look into complaints—and thus let the PPD Act come into use, i.e. objections apply.

In cases where the agency considers itself competent, and thus provides the data subjects with the opportunity to invoke the right of objection and consequently the right to be forgotten, the consideration of freedom of information and freedom of expression still plays a considerable role. Thus, the fact that certain data processing is covered by the PPD Act does not necessarily mean that restrictions do not apply to the right of objection, when the freedom of expression and freedom of information are taken into consideration.

There are no known examples in the Danish supervisory practices where such considerations have had any influence in matters relating to older information with high searchability. There are, however, several cases where the Danish DPA has assessed that the DPA was applicable, but an objection could not be met because of the consideration of the freedom of expression and freedom of information. An example can be found in a case from 2007, where the Danish DPA dealt with a complaint about the disclosure of personal information on, among other things, affiliation with political right-wing organisations in Denmark.[24] The publication

[23]White paper no. 1565, GDPR – and the legal framework of Danish legislation, p. 948.

[24]DPA file number 2007-229-0002, available at: https://www.datatilsynet.dk/tilsyn-og-afgoerelser/historiske-afgoerelser/2008/okt/afgoerelse-i-klagesag-om-offentliggoerelse-paa-redox-hjemmeside/.

was made on a website run by a Danish group called Reddox.[25] The individual, whom the information concerned, wanted the information removed from the website. The DPA found the PPD Act to apply, and in the role of controller, Reddox was responsible for observing the rules of the PPD. The DPA, however, did not agree with the requirement for removal of the information. One of the arguments of the DPA for not meeting the objection was that *"the freedom of information and freedom of expression must be considered, so that the PPD Act does not unduly impose restrictions on these freedom rights..."*

As also highlighted above, the interests of the data subjects will carry more weight in Danish law, if other controllers further enhance the searchability of the published personal data. Such other controllers can therefore be obliged to meet the objections raised and decrease searchability, cf. the ruling of the EU Court of Justice in Google Spain and the description of Danish practice under question 1.

2.2 Research and Statistics

As mentioned initially, the consideration for research and statistics in Danish law will also allow for a limitation in the right of objection and thus the right to be forgotten. This is stated in the rights guide mentioned in Question 1:

> In the case of processing for statistical or scientific purposes, an objection must, as a general rule, also not be accepted.

This rather comprehensive exception has its justification in that the GDPR and the PPD contains a special rule on processing for research and statistical purposes, which ensures a special protection of the data subjects under such circumstances. Once information has been processed for the purpose of carrying out statistical or scientific studies of significant social importance, the information cannot later be processed for other purposes, cf. subsection 2 of section 10 of the PPD Act. In the case of sensitive information, which is either (1) leaving the territory of the GDPR, (2) concerning biological material, or (3) meant for publication in scientific journals et.al., the information cannot be disclosed without prior permission by the DPA, cf. subsection 3 of section 10 of the PPD Act.

2.3 Documentation of the Public Administration Activities

The right of objection as laid down in the GDPR and DDPA, the former PPD Act, interacts with other legislations and unwritten principles of administrative law, based

[25]Redox describes themselves as a "left-wing, antifascist research group" that "digs deep into activities and structures of the extreme right-wing and then publish its findings", see https://redox.dk/om.

on the assumption that national law can impose an obligation to choose correction, addition or blocking prior to erasure, and that such an obligation is in accordance with the GDPR.[26]

Danish public authorities are as a main rule obliged not to erase any data due to documentary purposes, access options, and later archiving, in particular for historical research. This follows partly from unwritten administrative principles, partly from the obligation of public authorities to keep records of written proceedings,[27] and finally from the Danish Public Records Act.[28]

Therefore, erasure of otherwise correct and legally processed personal data is not really an option in the public sector. Correct personal data will not be erased, no matter how old it may be. Nor if the information has become incorrect or misleading will it be erased. Instead, personal data will be corrected by adding new information (files or remarks) specifying that the data subject has objected, or the data is found to be incorrect and thus no longer to be used by caseworkers.[29]

With regards to the very extensive public information databases in Denmark, incorrect information will of course be erased in the actively searchable part of such systems. However, the older versions of public databases are usually stored. This storage is done partly for the purpose of documentation (e.g. to document what information has previously been disclosed) and for archive purposes, i.e. for later historical research. The erasure of incorrect information in the user interface of public information systems is, however, more due to the interests of the quality of public data than the wish to provide the data subjects with a right to be forgotten.

In Danish law, the purpose of documentation of activities carried out by the public authorities often extends to include activities carried out by private individuals with public funding. This can be seen e.g. in healthcare.[30] The majority of healthcare professionals are covered by the Danish Authorisation Act,[31] and an accompanying executive order[32] listing detailed requirements for the content and

[26]Bill no. 147 of 31 May. 2000: the special notes to section 37.

[27]Law no. 606 of 12/06/2013 – the Danish Publicity and Freedom of Information Act, available at: https://www.retsinformation.dk/Forms/R0710.aspx?id=152299 (section 15).

[28]Executive Order no. 1201 of 28/09/2016 – Executive Order of the Danish Act on Public Records, available at: https://www.retsinformation.dk/Forms/R0710.aspx?id=183862.

[29]White paper nr. 1565, GDPR – and the legal framework of Danish legislation, pp. 329-330.

[30]The Danish hospital services are mainly organised as public authorities, but a number of supplementary healthcare services are provided by private operators such as doctors, dentists and chiropractors. However, the vast majority of health services are tax-financed in Denmark, whether they are carried out by public authorities or private companies.

[31]Executive Order no. 1356 of 23/10/2016 - Act on approval of health professionals and health professional activity, available at: https://www.retsinformation.dk/Forms/R0710.aspx?id=183809. Latest change was with order no. 990 of 18/08/2017, available at: https://www.retsinformation.dk/Forms/R0710.aspx?id=192522.

[32]Executive Order no. 1090 of 28/07/2016 – Act on authorised health persons medical records (record keeping, storage, disclosure and transfer, etc.), available at: https://www.retsinformation.dk/Forms/R0710.aspx?id=183578. Changed with order no. 530 of 24/05/2018, available at: https://www.retsinformation.dk/Forms/R0710.aspx?id=201378.

scope of the record keeping, cf. article 6 of the Act. This states that any information, once entered in a record, data must be kept for a specified period, cf. sections 14 and 15 of the Executive Order. There is no option to erase the data—only to add new information to complement the previously entered data. The latter follows from section 14 of the Executive Order, which reads as follows:

> Article 14. Information in the patient records shall not be erased or made illegible.
>
> Section 2. If it is necessary to correct or add to the patient record, it shall be done in such a way that the original text is preserved. It shall be stated who made the correction or addition and when.
>
> Section 3. In an electronic patient record, the original version of the information that has been modified by correcting or adding shall remain available.

In other words, data subjects in Denmark cannot request (a complete) erasure of information in patient records—regardless of their correctness or age—within the fixed storage periods.

2.4 *Intelligence Work*

Neither the Data Protection Directive nor the GDPR regulates the processing of personal data as part of (national) intelligence work. This can be seen in article 13, section 2 in DDPA (this was in subsection 11 of section 2 of the PPD Act). It states that:

> (2) This Act shall not apply to the processing of data which is performed on behalf of the intelligence services of the police and the national defence.

The Danish security service consists of the Danish Security and Intelligence Service (PET), the Danish Defence Intelligence Service (FE) and the Center for Cyber Security (CFCS). Legislation related to these security services does not include rules on the right of objection or the right to be forgotten.

However, section 1 of article 9 of the Act on the Danish Security and Intelligence Service (the PET Act) states that PET must erase information about natural and legal persons if the data are provided as part of studies or investigations aimed at said persons if the studies or investigations have not provided new information within the past 15 years.[33] In sections 1-2 of article 6 of the Danish FE Act, the same is stated relating to information about natural and legal persons residing in Denmark and raw data provided as part of intelligence work. In both cases, however, erasure may in some cases be omitted if essential consideration for the performance of intelligence services makes it necessary, cf. section 2 of article 9 of the PET Act and section 3 of article 6 of the FE Act. Article 14 of the CFCS Act contains a time limit as a manifestation of the general storage limitation principle.

In summary, according to Danish law, the data subjects do not have the right to object or request the erasure of correct information processed by the Danish intelligence services—not even data of older origin. However, an indirect request system

[33] Hereinafter the PET Act.

is in place, allowing data subjects to make a request to the Danish Intelligence Oversight Board. The Board then investigates whether the data processing in question violates relevant regulations. The Board will notify the data subject in question.

2.5 *Law Enforcement Authorities' Processing of Personal Data*

For Danish law enforcement authorities, the Act on data protection by law enforcement authorities applies.[34] Article 1 of the law states that the law enforcement authorities in Denmark is regarded as the police, the public prosecutor's office, including the military prosecutor, the Prison Services, the Independent Police Complaints Authority and the courts.

The law is an implementation of the EU Directive on data protection within law enforcement,[35] and as such regulates the processing of personal data by these authorities as data controllers for the purposes of preventing, investigating, detecting or prosecuting criminal offences or enforcing criminal sanctions, including protection against or prevention from threats to public security.[36]

The general principle of storage limitation is laid down in both section 6 of article 4 of the Danish law on law enforcement authorities[37] and article 4 of the underlying directive.

> The more specific regulation aimed at implementing the time-limitation principle have been revised following the adoption of the new act on law enforcement authorities during 2017. Among others the regulation of Danish Criminal Register, to mention one example. See Executive Order no. 1079 of 20 September 2017 concerning the processing of personal data in the Danish Criminal Register.[38] Regarding the courts, it is stated in section 1 of article 221 of the Danish Administration of Justice Act that the court may, at any time in accordance

[34]Act no. 410 of 27/07/2017 – Act on processing of personal data by law enforcement authorities – entered into force on 1 May 2017, available at: https://www.retsinformation.dk/Forms/R0710.aspx?id=189891.

[35]Directive (EU) 2016/680 of the European Parliament and of the Council of 27 April 2016 on the protection of natural persons with regard to the processing of personal data by competent authorities for the purposes of the prevention, investigation, detection or prosecution of criminal offences or the execution of criminal penalties, and on the free movement of such data, and repealing Council Framework Decision 2008/977/JHA, available at: http://eur-lex.europa.eu/legal-content/ENG/TXT/PDF/?uri=CELEX:32016L0680&from=DA.

[36]Provided processing is done in whole or in part by means of automatic data processing or information are or will be contained in a manual directory.

[37]Act no. 410 of 27/04/2017 – Act on processing of personal data by law enforcement authorities, available at: https://www.retsinformation.dk/Forms/R0710.aspx?id=189891.

[38]Executive Order no. 1079 of 20 September 2017 concerning the processing of personal data in the Danish Central Crime, available at: https://www.retsinformation.dk/Forms/R0710.aspx?id=192722.

> with or upon application, correct typographical errors which have occurred in terms of words, names or numbers, mere miscalculations and errors and omissions relating only to the form. Other errors and omissions may also be corrected if the parties do not oppose to it. The court may also, under a number of conditions, correct information in case-law but not otherwise make changes in the reasons or results of cases before the courts.

The Act on law enforcement authorities contains a further right of objection in section 17, but this right does not encompass correct information processed in accordance with the personal data principles and processing conditions in Chapter 3 of the Act. In other words, the right of objection described above in Question 1 does not apply to the processing of *correct* information with the Danish law enforcement authorities. It may be assumed that the provisions of the Act will be administered in such a way that data subjects in Denmark can have older information erased or have reduced searchability (only) if this is (already) required by section 6 of article 4 of the Act on time limitation, or when the authorities are required to erase data due to other regulations.

Noting that the EU data protection reform has not caused major changes in Danish regulation. On the exceptions to the right to be forgotten, the above-described white paper states i.a, regarding the GDPR[39]:

> 4.7.3.3. Exceptions – section 3 of article 17
>
> Section 3 of article 17 provides a number of exceptions to the right of erasure for data subjects and the controller's obligation to erase without undue delay, as well as expectations to the controller's obligation to notify the subject and thus from the "right to be forgotten".
>
> Information may pursuant to litra (a) be retained to the extent necessary for the exercise of the right to freedom of expression and information [. . .].
>
> Information may be retained in accordance with litra (b) to the extent necessary to comply with a legal obligation requiring processing under EU or national law of the controller or to perform a task in the interests of society or as the subject of public authority, which has been imposed on the controller.
>
> It must therefore be assumed that Article 17 does not contain an independent right to be forgotten in the public sector. This is because it will often be necessary for public authorities to be able to document the basis on which a ruling or decision was based. Public authorities in particular should exhibit considerable reluctance to erase information, which at one time formed part of the basis on which a decision was made. Therefore, when public authorities correct incorrect or misleading information, it will often be necessary to note the correction (the correct information) of the case without removing the information already provided. This may be in the form of an added note.
>
> An actual erasure may be required more often if information is included in a register or other information system from which data is provided to other controllers. It must, however, also be accepted in the circumstances here that, through the continued storage of a copy of the register in its previous version or in any other way, it may be documented which information was previously disclosed.
>
> It should also be noted that, in any case, a continued storage of archives could be considered. See section 10.6. on the framework of sections 1, 3 and 4 of Article 89 on archiving purposes

[39]White paper no. 1565, GDPR – and the legal framework of Danish legislation, p. 330 et seq.

in the public interest [GDPR Article 89 allows derogations relating to processing for archiving purposes in the public interest, scientific or historical research purposes or statistical purposes]. It follows from litra (c) that information can be preserved to the extent that processing is necessary for public interest in public health. It follows from [Article 89] litra (d) that information can be preserved to the extent that processing is necessary for archival purposes in the public interest, scientific or historical research purposes or statistical purposes. Finally, it follows from litra (e) that information can be preserved to the extent necessary for legal requirements to be established, enforced or defended.

The exceptions in section 3 of article 17 are very considerable. It must therefore be assumed that there will be only a limited space for the data subject to utilise the right to be forgotten in the public sector.

In other words, it is to be expected that the described legal position will largely remain in force after 25 May 2018.

3 What Are, in Your Law, the Legal Remedies Available to Enforce the Right to Be Forgotten?

It is possible for the Danish courts to grant compensation from the liable data controllers to the data subjects pursuant to article 40 in the DDPA, in comparison with article 82, section 2 in the GPDR. The courts may also impose fines on the data controller (and to some extent the data processor) pursuant to article 41 of the DDPA, as well as deprive natural persons as well as companies the right to act as data processors.

In accordance with paragraph 3, article 8 of the EU Charter[40] and the Personal Data Directive, data subjects also have the right to appeal to an independent supervisory body, cf. section 39 of the current DDPA.

As far as the responses to Google Spain are concerned, the DPA has officially referred citizens to submit erasure requests to the controllers (the search engines). If the controller refuses to remove links, the data subject may then appeal to the DPA. About the DPA assessment, the agency has stated:

In assessing whether a search result is to be removed (erased), the DPA will initially decide whether the PPD Act applies. According to the "Google ruling", this will be the case when the company behind the search engine is established in Denmark or has established a branch or subsidiary in Denmark, which advertises aimed at residents in this country.

If the PPD Act applies, the DPA will assess whether the data processing conditions ... are met. If the DPA finds that a search engine provider's processing of personal data can take place within the framework ... the DPA will also decide whether the rights of data subjects ... can nevertheless lead to erasure. The DPA will ultimately decide whether the data

[40]Charter of fundamental Rights of the European Union, article 8 (3): "Compliance with these rules shall be subject to control by an independent authority". Available at: http://eur-lex.europa.eu/legal-content/EN/TXT/HTML/?uri=CELEX:12012P/TXT&from=DA.

> subject's objection is justified and, in that case, order the search engine to erase the relevant search results.[41]

In relation to the media, it is possible to file a complaint to the Danish Press Council and the council can assess whether the processing constitutes a violation of the ethics standards—but unlike the DPA, the Commission cannot order an erasure.[42]

The Danish Court Administration supervises the processing of personal data when the courts are data controllers, cf. article 38 of the DDPA and article 38 of the Danish Act on Law Enforcement Authorities. It follows from the first subsection of this provision that the Danish Court Administration oversees the processing of personal data outside of the jurisdiction of the judicial functions, i.e. the administrative functions. The courts oversee other forms of processing of personal data in the judicial organisations, c.f. subsection 2 of article 38 of the Act. It follows from article 42 of the Act on Law Enforcement Authorities that the Danish Court Administration may impose orders on processing activities or temporarily or permanently limit, including prohibit, processing of personal data.

Supervision of the Danish security service is not placed with the DPA but with the Danish Intelligence Oversight Board, which was established on 1 January 2014, see section 13 of the Danish FE Act, section 16 of the Danish PET Act and section 19 of the CFCS Act.[43]

> The Danish Intelligence Oversight Board is a specific independent inspection body consisting of five members appointed by the Minister of Justice in negotiation with the Minister for Defense. The chairman, a Danish high court judge, is to be appointed on the recommendation of both the Presidents of the Eastern High Court and the Western High Court in Denmark, while the other members have been appointed following discussions with the Danish Parliamentary Committee on the Intelligence Services.

The Intelligence Oversight Board may order PET and FE to erase data, but the supervisory authority does not have the power to order the intelligence services to take specific measures in relation to data processing.[44]

---

[41]DPA, Annual Report 2016 p. 13, available at: https://www.datatilsynet.dk/media/6515/datatilsynets_aarsberetning_2016_web.pdf.

[42]The Danish Press Council, however, does not have the jurisdiction to order a media to erase or otherwise prevent the availability of the content in question.

[43]Executive Order on the Danish Security and Intelligence Service (PET), https://www.retsinformation.dk/Forms/R0710.aspx?id=186190, Executive Order on PET's processing of information on natural and legal persons etc, https://www.retsinformation.dk/Forms/R0710.aspx?id=164082, Executive Order on the Danish Defence Intelligence Service (FE), https://www.retsinformation.dk/Forms/R0710.aspx?id=176852 and the Act on Center for Cyber Security, https://www.retsinformation.dk/Forms/R0710.aspx?id=163853.

[44]The supervisory authority may, however, issue statements to the intelligence services and the CFCS, in which the supervision expresses its opinion e.g. as to whether the intelligence services and the CFCS comply with the rules on data information. If an intelligence service or the CFCS, exceptionally decide not to follow a recommendation in a statement from the supervisory authority, the intelligence service or CFCS shall inform the supervisory authority thereof and forthwith submit the matter to the relevant minister for a decision.

In addition, there are a number of more special supervisory arrangements in Denmark. In the health sector, for example, the ombudsman for patients addresses certain cases, such as the disclosure of data concerning health, according to the rules of the Danish Health Act. The ombudsman for patients also supervises some issues regarding erasure of data in patient records.[45] Another example is, if a directory inquiry service, such as the Yellow Pages, Krak, Eniro etc publish incorrect telephone numbers, addresses etc, the Danish Energy Agency is the supervisory authority.[46]

In Denmark, there are currently very few lawsuits related to data protection regulation. As far as the authors know, the few lawsuits during the former legislative regime was primarily related to employment.[47] The GDPR, however, stipulates the right to lodge a complaint with a supervisory authority (article 77), the right to an effective judicial remedy against a supervisory authority (article 78), the right to an effective judicial remedy against a controller or processor (article 79), representation of data subjects (article 80), suspension of proceedings (article 81), the right to compensation and liability (article 82), as well as the general conditions for imposing administrative fines (article 83) and other penalties (article 84). Particularly article 83 constitute a major change of Danish data protection regulation, since the provision implies a significant increase in the size of fines imposed.

4 As a Follow-up to the Previous Question, Does Your Law Allow the Plaintiff to Receive Material or Immaterial Damages? If Yes, Is Such a Remedy Realistic in Practice?

Section 40 of the PPD Act stipulates that any person suffering from material or immaterial damages resulting from illegal processing of data, or any other act in contravention to the PPD and the GDPR is entitled to compensation according to Article 82 of the GDPR.

> Danish law operates with a number of distinctions. A distinction is made between integrity violations (also called material damage) and non-integrity violations (also referred to as immaterial damage). Integrity violations encompasses any damage to the human body, or things, animals or real estate, caused by physical means. All other damages are considered non-integrity violations. The distinction is therefore whether the damage is caused by physical or non-physical means. Non-integrity violations include violations of both author and artist rights, patents, trademarks etc, unlawful conduct during trade, such as improper marketing (financial damage), as well as violations of privacy, defamation, violations of name etc. (non-financial damage). In other words, compensation in connection with the term immaterial damage in Denmark involves a financial loss, e.g. lost turnover. In addition, a distinction is made between compensation for financial damage and compensation for

[45]Further information https://stps.dk/da/borgere/rapporter-en-utilsigtet-haendelse/.

[46]The Danish Energy Agency's website: http://www.ens.dk/.

[47]Blume (2016), p. 169 et seq.

non-financial damage – usually referred to as "remuneration". Elsewhere in legal literature on the topic of liability, the concept of damage is further divided. Firstly into damage to persons or property as well as damages that occur because of damage to persons or property (integrity violations), and secondly into damage occurring without associated damage to persons or property, so-called plain (or "clean") property damage (non-integrity violations).

In the recent white paper by the Danish Ministry of Justice, mentioned in the introduction, it is assumed that only property damage was covered by the then current version of the PPD—i.e. the provision only covered financial damage. In all other cases, claims must be advanced using a different, widely applicable, rule on remuneration available in section 26 of the Danish Liability and Compensation Act.[48] The latter provision stipulates, that the person responsible for an unlawful violation of another person's freedom, peace, honour or body shall pay a compensation for moral damages.

It follows from Danish tort law that the initial burden of proof lies with the data subject (the sufferer), i.e. the data subject must prove that there is a causal link between the unlawful data processing and loss, and that the financial loss is a foreseeable consequence of the unlawful processing, and also that a loss has in fact been suffered.[49] However, it is the responsibility of the controller to prove that he has not acted culpably. As regards a compensation for moral damages (non-financial damages), the issue of fault is also covered by the data subject's (the sufferer) burden of proof, according to section 26 the Danish Liability and Compensation Act. According to general Danish rules, if more than one party is liable for the same damage, the general rule is that they shall be jointly and severally liable towards the sufferer. In other words, the sufferer may choose to have the entire compensation paid from any one person causing losses. All persons causing losses will then be released from the claim of the injured party when the injured party has received the compensation amount. The breakdown of claims between the persons causing losses is then governed by subsection 1 of section 25 of the Danish Liability and Compensation Act, according to which the mutual distribution of the compensation burden between several jointly liable parties is made according to what may be considered reasonable in light of the nature of the liability and other circumstances. In other words, when more than one party is liable, the one party compensating the data subject may, in general, seek recourse with the other parties liable.

Compensation for moral damages in accordance with section 26 of the Danish Liability and Compensation Act requires fulfilment of several other conditions not covered by the provisions of the PPD Act, including that the violation must have been serious and that the tort in question must be considered capable of violation of integrity from an objective perspective. Contrary to the general conditions for granting compensation, however, it is not required to prove a financial loss.

There are no examples in Danish case law of compensation being granted in accordance with the PPD Act prior to the implementation of the GDPR. There are, however, examples of compensation under section 26 of the Danish Liability and Compensation Act for violation of the PPD Act—but not in relation to the right of objection and the right to be forgotten.

---

[48]White paper no. 1565, GDPR – and the legal framework of Danish legislation, p. 902.

[49]Waaben and Nielsen (2015), p. 654 et seq.

E.g. in ruling no. U 2008.727/2S, Sø- og Handelsretten (the Copenhagen Maritime and Commercial Court) considered whether an employer's video-surveillance of an employee was eligible for compensation for moral damages according to section 26 of the Danish Liability and Compensation Act. The store employee, CL, was subject to video-surveillance by his employer from the private residence of the employer for half an hour to forty-five minutes. The surveillance was not motivated by work or safety reasons. The surveillance led to a collection of information (images) of CL for purposes which CL was not aware of. It was therefore in breach of subsections 1 and 2 of section 5 of the PPD Act concerning good data processing practices which specifies that the collection of information must be for specified, explicit and legitimate purposes. Furthermore, the installation of surveillance was in violation of an agreement on control measures concluded between the Danish Confederation of Trade Unions (LO) and the Danish Employers' Confederation (DA). The tort thus justified a compensation for moral damages for CL according to section 26 of the Danish Liability and Compensation Act. The fact that the Danish Labour Court had fined the employer for violation of the agreement between LO and DA on control measures could not lead to acquittal. The nature and scope of the surveillance resulted in a compensation for moral damages of DKK 25,000 (approx. EUR 3,333.33). Another example is from 2011. In this case, the Danish Supreme Court ruled that a municipality's (H) disclosure of information about a suspected alcohol abuse as part of a potential employer's (S) obtaining of reference information was an unlawful disclosure of personal data pursuant to subsection 1 of section 7 of the PPD Act.[50] The Supreme Court did not find that A would have recruited S if the information had not been disclosed, which is why A did not find it necessary to claim compensation as a result of the unlawful disclosure. Pursuant to section 26 of the Danish Liability and Compensation Act, A was, however, granted a compensation for moral damages of DKK 25,000 (approx. EUR 3,333.33) from H.

Denmark needed to adapt national law to the GDPR. In many contexts, this did not lead to changes in existing law—and in other cases changes were made to the then current legislation. On this topic, the committee wrote the following in the white paper:

Section 1 of article 82 of the Regulation extends current legislation, in that the processor can also be held liable for damage. In addition, the wording of the regulation clarifies that any person shall have the right to compensation for material and immaterial damage, which, however, must be assumed to correspond to current legislation. In addition, section 2 of article 82 of the Regulation introduces a liability for processors under certain circumstances, in particular if the processor has not complied with obligations of the Regulation.[51]

The GDPR did therefore extend the legislation in that the processor can now be held accountable. If the interpretation found in the white paper is true, however, not much else has changed. It is included implicitly in the introductory remarks to this section that the compensation for unlawful processing of personal data—including lack of respect for the right to be forgotten—until now has been a largely unrealistic means of ensuring compliance with the regulation. Judicial reviews will continue to incur large costs in Denmark. Therefore, it is not likely that significant additional liability suits will occur now that the GDPR has been implemented. The potential amount of compensation will most likely not be sufficiently high to make data subjects run the risk of starting a process.

[50]U.2011.2343.H.

[51]White paper no. 1565, GDPR – and the legal framework of Danish legislation, p. 917 et seq.

5 In General, How Do You Assess the Implementation of the Right to Be Forgotten in Your Law? Is It Effective? Is It Used in Practice? Are There Particular Obstacles in the Implementation of This Right?

It is probably a characteristic of Danish law that there is more special legislation than in other countries. Also, there seems to be a tendency for data protection law to not exert any particular influence in the drafting of special legislation. This is most likely the reason why the question cannot be fully answered. An implementation took place in the wake of Google Spain in terms of search engines. The legal status, however, remains unchanged.

6 How Did Courts and Commentators in Your Country Welcome the ECJ Ruling on Google v González?

The Google Spain judgment led to a certain degree of media coverage in Denmark, but has not, as such, had any significance for practice or legislation. Danish courts generally do not make *obiter dictum* statements—and have not in any specific cases had the opportunity to comment on the right to be forgotten. The DPA has, however, issued a communication on the detailed procedure, see question 3.

7 For Those Who Are from a Country That Is Not Part of the European Union, Did Your Courts Follow the ECJ Ruling on the Right to Be Forgotten? Is it Likely That They Will Follow It?

Denmark is a member.

8 Did Your Law Already Grant a Similar Right to Be Forgotten to the One Stated in the ECJ Ruling?

If Google Spain is in fact interpreted as has been communicated in Denmark: Yes.

9 To Implement the ECJ Ruling, Google Has Created a Form in Which Anyone Interested Can Submit a Request to Have Information About Him-or Herself Be Delisted. Based on This Request, Google Will Weigh the Private Interest of the Petitioner and the Public Interest to Be Informed. Google Does not Disclose the Ways in Which it Deals with Requests. In Particular, Google Does not Fully Disclose the Category of Requests That Are Excluded or Accepted, the Proportion of Requests and Successful De-listings and, Among Others, the Reason for the Denial of Delisting. Do You Think That Google Should Be More Transparent About the Ways It Uses to Implement the Right to Be Forgotten?

Yes, transparency and justification must be regarded as fundamental procedural guarantees—and should also apply here.

10 Is the Procedure Prepared by Google Used in Your Country?

Yes.

11 Is There Any Upcoming Legal Reform in Your Country Whose Purpose Is to Reinforce or Modify the Right to Be Forgotten?

The GDPR was implemented in May 2018, and although it caused some adjustments, there were no major changes. There are currently no upcoming reforms on the agenda.

12 In Your Opinion, What Should Be the Next Step in the Protection of the Right to Be Forgotten? Do You Think That One Must Go Further and Strengthen the Right to Be Forgotten? Do You Think That the European Union Should Modify or Adapt Its Legislation on the Right to Be Forgotten?

As is stated implicitly in the above, the regulation of the right to be forgotten in the GDPR is not particularly radical, as the provisions in article 17 are more or less perforated due to the many exceptions. Since Denmark is one of the countries in the world where the government holds the most extensive records of the population, it should be considered whether a clearer legal situation could be established as regards the right to be forgotten with the authorities. This, however, only applies if such a right is interpreted as a right to make old and no longer relevant information less searchable for the public authorities in connection with exercising authority or distribution to the public.

References

Act no. 410 of 27/04/2017 – Act on processing of personal data by law enforcement authorities. Available at: https://www.retsinformation.dk/Forms/R0710.aspx?id=189891

Act no. 410 of 27/07/2017 – Act on processing of personal data by law enforcement authorities – entered into force on 1 May 2017. Available at: https://www.retsinformation.dk/Forms/R0710.aspx?id=189891

Act No. 429 of 31 May 2000 as amended by section 7 of Act No. 280 of 25 April 2001, section 6 of Act No. 552 of 24 June 2005, section 2 of Act No. 519 of 6 June 2007, section 1 of Act No. 188 of 18 March 2009, section 2 of Act No. 503 of 12 June 2009, section 2 of Act No. 422 of 10 May 2011, section 1 of Act No. 1245 of 18 December 2012 and section 1 of Act No. 639 of 12 June 2013

Bill no. 147 of 31 May. 2000 (the Folketing Hansard 1999/00): the special notes to section 2 & the special notes to section 37

Blume P (2014) Overvågning: Kan persondataretten gøre nytte? Tidsskrift for Informationsvidenskab og Kulturformidling, no. 2/3, year 3, p 74

Blume P (2016) Persondataforordningen. Juristen, no. 4, p 169

Charter of fundamental Rights of the European Union, article 8 (3)

Directive (EU) 2016/680 of the European Parliament and of the Council of 27 April 2016 on the protection of natural persons with regard to the processing of personal data by competent authorities for the purposes of the prevention, investigation, detection or prosecution of criminal offences or the execution of criminal penalties, and on the free movement of such data, and repealing Council Framework Decision 2008/977/JHA, available at: http://eur-lex.europa.eu/legal-content/ENG/TXT/PDF/?uri=CELEX:32016L0680&from=DA

Directive 95/46/EC of the European Parliament and of the Council of 24 October 1995 on the protection of individuals with regards to the processing of personal data and on the free movement of such data, available at: http://eur-lex.europa.eu/LexUriServ/LexUriServ.do?uri=CELEX:31995L0046:EN:HTML

DPA, Annual Report 2016 p. 13, available at: https://www.datatilsynet.dk/media/6515/datatilsynets_aarsberetning_2016_web.pdf
DPA file 2011-222-0094, available at: https://www.datatilsynet.dk/tilsyn-og-afgoerelser/historiske-afgoerelser/2012/jun/offentliggoerelse-af-personoplysninger-paa-hjemmeside-ii/
DPA file number 2002-082-0075, available at: https://www.datatilsynet.dk/tilsyn-og-afgoerelser/historiske-afgoerelser/2004/mar/spoergsmaal-om-persondatalovens-anvendelse-paa-faglitteratur/
DPA file number 2004-313-0247, available at: https://www.datatilsynet.dk/tilsyn-og-afgoerelser/historiske-afgoerelser/2005/jun/klage-over-offentliggoerelse-af-navn-og-adresse-paa-kommunes-hjemmeside/
DPA file number 2006-321-0457, 2007-321-0039 and 2008-321-0134, available at: https://www.datatilsynet.dk/tilsyn-og-afgoerelser/historiske-afgoerelser/2006/okt/offentliggoerelse-af-oplysninger-om-modstandsfolk-paa-internettet-i/, https://www.datatilsynet.dk/tilsyn-og-afgoerelser/historiske-afgoerelser/2007/sep/offentliggoerelse-af-oplysninger-om-modstandsfolk-paa-internettet-ii/ and https://www.datatilsynet.dk/tilsyn-og-afgoerelser/historiske-afgoerelser/2009/apr/offentliggoerelse-af-oplysninger-om-modstandsfolk-paa-internettet-iii/
DPA file number 2007-229-0002, available at: https://www.datatilsynet.dk/tilsyn-og-afgoerelser/historiske-afgoerelser/2008/okt/afgoerelse-i-klagesag-om-offentliggoerelse-paa-redox-hjemmeside/
DPA file number 2011-215-0874, available at: https://www.datatilsynet.dk/tilsyn-og-afgoerelser/historiske-afgoerelser/2012/jun/offentliggoerelse-af-personoplysninger-paa-hjemmeside-i/
Executive Order no. 1079 of 20 September 2017 concerning the processing of personal data in the Danish Central Crime, available at: https://www.retsinformation.dk/Forms/R0710.aspx?id=192722
Executive Order no. 1090 of 28/07/2016 – Act on authorised health persons medical records (record keeping, storage, disclosure and transfer, etc.), available at: https://www.retsinformation.dk/Forms/R0710.aspx?id=183578. Changed with order no. 530 of 24/05/2018, available at: https://www.retsinformation.dk/Forms/R0710.aspx?id=201378
Executive Order no. 1201 of 28/09/2016 – Executive Order of the Danish Act on Public Records. Available at: https://www.retsinformation.dk/Forms/R0710.aspx?id=183862
Executive Order no. 1356 of 23/10/2016 - Act on approval of health professionals and health professional activity. Available at: https://www.retsinformation.dk/Forms/R0710.aspx?id=183809
Executive Order on PET's processing of information on natural and legal persons etc, https://www.retsinformation.dk/Forms/R0710.aspx?id=164082
Executive Order on the Danish Defence Intelligence Service (FE), https://www.retsinformation.dk/Forms/R0710.aspx?id=176852 and the Act on Center for Cyber Security, https://www.retsinformation.dk/Forms/R0710.aspx?id=163853
Executive Order on the Danish Security and Intelligence Service (PET), https://www.retsinformation.dk/Forms/R0710.aspx?id=186190
Guide no. 126 of 10/07/2000 on data subjects' rights under the rules in chapters 8-10 of the Act on processing of personal data, see: https://www.retsinformation.dk/Forms/R0710.aspx?id=852
Jakobsen SS, Schaumburg-Müller S (2015) Retten til at blive glemt: og forholdet til medierne og informationsfriheden. Juristen 5(5):176–186
Jakobsen SS, Schaumburg-Müller S (2016) Mediejura for journalister og andre mediearbejdere. Jurist- og Økonomforbundets Forlag, Copenhagen, p 255, 263 et seq
Law no. 606 of 12/06/2013 – the Danish Publicity and Freedom of Information Act., available at: https://www.retsinformation.dk/Forms/R0710.aspx?id=152299 (section 15)
White paper no. 1565, GDPR – and the legal framework of Danish legislation, p 329-330, 330, 902, 917, 948, 958 et seq
Waaben H, Nielsen KK (2015) Lov om behandling af personoplysninger. Jurist- og Økonomforbundets Forlag, Copenhagen, p 121, 538, 654

Finland: The Right to Be Forgotten

Anette Alén-Savikko

Abstract The Finnish Data Protection Ombudsman (DPO) and courts have handled cases concerning the 'right to be forgotten' as rectification matters. Thus, the personal data in question must have been unnecessary, false, incomplete or outdated in relation to the purpose of processing. This chapter deals with the Finnish legal state prior to the General Data Protection Regulation (EU) 2016/679 (GDPR) which currently applies, alongside the supplementary national Data Protection Act (1050/2018).

1 How Is the "Right to Be Forgotten" Protected Under Finnish Law?

With regard to the Court of Justice of the European Union (CJEU) *Google Spain* case (C-131/12), the right to be forgotten is linked to data protection law. The Personal Data Act (523/1999; PDA)[1] is the Finnish general data protection law which for its part includes provisions implementing the EU Data Protection Directive (95/46/EC; DPD). The Finnish Data Protection Ombudsman (DPO) handles referrals related to the 'right to be forgotten' as rectification matters, that is, in the context of a data subject's right to have data corrected or erased; the personal data in question must therefore be unnecessary, false, incomplete or outdated in relation to the purpose of processing (§§ 9, 29 PDA). The DPO has taken into account national law, CJEU (and European Court of Human Rights; ECtHR) case law, as well as WP29 guidelines in particular.[2] Overall, DPO documents and subsequent court cases on the removal of links from Google search results have referred to Sections 2–3, 3 (4), 6, 8–9, 9(1), 29, 29(1), and/or 40, 40(2) of the Personal Data Act whereby it has

[1]Unofficial translation by the Ministry of Justice.

[2]See information given by the Office of the DPO: Tietosuojavaltuutetun toimisto (2017); see also Working Party (2014).

A. Alén-Savikko (✉)
Faculty of Law, University of Helsinki, Helsinki, Finland
e-mail: anette.alen@helsinki.fi

F. Werro (ed.), *The Right To Be Forgotten*, Ius Comparatum – Global Studies in Comparative Law 40, https://doi.org/10.1007/978-3-030-33512-0_5

been a question of assessing whether the search results are erroneous, unnecessary, incomplete or obsolete and whether the controller is obliged to remove such results.[3]

In its practice, the DPO has been reluctant to order removals in cases concerning professional, public or political activity in particular. Also, criminal activity has been many times under scrutiny with notes on ia severity of crimes and time passed. In two court cases before the Helsinki Administrative Court, the appellant was the controller or company operating in Finland: The first case (docket no 06038/16/1204, 20.1.2017) concerns the designation of a company as a controller. Google Finland Oy claimed that, although the request for removal of links was correctly dismissed by the DPO (docket no 1869/533/2015, 10.5.2016[4]), Google Finland Oy could not be designated a "controller" together with Google Inc. in the DPO decision. Two separate legal entities cannot amount to one party and the data protection obligations only concern Google Inc. directly. The appellant demanded the decision be repealed concerning Google Finland Oy or returned before the DPO. The DPO stated that it has not suggested an independent position for Google Finland Oy. Google Search is operated by Google Inc., while Google Finland Oy is the local place of business (§ 4(1)[5] PDA) with which data subjects in Finland may be in

[3]See DPO docket no 2351/533/2015, 15.12.2016, docket no 2199/533/20114, 25.11.2016, docket no 1891/533/2015, 21.10.2016, docket no 1869/533/2015, 10.5.2016, docket no 2024/533/2014, 5.1.2016, docket no 3190/533/2014, 19.2.2016, docket no 1610/533/2015, 4.2.2016, docket no 1374/533/2015, 3.12.2015, docket no 1924/533/2014, 29.4.2015, 1940/533/2014, 3.7.2015, docket no 2181/533/2014, 15.9.2015; see also Helsinki Administrative Court docket no 06038/16/1204, 20.1.2017 and Helsinki Administrative Court (2016) which was appealed; see below Supreme Administrative Court (2018). Some cases were apparently pending in other courts during 2017.

[4]The case before the DPO concerned a request to remove several url-search results from Google Search. The applicant claimed that the information made it impossible to move on and apply for a job, while it also leads to prank calls and various threats. Moreover, the information causes suffering, insecurity, and problems for the applicant and the applicant's family. The controller had refused to remove the results; it stated that the information was actual and appropriate, while the availability was justified by the public interest. Any defamation claims must be made separately. Before the DPO, the applicant added that the information had caused damage and that the applicant had already been punished for the crimes committed by the applicant. Moreover, some information was sensitive and the fundamental rights of the applicant, including privacy, were at stake. For instance, the photo, date of birth, and place of residence of the applicant were no longer relevant information. Some links concerned journalistic content, while others were online discussion forums. Comments by the public accompanying news publications were especially inappropriate. The DPO assessed the forums (discussions on sex offenders) as well as online publications and stated that all information was linked to the applicant's criminal activity. The information and documents regarding the crimes were also accurate and the conviction had force of law. The DPO assessed only Google's activity, not that of the original publishers. Referring to CJEU praxis and the WP29 guidelines the DPO did not see a reason to order removal; interference with the data subject's fundamental rights was justifiable taking into account "the role played by the data subject in public life" and the interest of the general public to have access to the information. This role bases, among others, on the media exposure due to the data subject's activity.

[5]§ 4(1) PDA: "This Act applies to processing of personal data where the controller is established in the territory of Finland or otherwise subject to Finnish law." (Unofficial translation by the Ministry of Justice).

contact. Therefore, the status as a party in the case is justified. The court dismissed the appeal; it stated that Google Finland Oy is the local place of business pursuant to Section 4(1) of the Personal Data Act. It has not been designated as controller, nor has it been obliged to any action.

The second case (docket no 01135/16/1204, 8.12.2016) concerns removal of links to information about a crime and punishment as well as information on the medical condition of the convicted person. The DPO (docket no 1374/533/2015, 3.12.2015[6]) had ordered removal of search results. Google Inc. and Google Finland Oy appealed, referring to the severity and actuality of the crime, as well as to the public interest in discussing the length of punishments in Finland. The appellants also referred to art 10 ECHR, while noting that information had also been made public by the mother of the person in question. Moreover, they noted that both companies cannot be addressed by the same decision; therefore the decision should at least be repealed with regard to Google Finland Oy. The DPO noted that health data was not given enough weight by the controller (diminished accountability in the criminal case). The court was of the same opinion as the DPO and found the balancing of privacy and free speech correct: the court noted, among others, that health documents are confidential under secrecy laws. Taking into account privacy as a fundamental right as well as the content of the linked information, the court found the search results unnecessary and subject to removal. Privacy of the data subject outweighed the public's right to information. There was no public role due to the criminal activity or the information given by third parties. For its part, Google Inc. was the controller and Google Finland Oy was the local place of business. The appeal was dismissed.

In 2018, the Supreme Administrative Court ruled on the matter.[7] Data related to crime and punishment as well as health data were both among the special categories of data ("sensitive data"; see Sec 11 PDA). The interest of the public did not prevail over privacy and data protection with regard to health related information of a

[6]The applicant had requested the removal of search results from Google Search, but the controller refused to remove the results due to the legitimate interest of the public, taking into account the nature of the crime and actuality thereof. The background involved media coverage of a crime committed by the applicant; even though the applicant was found to have diminished accountability one paper had published the name of the applicant as well as some health related information. The applicant stated that this was contrary to established media practice and led to stigmatization. The applicant had also served most of the punishment. Moreover, health information is sensitive data whereby the interest of the public cannot prevail. The DPO ordered the controller to remove the search results. One link led to an online discussion forum where the crime, the applicants name and health was discussed, while another link led to news and media coverage. The information concerned crime and punishment on one hand and health on the other hand. The DPO referred to CJEU praxis and WP29 guidelines whereby the crime and time lapsed could be taken into account in assessing the issue. The serious crime, with a severe punishment, was actual. However, the health related information made the issue wider than this; it was not incorrect, but nonetheless unnecessary. Overall, the DPO noted that national legislation restricts access to health data in many ways, while journalistic practice also involves restrictions with regard to sensitive data in crime news.

[7]See ECLI:FI:KHO:2018:112.

convicted person. Moreover, the websites in question were searchable without using the person's name, which meant that societal debate was practically unrestricted. The crime was severe and quite recent, and, consequently, the person's role in public life had relevance for weighing fundamental rights. However, the fact that health data situated at the core of privacy had to be taken into account in the balancing act as well.

The provisions in Sections 2–3 of the Finnish Personal Data Act include the scope of application and secondary nature of the act as well as the definitions of terms and concepts respectively, while Section 6[8] provides the requirement for advance planning any processing of personal data. The purposes, sources and recipients must also be defined in advance whereas the purposes must be defined so as to clearly indicate the corresponding tasks of the controller (§ 6). Section 8 includes the general prerequisites for processing. According to Section 8(1):

Personal data shall be processed only if:

(1) the data subject has unambiguously consented to the same;
(2) the data subject has given an assignment for the same, or this is necessary in order to perform a contract to which the data subject is a party or in order to take steps at the request of the data subject before entering into a contract;
(3) processing is necessary, in an individual case, in order to protect the vital interests of the data subject;
(4) processing is based on the provisions of an Act or it is necessary for compliance with a task or obligation to which the controller is bound by virtue of an Act or an order issued on the basis of an Act;
(5) there is a relevant connection between the data subject and the operations of the controller, based on the data subject being a client or member of, or in the service of, the controller or on a comparable relationship between the two (connection requirement);
(6) the data relate to the clients or employees of a group of companies or another comparable economic grouping, and they are processed within the said grouping;
(7) processing is necessary for purposes of payment traffic, computing or other comparable tasks undertaken on the assignment of the controller;
(8) the matter concerns generally available data on the status, duties or performance of a person in a public corporation or business, and the data is processed in order to safeguard the rights and interests of the controller or a third party receiving the data; or
(9) the Data Protection Board has issued a permission for the same, as provided in section 43(1).[9]

Section 9 PDA includes the principles related to data quality, that is, necessity and accuracy, as follows:

---

[8] § 6 PDA (Defined purpose of processing): "It must be appropriate and justified to process personal data in the operations of the controller. The purpose of the processing of personal data, the regular sources of personal data and the regular recipients of recorded personal data shall be defined before the collection of the personal data intended to be recorded in the file or their organisation into a personal data file. The purpose of the processing shall be defined so that those operations of the controller in which the personal data are being processed are made clear." (Unofficial translation by the Ministry of Justice).

[9] Unofficial translation by the Ministry of Justice.

(1) *The personal data processed must be necessary for the declared purpose of the processing (necessity requirement).*
(2) The controller shall see to that no erroneous, incomplete or obsolete data are processed (accuracy requirement). This duty of the controller shall be assessed in the light of the purpose of the personal data and the effect of the processing on the protection of the privacy of the data subject.[10] (Italics by the author)

For its part, Section 29 regulates rectification so that:

(1) *The controller shall, on its own initiative or at the request of the data subject, without undue delay rectify, erase or supplement personal data contained in its personal data file and erroneous, unnecessary, incomplete or obsolete as regards the purpose of the processing. The controller shall also prevent the dissemination of such data, if this could compromise the protection of the privacy of the data subject or his/her rights.*
(2) If the controller refuses the request of a data subject of the rectification of an error, a written certificate to this effect shall be issued. The certificate shall also mention the reasons for the refusal. In this event, the data subject may bring the matter to the attention of the Data Protection Ombudsman.
(3) The controller shall notify the rectification to the recipients to whom the data have been disclosed and to the source of the erroneous personal data. However, there is no duty of notification if this is impossible or unreasonably difficult.[11] (Italics by the author)

Finally, Section 40 includes provisions on the measures the DPO undertakes:

(1) The Data Protection Ombudsman shall promote good processing practice and issue directions and guidelines so as to achieve a situation where unlawful conduct is not continued or repeated. Where necessary, the Data Protection Ombudsman shall refer the matter to be dealt with by the Data Protection Board, or report it for prosecution.
(2) *The Data Protection Ombudsman shall decide matters brought to his/her attention by data subjects on the basis of sections 28 and 29. The Ombudsman may order a controller to realise the right of access of the data subject or to rectify an error.*
(3) The Data Protection Ombudsman may issue more detailed guidelines on how personal data is to be secured against unlawful processing.[12] (Italics by the author)

According to the preparatory legislative work leading to the Personal Data Act,[13] Sections 2–3, 3(4), 6, 8–9, 9(1), 29, 29(1), and 40 of the Personal Data Act implement the Data Protection Directive as follows: Section 2(2) corresponds to the scope of application provided in Article 3(1) DPD, while Section 2(3) corresponds to Article 3(2) (household exemption). Section 2(5) on derogations for purposes of journalistic, artistic and literary expression was in line with Article 9 of the directive.[14] Definitions in Section 3 of the act, including that of a controller (§ 3(4)), were enacted in light of Article 2 of the directive.[15] Section 6 implements Article 6(1)(a) alongside Section 5 on duty of care and lawfulness of processing.[16]

[10]Unofficial translation by the Ministry of Justice.

[11]Unofficial translation by the Ministry of Justice.

[12]Unofficial translation by the Ministry of Justice.

[13]Government (1998).

[14]Government (1998), pp. 32–34.

[15]Government (1998), pp. 34–36.

[16]Government (1998), p. 38.

Section 8 of the act bases on Article 7 of the directive; Article 7(d)-(f), including legitimate interest ground, was, however, implemented in Finland so as to require the permission of the Data Protection Board.[17] For its part, Section 9 of the Personal Data Act implements the requirements laid down in Article 6(1)(d) and 6(2) of the directive; Section 9(1) in particular includes the necessity requirement and paragraph 2 the accuracy requirement.[18] Section 29 on rectification corresponds to Article 12 (b)-(c) of the directive, while Section 40 includes provisions implementing the requirements for supervisory authorities' powers laid down in Article 28(3) and those for remedies laid down in Article 22.[19]

The right to be forgotten has links already to several provisions of the Data Protection Directive—despite the fact that the right is explicitly included in the General Data Protection Regulation (EU) 2016/679 (GDPR).[20] The current Finnish Personal Data Act implements the Directive. With regard to the Directive, "the right to be forgotten", as enshrined in the praxis of the CJEU, especially bases on Articles 6, 12(b) and 14(1)(a).[21] Article 6 includes provisions on the general principles, such as lawfulness of processing and adequacy, relevance and accuracy of personal data. Article 12 regulates right of access to data, while the obligation to guarantee for data subjects the right to rectification, erasure or blocking of data, particularly in cases of incompleteness or inaccurateness of the data, is laid down in Article 12(b). According to Article 14(1)(a), Member States must also grant data subjects the right to "at least in the cases referred to in Article 7 (e) and (f), to object at any time on compelling legitimate grounds relating to his particular situation to the processing of data relating to him, save where otherwise provided by national legislation". In addition, the right to be forgotten is linked to Articles 7 and 8 of the EU Charter of Fundamental Rights (CFR).[22]

In the Finnish Personal Data Act Section 30 on the right to prohibit processing provides for a data subject's right to prohibit processing of personal data "for purposes of direct advertising, distance selling, other direct marketing, market research, opinion polls, public registers or genealogical research".[23] The right to prohibit processing in the context of direct marketing bases on Article 14(1)(b) of the directive. However, the act does not contain a general provision on the right to object pursuant to Article 14(1)(a). Processing on the basis of Article 7(e)-(f) as well the data subjects' rights would already be assessed in the context of Section 8 and Chapter 4 (processing for special purposes, including direct marketing) of the act.[24]

[17]Government (1998), pp. 38–41. Cf. C-468/10 and C-469/10 *ASNEF and FECEMD* ECLI:EU:C:2011:777.

[18]Government (1998), p. 42.

[19]Government (1998), pp. 63–64, 72.

[20]The right to erasure (the right to be forgotten) is enshrined in Article 17 GDPR.

[21]C-131/12 *Google Spain and Google* ECLI:EU:C:2014:317, paras 88, 93, 99.

[22]C-131/12, paras 74, 91, 97, 99.

[23]Unofficial translation by the Ministry of Justice.

[24]Government (1998), pp. 64–65.

In Finland, the national level constitutional ground for data protection is privacy as enshrined in Section 10 of the Constitution of Finland (731/1999).[25] The Data Protection Directive also notes fundamental rights, in particular the right to privacy, in its Article 1.[26] Thereby, the Finnish Personal Data Act aims to protect fundamental rights related to privacy in the context of processing personal data.[27] The right to privacy in Section 10 of the Constitution explicitly covers private life, domestic peace and honour, while more detailed provisions on personal data protection are to be laid down in law (§ 10(1)). Confidentiality of communication is also protected (§ 10(2)). The protection of privacy also entails the right to self-determination and the right to be informed and decide on the use of personal data.[28]

With regard to privacy and private life, the right to be forgotten in data protection legislation is different from the criminalization of unlawful dissemination of information violating personal private life (§ 8, Ch. 24 Penal Code 39/1889).[29] The provision applies to unjustified dissemination, via mass media or otherwise widely, of information, an insinuation, or an image of the private life of another person, in a manner apt to cause damage, suffering, or contempt (§ 8(1)) with an exception for persons in politics, business, in public office or position or in a comparable position and where necessary for public interest (§ 8(2)). The aggravated form (§ 8 a) applies where there is great suffering or particularly vast damage as well as an overall seriousness.[30]

2 What Are the Limits to the Right to Be Forgotten Under Finnish Law?

National provisions as explained above are interpreted in light of EU law. As noted, Article 14(1)(a) DPD has not been implemented in Finnish law as a general provision for processing based on Article 7(e)-(f) of the directive. In the initial discussions around the Google Spain case, dr. Riitta Ollila pointed to this defect in a blog (in the position of the vice chairman of the Council for Mass Media) as a possible source for challenges in applying the right to be forgotten in Finland.[31]

[25]Unofficial translation by the Ministry of Justice.

[26]Cf. GDPR and the right to protection of personal data as enshrined in Article 8 of the CFR and Article 16(1) of the Treaty on the Functioning of the European Union (TFEU) (GDPR, recitals 1, 4).

[27]Government (1998), p. 30.

[28]Government (1993), pp. 52–53; (1998), p. 30. For a comprehensive account on privacy in Finland, see Neuvonen (2014).

[29]Unofficial translation by the Ministry of Justice.

[30]For a comprehensive account on crimes related to freedom of expression, see Tiilikka (2007, 2008). See also Alén-Savikko (2016), pp. 247–249.

[31]Ollila (2014b); see also Ollila (2014a), pp. 818–819 where similar remarks are made with regard to the scope and acceptability of the requests for removal.

However, Helsinki Administrative Court has based its ruling on the necessity requirement and rectification (§§ 9, 29 PDA).[32] Similarly, the Supreme Administrative Court has referred to Sections 9 and 29 PDA.[33]

3 What Are the Legal Remedies Available to Enforce the "Right to Be Forgotten" in Finnish Law?

National substantive provisions as explained above are interpreted in light of EU law. With regard to the process, first, the data subject must contact the controller, then the Data Protection Ombudsman may be contacted and, finally, parties may appeal to the Administrative Court and even to the Supreme Administrative Court (leave to appeal). In general, when a referral is made, the DPO should be informed on the initial contact to the controller and the negative outcome (refusal attached), while also the links in question should be specified.[34] Section 29 of the Personal Data Act regulates rectification so that:

(1) *The controller shall, on its own initiative or at the request of the data subject, without undue delay rectify, erase or supplement personal data contained in its personal data file and erroneous, unnecessary, incomplete or obsolete as regards the purpose of the processing. The controller shall also prevent the dissemination of such data, if this could compromise the protection of the privacy of the data subject or his/her rights.*

(2) If the controller refuses the request of a data subject of the rectification of an error, a written certificate to this effect shall be issued. The certificate shall also mention the reasons for the refusal. In this event, the data subject may bring the matter to the attention of the Data Protection Ombudsman.

---

[32]Helsinki Administrative Court (2016). The case concerns removal of links to information about a crime and punishment as well as information on the medical condition of the convicted person. The DPO (docket no 1374/533/2015, 3.12.2015) had ordered removal of search results. Google Inc. and Google Finland Oy appealed, referring to the severity and actuality of the crime, as well as to the public interest in discussing the length of punishments in Finland. The appellants also referred to art 10 ECHR, while noting that information had also been made public by the mother of the person in question. Moreover, they noted that both companies cannot be addressed by the same decision; therefore the decision should at least be repealed with regard to Google Finland Oy. The DPO noted that health data was not given enough weight by the controller (diminished accountability in the criminal case). The court was of the same opinion as the DPO and found the balancing of privacy and free speech correct: the court noted, among others, that health documents are confidential under secrecy laws. Taking into account privacy as a fundamental right as well as the content of the linked information, the court found the search results unnecessary and subject to removal. Privacy of the data subject outweighed the public's right to information. There was no public role due to the criminal activity or the information given by third parties. For its part, Google Inc. was the controller and Google Finland Oy was the local place of business. The appeal was dismissed. The case was appealed. See Supreme Administrative Court (2018).

[33]Supreme Administrative Court (2018).

[34]See DPO Office (2017).

(3) The controller shall notify the rectification to the recipients to whom the data have been disclosed and to the source of the erroneous personal data. However, there is no duty of notification if this is impossible or unreasonably difficult.[35] (Italics by the author)

Section 40 includes provisions on the measures the DPO undertakes:

(1) The Data Protection Ombudsman shall promote good processing practice and issue directions and guidelines so as to achieve a situation where unlawful conduct is not continued or repeated. Where necessary, the Data Protection Ombudsman shall refer the matter to be dealt with by the Data Protection Board, or report it for prosecution.
(2) *The Data Protection Ombudsman shall decide matters brought to his/her attention by data subjects on the basis of sections 28 and 29. The Ombudsman may order a controller to realise the right of access of the data subject or to rectify an error.*
(3) The Data Protection Ombudsman may issue more detailed guidelines on how personal data is to be secured against unlawful processing.[36] (Italics by the author)

Section 46 provides the means for the DPO to impose a threat of a fine to reinforce the duty to provide access to data (§§ 39(1) and 39(3)) and a decision based on Section 40(2). Section 45 includes provisions on appeal (as amended by Act 901/2015[37]):

(1) *The decision of the Data Protection Ombudsman and the Data Protection Board may be appealed to the Administrative Court in accordance with the Administrative Judicial Procedure Act (586/1996).*[38] The Data Protection Ombudsman may appeal the decision of the Data Protection Board made pursuant to section 43.[39]
(2) *The decision of the Administrative Court may be appealed only if the Supreme Administrative Court grants leave to appeal.*[40]
(3) It may be ordered in a decision of the Data Protection Board that it is to be complied with regardless of appeal, unless otherwise ordered by the appellate authority.[41] (Italics by the author)

Appeals are regulated in the Administrative Judicial Procedure Act as follows:

§ 4 An administrative decision may be challenged by an appeal as provided in this Act.[42]

§ 9 Appeal against a decision of an Administrative Court shall be lodged in the Supreme Administrative Court.[43]

[35] Unofficial translation by the Ministry of Justice.

[36] Unofficial translation by the Ministry of Justice.

[37] See Government (2014) whereby the system of leave to appeal was widened to cover more matters than before, including in the field of data protection.

[38] Unofficial translation by the Ministry of Justice.

[39] Translation by the author.

[40] Translation by the author.

[41] Unofficial translation by the Ministry of Justice.

[42] Unofficial translation by the Ministry of Justice.

[43] Unofficial translation by the Ministry of Justice.

§ 13 (as amended by Act 891/2015)

(1) *The law shall contain specific provisions on situations where decisions of an authorities referred to in sections 7–9 may not be appealed or where leave to appeal is required for an appeal before the Supreme Administrative Court.*
(2) Where a leave to appeal is required for appealing an Administrative Court decision before the Supreme Administrative Court, the leave must be granted if:

1) it is important to have the matter submitted before the Supreme Administrative Court due to application of the law in similar cases or uniformity of praxis;
2) there is a particular reason, due to a manifest error, to have the matter tried before the Supreme Administrative Court; or
3) there is some other weighty reason for granting the leave.

(3) The leave may also be granted so as to cover only a part of the Administrative Court decision under appeal.
(4) If an appeal on a decision in the main issue is prohibited or subject to a leave to appeal, the same restriction applies to an appeal in a related issue.
(5) The restrictions of the right to appeal laid down in this section do not concern appeals on Administrative Court decisions in matters of administrative litigation if not otherwise provided in the law.[44] (Italics by the author)

There are also criminal law dimensions to data protection in Finland.[45] Section 48 (1) of the Personal Data Act refers to the provisions of the Penal Code (39/1889; PC[46]) as follows:

The penalty for a personal data offence is provided in chapter 38, section 9 of the Penal Code (39/1889) and for breaking into a personal data file in chapter 38, section 8 of the Penal Code. The penalty for a violation of the secrecy obligation referred to in section 33[47] is provided in chapter 38, section 1 or 2 of the Penal Code, unless the act is punishable under chapter 40, section 5 of the Penal Code or a more severe penalty is provided in another Act.[48]

[44]Translation by the author.

[45]For more, see Pitkänen et al. (2013), pp. 287–307.

[46]Unofficial translation by the Ministry of Justice.

[47]§ 33: "Anyone who has gained knowledge of the characteristics, personal circumstances or economic situation of another person while carrying out measures relating to data processing shall not disclose the data to a third person against the provisions of this Act." (Unofficial translation by the Ministry of Justice).

[48]Unofficial translation by the Ministry of Justice.

The referred provisions thus include data protection offense (§ 9, Ch. 38[49]), computer break-in (§ 8, Ch. 38[50]), secrecy offense and violation (§§ 1-2, Ch. 38[51]), as well as breach of official secrecy (§ 5, Ch. 40[52]). There is Supreme

[49]§ 9: "A person who intentionally or grossly negligently (1) processes personal data in violation of the provisions of the Personal Data Act (523/1999) on the exclusivity of purpose, the general prerequisites for processing, the necessity and integrity of data, sensitive data, identification codes or the processing of personal data for specific purposes, or violates a specific provision on the processing of personal data, (480/2001) (2) by giving false or misleading information prevents or attempts to prevent a data subject from using his or her right of inspection, or (3) conveys personal data to states outside the European Union or the European Economic Area in violation of Chapter 5 of the Personal Data Act, *and thereby violates the privacy of the data subject or causes him or her other damage or significant inconvenience*, shall be sentenced for a data protection offence to a fine or to imprisonment for at most one year." (Unofficial translation by the Ministry of Justice; Italics by the author) Section 9(a) includes provisions on identity theft.

[50]§ 8: "(1) A person who by using an access code that does not belong to him or her or by otherwise breaking a protection unlawfully hacks into an information system where information or data is processed, stored or transmitted electronically or in a corresponding technical manner, or into a separately protected part of such a system, shall be sentenced for a computer break-in to *a fine or to imprisonment for at most two years*. (2) Also a person who, without hacking into the information system or a part thereof, (1) by using a special technical device or (2) otherwise by by-passing the system of protection in a technical manner, by using a vulnerability in the information system or otherwise by evidently fraudulent means unlawfully obtains information or data contained in an information system referred to in subsection 1, shall be sentenced for a computer break-in. (3) An attempt is punishable. (4) This section applies only to acts that are not subject to an equally severe or more severe penalty provided elsewhere in the law." (Unofficial translation by the Ministry of Justice; Italics by the author) The aggravated form is regulated in Section 8a, while Section 8 (b) includes provisions on an offence involving a system for accessing protected services.

[51]§ 1: "A person who in violation of a secrecy duty provided by an Act or Decree or specifically ordered by an authority pursuant to an Act (1) discloses information which should be kept secret and which he or she has learnt by virtue of his or her position or task or in the performance of a duty, or (2) makes use of such a secret for the gain of himself or herself or another shall be sentenced, unless the act is punishable under Chapter 40, section 5, for a secrecy offence to *a fine or to imprisonment for at most one year*. (Unofficial translation by the Ministry of Justice; Italics by the author) § 2: "(1) If the secrecy offence, in view of the significance of the act as concerns the protection of privacy or confidentiality, or the other relevant circumstances, is petty when assessed as a whole, the offender shall be sentenced for a secrecy violation to a fine. (2) Also a person who has violated a secrecy duty referred to in section 1 and it is specifically provided that such violation is punishable as a secrecy violation, shall also be sentenced for a secrecy violation." (Unofficial translation by the Ministry of Justice; Italics by the author).

[52]§ 5: "(1) If a public official intentionally, while in service or thereafter, unlawfully (1) discloses a document or information which pursuant to the Act on the Openness of Government Activities (621/1999) or another Act is to be kept secret of not disclosed, or (2) makes use of the document or information referred to in paragraph (1) to the benefit of himself or herself or to the loss of another, shall, unless a more severe penalty has been provided elsewhere in law for the act, be sentenced for breach of official secrecy to a fine or to imprisonment for at most two years. A public official may also be sentenced to dismissal if the offence demonstrates that he or she is manifestly unfit for his or her duties. (2) If a public official commits the offence referred to in subsection 1 through negligence, and the act, in view of its harmful and damaging effects and the other relevant circumstances, is not of minor significance, he or she shall, unless a more severe penalty has been provided elsewhere in law for the act, be sentenced for negligent breach of official secrecy to *a fine or to imprisonment for at most six months*." (Unofficial translation by the Ministry of Justice; Italics by the author).

Court praxis regarding the provisions.[53] For example, two cases concerning data protection offence provide examples on the additional requirements alongside acting contrary to data protection law in a manner specified in § 9, Ch. 38 PC; violation of privacy of the data subject or damage or significant inconvenience to the data subject is required. The first case (KKO 1998:85; vote) concerned transfer by a company X of its newspaper subscriber register data to other companies whereby the court found violation of privacy. However, due to mitigating circumstances (time lapsed, direct marketing purposes, action after being contacted by the DPO), the representatives of company X were not sentenced to punishment. The other case (KKO 1999:127) concerned an insurance salesman who switched jobs and took advantage of a customer register from the previous employer. No evidence was found to support violation of privacy, while no accusations were made with regard to damage or inconvenience. Charges were dismissed.[54]

For its part, Section 48(2) of the Personal Data Act includes a provision on personal data violation as follows:

> A person who intentionally or grossly negligently and contrary to the provisions in this Act:
>
> (1) fails to comply with the provisions on the definition of the purpose of the processing of the personal data, the drawing up of the description of the file, the information on data processing, the rectification of the file, the right of the data subject to prohibit the processing of data or the notification of the Data Protection Ombudsman;
> (2) provides false or misleading data to a data protection authority in a matter concerning a personal data file;
> (3) breaks the rules or regulations on the protection and destruction of personal data files; or
> (4) breaks a final order issued by the Data Protection Board on the basis of section 43(3), thus compromising the protection of the privacy of the data subject or his/her rights,
>
> shall be sentenced for a personal data violation to *a fine*, provided that a more severe penalty is not provided in another Act.[55] (Italics by the author)

This provision on personal data violation is secondary, and, thus, the applicability of other provisions must be assessed first.[56] Criminal law applies to the most severe acts following the principle of *ultima ratio*; it provides protection for interests directly deriving from fundamental rights.[57]

According to the report of the Working Group, set up by the Ministry of Justice to prepare a proposal concerning national leeway and assess the need for a general legislative act, the existing provisions on personal data offence (§ 9, Ch. 38 PC) would be replaced by introducing provisions on data protection offence. According to the report, an all-encompassing provision is no longer necessary, while the sanction system undergoes a radical change with the GDPR. Criminal provisions should only cover areas not covered by administrative fines pursuant to the

[53]For more, see Pitkänen et al. (2013), pp. 290–307.

[54]For more, see Pitkänen et al. (2013), pp. 290–291.

[55]Unofficial translation by the Ministry of Justice.

[56]Pitkänen et al. (2013), p. 289.

[57]Pitkänen et al. (2013), pp. 287–289.

Regulation. Thereby, it would cover situations where a person is not acting as a controller or a processor. Corporate criminal liability would not be applicable. An example is snooping out of curiosity without a legal basis for processing (e.g., in hospitals). The provision would also apply to a person acting against data security, such as someone disposing of personal data documents without taking into account security regulations.[58] Currently, the GDPR and the national Data Protection Act (1050/2018) apply.

4 Does Finnish Law Allow the Plaintiff to Receive Material or Immaterial Damages? How Realistic Is the Remedy in Practice?

Section 47 of the Personal Data Act includes provisions on the liability for damages whereby

> (1) The controller is liable to compensate for the economic and other loss suffered by the data subject or another person because of processing of personal data in violation of the provisions of this Act.

However, otherwise the provisions of the Tort Liability Act (412/1974; TLA)[59] apply, more specifically Sections 2–3 of Chapter 2, Sections 4 and 6 of Chapter 3 and Chapters 4, 6 and 7 (§ 47(2)). This means that pursuant to the aforementioned Section 47(1) of the Personal Data Act, liability for damages is wider than otherwise since no intent or negligence is required.[60]

The Tort Liability Act is not applicable to contractual liability or liability for damages within the scope of a special law, unless the law provides otherwise (§ 1 TLA). In case a contractual relation exists, contractual liability applies.[61]

Illegal processing of personal data or processing violating the law typically involves so called pure economic loss, that is, loss without a connection to personal injury[62] or damage to property, or it involves suffering. The specific requirement for compensation included in Chapter 5[63] of the Tort Liability Act are, however, not

[58]Ministry of Justice (2017), pp. 173–174.

[59]Unofficial translation by the Ministry of Justice.

[60]Pitkänen et al. (2013), p. 278.

[61]Pitkänen et al. (2013), p. 278.

[62]Damages for personal injury pursuant to Section 2 of Chapter 5 of the Tort Liability Act cover compensation for medical costs, loss of income as well as pain and harm, among others. See also Alén-Savikko (2016), p. 250.

[63]Damages only cover compensation for pure economic loss where it is a question of exercise of public authority, punishable acts, or particularly weighty reasons (§ 1, Ch. 5 TLA). With regard to compensation for suffering, for instance, a punishable act targeting liberty, peace, honor, or privacy is required or human dignity must be seriously infringed (§ 6, Ch. 5 TLA). For more, see Tiilikka (2008), pp. 308–324; Tiilikka (2007), pp. 312–314; Government (2003), pp. 16–17. See also Alén-Savikko (2016), p. 251.

applied in the context of personal data processing. It suffices to act against the Personal Data Act. This means a wider range of compensable loss than generally in situations involving pure economic loss.[64] Pure economic loss may occur, for example, where a data subject loses income due to a data leakage, where banking credentials are leaked due to insufficient security measures or where false data lead to higher interest rate for a credit.[65] Suffering, in turn, refers to feelings of humiliation, fear, despair or the like. This may occur, among other, in a situation where personal photos are leaked from a cloud service to the open internet.[66] Personal injury, such as a psychological injury, is compensable separately—but this would be exceptional in the case of personal data processing.[67] In general, liability requires causality and the one who has suffered loss must show it.[68] With regard to suffering, the underlying act must be such as to typically cause suffering (the nature of the act is enough with no further proof necessary); thus, subjective experience is in principle irrelevant (even if individual features might be taken into account in the amount of compensation), but it would also be difficult otherwise to show causality and amount of suffering, and thus, get compensation.[69]

The GDPR includes provisions on effective judicial remedies (Art. 79 GDPR), while Article 82[70] GDPR in particular includes provisions on the right to compensation and liability (cf. Arts. 22–23 DPD). The report of the Working Group, set up by the Ministry of Justice to prepare a proposal concerning national leeway and assess the need for a general legislative act, notes that Article 82 GDPR allows for no

[64]Pitkänen et al. (2013), pp. 278–279.

[65]Pitkänen et al. (2013), p. 279.

[66]Pitkänen et al. (2013), pp. 279–280.

[67]Pitkänen et al. (2013), p. 280.

[68]For burden of proof, see Ch. 17 Code of Judicial Procedure (4/1734).

[69]Pitkänen et al. (2013), p. 281. See e.g. Supreme Court (2000, 2005).

[70]"1. *Any person who has suffered material or non-material damage as a result of an infringement of this Regulation shall have the right to receive compensation from the controller or processor for the damage suffered.* 2. Any controller involved in processing shall be liable for the damage caused by processing which infringes this Regulation. A processor shall be liable for the damage caused by processing only where it has not complied with obligations of this Regulation specifically directed to processors or where it has acted outside or contrary to lawful instructions of the controller. 3. A controller or processor shall be exempt from liability under paragraph 2 if it proves that it is not in any way responsible for the event giving rise to the damage. 4. Where more than one controller or processor, or both a controller and a processor, are involved in the same processing and where they are, under paragraphs 2 and 3, responsible for any damage caused by processing, each controller or processor shall be held liable for the entire damage in order to ensure effective compensation of the data subject. 5. Where a controller or processor has, in accordance with paragraph 4, paid full compensation for the damage suffered, that controller or processor shall be entitled to claim back from the other controllers or processors involved in the same processing that part of the compensation corresponding to their part of responsibility for the damage, in accordance with the conditions set out in paragraph 2. 6. Court proceedings for exercising the right to receive compensation shall be brought before the courts competent under the law of the Member State referred to in Article 79(2)." (Italics by the author) See also recital 146.

national leeway.[71] However, some issues, such as causality and burden of proof, appear to stay dependent on national frameworks, and achieving the objective of full compensation seems uncertain; there is also uncertainty with regard to data protection damage being apt for claims.[72] Recital 75 in the preamble of the Regulation states that the damage may be "physical, material or non-material damage, in particular: where the processing may give rise to discrimination, identity theft or fraud, financial loss, damage to the reputation, loss of confidentiality of personal data protected by professional secrecy, unauthorised reversal of pseudonymisation, or any other significant economic or social disadvantage", which arguably covers pure economic loss as well.[73] With regard to intent or negligence, the Regulation leaves some open questions in terms of the strictness of liability, while national traditions differ and continue to play a role.[74] The GDPR includes explicit provisions on processor liability (see Arts. 28, 82 GDPR), and data subjects may rely directly on the Regulation for their rights (instead of contracts). This may be deemed a significant improvement for data subjects.[75] For its part, CJEU praxis in other areas might include relevant findings for data protection contexts as well.[76]

5 How Do You Assess the Implementation of the "Right to Be Forgotten" in Finnish Law? Is It Effective? Is It Used in Practice? Are There Particular Obstacles in the Implementation of This Right?

National provisions are explained in Q1. They must be interpreted in light of EU law. Thus, even if the right to be forgotten was not as such implemented in Finnish law, the provisions underlying the CJEU ruling have been transposed into national law. Some leeway has, however, been used at the national level when it comes to Article 7(d)-(f) and Article 14(1)(a) DPD (see Q1/Q2).

The right to be forgotten is used in practice. Finnish people began to utilize the opportunity to request the removal of links from Google's search results. According to news coverage, by the end of June 2014, around 800 requests (concerning almost 3000 web addresses) were made from Finland. This was a lot in relation to population and compared to some other European countries.[77] The DPO was also later contacted. By the end of November 2014, a couple of dozen referrals to the DPO had

[71]Ministry of Justice (2017), p. 63.

[72]Wennäkoski (2017), pp. 68–74, 76–77, 79–80, 91–93.

[73]Wennäkoski (2017), p. 73.

[74]Wennäkoski (2017), pp. 74–79.

[75]Wennäkoski (2017), pp. 84–85, 92.

[76]See e.g., C-367/15 and C-479/93 for CJEU views on showing damage and imposing national requirements respectively. See also Wennäkoski (2017), pp. 79–80.

[77]Naalisvaara (2014) and Brunila (2014).

been made. The first orders by the DPO were issued in December 2015; one of these decisions concerning links to information about a crime and a convicted person was appealed by Google, as noted above. By January 2016, a dozen decisions had been made by the DPO, while referrals kept coming in.[78] According to information from the Office of the DPO over 30 referrals were made during 2016, whereas early 2017 saw almost 70 referrals pending at the office. Most of these concern requests denied by search engines. By April 2017, 12 decisions[79] had been made by the DPO out of which 8 were dismissals, 2 were orders for partial removal, and 2 were orders for full removal. Some decisions have been appealed.[80]

Currently, Article 17 of the GDPR applies.

6 How Did Courts and Commentators in Finland Welcome the ECJ Ruling on *Google v González*?

The ruling evoked some discussion among legal scholars. The ruling was both analyzed and criticized. As noted above, Article 14(1)(a) DPD has not been implemented in Finnish law as a general provision for processing based on Article 7(e)-(f) of the directive. In the initial discussions around the Google Spain case, dr. Riitta Ollila pointed to this defect.[81] Later, she also analyzed the case in an article on data protection as an EU fundamental right.[82] She noted the fact that national legislative solutions may differ in some regards (Art. 14(1)(a)) which might affect the scope and acceptability of the requests for removal. Moreover, she pointed to the means available pursuant to the Act on the Exercise of Freedom of Expression in the Mass Media (460/2003; Freedom of Expression Act; FEA) in situations of criminal content (defamation, dissemination of information violating personal private life, etc.).[83]

Moreover, the flexibility of the Data Protection Directive in light of the developing technology was noted alongside the need to provide a (more) unified legal framework in the EU and take into account the global implications of the online environment.[84] For instance, Senior Councelor Timo Ruikka (DLA Piper Finland Oy) describes the experience of dealing with Google regarding requests for removal as follows:

78 Rautio (2016a, b), Junttila (2014) and Karhula (2016).

79 See n 3 supra.

80 See DPO Office (2017).

81 Ollila (2014b).

82 Ollila (2014a), pp. 818–820.

83 Ollila (2014a), pp. 818–819.

84 Virtanen (2014), pp. 26–27.

> Dealing with Google is an alienating experience in an Orwellian, if not outright a Kafka-like, manner. Not one of the Google documents includes the name of the person who handled the matter. It is quite possible that at least part of the messages are generated fully automated. The documents have numerous language errors pointing to machine-made translation. (. . .) The letter has not been signed by hand, but it has been send by "The Google Removals team" which remains unknown. The e-mails include the signature "Google-tiimi" [in Engl. Google team]. Therefore, the documents contribute to an uncertainty of whether the matter has even been considered appropriately on the side of Google, or that even the material decision would be made by a human being or human beings.[85]

The right to be forgotten was also analyzed in a wider communicational context and in relation to the dialectics of remembering and forgetting.[86] Criticism against the decision was mapped (including from the US), for instance, the resources required from relevant controllers, the EU "bubble" on the internet and the principles related to freedom of expression and information. The protection of an individual was emphasized as the rationale for the right to be forgotten instead of online censorship. However, the responsibilities of private entities deriving from the right to be forgotten were also deemed conducive to increase or reinforce their power in the online environment; the execution of the right is largely left within the hands of private corporations who thus shape the online public sphere by their decisions and processes.[87]

The DPO also provide information on the ruling and its implications. The DPO's office informed the public about the ruling on its website.[88] It also further informed the public on the core elements of the ruling as well as the fact that the controller must be contacted first, not directly the DPO, while promising to update its guidelines on the removal of information from online search engines.[89] The DPO's office also announced that Google had set up a mechanism to handle the requests for removal and referred to the Finnish version of the online form.[90] The DPO, Reijo Aarnio, was interviewed in the professional magazine of the Association of Finnish Lawyers and one part of the interview concerned the CJEU ruling on the 'right to be forgotten'; the DPO praised the ruling in many regards, for instance, where Google's role as a controller and the territorial scope of the directive were concerned.[91]

With regard to news coverage in general, the ruling was also noted, informing people of the right to be forgotten, while some requests for removal or Google's processes and decisions were especially highlighted. In some news coverage, the DPO was consulted, while some referred to other sources. Just to give a few examples, Google's Country Manager in Finland, Anni Ronkainen, was amazed by the ruling (Helsingin Sanomat, 5.7.2014), while the DPO was pleased by it (Helsingin Sanomat, 14.5.2014, hs.fi[92]). The DPO predicted that many people

[85]Ruikka (2016), p. 698. Translation by the author.

[86]Alén-Savikko (2015) referring to Mayer-Schönberger (2009).

[87]Alén-Savikko (2015), pp. 431–433. See also Powles and Chaparro (2015).

[88]DPO Office (2014a).

[89]DPO Office (2014b).

[90]DPO Office (2014c).

[91]Lähdevirta (2014), p. 34.

[92]Kerkelä and Peurakoski (2014).

would jump at the chance to have unwanted links removed and pointed to the importance of having a legal tool with which to operate in the matter as well as people being informed of their rights (13.5.2014, yle.fi[93]). News coverage also followed the introduction of the service (online form) for removal request created by Google, while also other measures were covered (e.g., plans to construct a committee of outside experts to advise on issues related to free flow of information and data protection etc.) (30.5.2014, yle.fi[94]).

7 For Those Who Are from a Country That Is Not Part of the European Union, Did Your Courts Follow the ECJ Ruling on the Right to Be Forgotten? Is It Likely That They Will Follow It?

Non-applicable.

8 Did Finnish Law Already Grant a Similar Right to Be Forgotten Than the One Stated in the ECJ Ruling?

National provisions are explained above. They are interpreted in light of EU law. The law was not changed in wording after the CJEU ruling. In its ruling on the right to be forgotten, Helsinki Administrative Court has referred to the necessity requirement and rectification (§§ 9, 29 PDA).[95] However, prior to the ruling, people were

[93]Nieminen (2014b).

[94]Parviainen (2014); see also Waters (2014).

[95]Helsinki Administrative Court (2016). The case concerns removal of links to information about a crime and punishment as well as information on the medical condition of the convicted person. The DPO (docket no 1374/533/2015, 3.12.2015) had ordered removal of search results. Google Inc. and Google Finland Oy appealed, referring to the severity and actuality of the crime, as well as to the public interest in discussing the length of punishments in Finland. The appellants also referred to art 10 ECHR, while noting that information had also been made public by the mother of the person in question. Moreover, they noted that both companies cannot be addressed by the same decision; therefore the decision should at least be repealed with regard to Google Finland Oy. The DPO noted that health data was not given enough weight by the controller (diminished accountability in the criminal case). The court was of the same opinion as the DPO and found the balancing of privacy and free speech correct: the court noted, among others, that health documents are confidential under secrecy laws. Taking into account privacy as a fundamental right as well as the content of the linked information, the court found the search results unnecessary and subject to removal. Privacy of the data subject outweighed the public's right to information. There was no public role due to the criminal activity or the information given by third parties. For its part, Google Inc. was the controller and Google Finland Oy was the local place of business. The appeal was dismissed. The case was appealed. See Supreme Administrative Court (2018).

advised by the DPO's office in a dedicated guide to turn to the publisher in order to have information removed from the internet after which the search engine could be contacted to speed the process of updating the results. The guide also pointed to the importance of approaching online sources with critique; people must remember that there are legitimate reasons to publish information on individuals, while there may be other people by the same name (it is advised that people check whom the information concerns) and search engines may make mistakes.[96] The DPO also noted that the 'right to be forgotten' only extended to removal of links from search results by the name of the person in question, while it did not imply a total removal of content from the internet. Initial publishers must be contacted in case a person seeks removal of content. Moreover, crimes related to the content of online messages or web pages (e.g., defamation) are out of the scope of the DPO.[97]

In its praxis, the Council for Mass Media (CMM) in Finland had outlined that removal of information from news archives is usually not possible.[98] However, with regard to good journalistic practice, the Guidelines for Journalists require speedy correction of false information.[99] In 2018, the CMM issued a resolution on removal of online content whereby the Council acknowledges the role of search engines and discussion on informational self-determination, including the "right to be forgotten". It refers to the development in the field of online content and search operations and therefore notes the need to revise its previous outlines on the issue of removal of online content.[100]

9 Do You Think That Google Should Be More Transparent About the Ways It Uses to Implement the "Right to Be Forgotten"?

Yes, more transparency is definitely needed. Even if the procedure at the first stage is left in the hands of private entities (companies), the procedure should be more transparent. Incumbent companies operating online and in the digital public sphere

---

[96]DPO Office (2010).

[97]See DPO Office (2017).

[98]Pitkänen et al. (2013), pp. 125–126; see also CMM (2009).

[99]CMM (2014), Sec 20: "Essentially incorrect information must be corrected without delay and so as to reach, to the highest extent possible, the attention of those who have had access to the incorrect information. The correction must be published on the editorial website of the media in question, as well as in the publication or broadcast in which the incorrect information was originally given." See also Ollila (2014a), p. 818.

[100]CMM (2018). The editor-in-chief may allow removal of content (incl. individual pictures and names) after publication if there are unreasonable consequences for private individuals, taking into account the right of the public to information. Publications should however be archived in original form as well. Moreover, the CMM notes that it is not competent to order removal and it may investigate publications beyond 3-months' time only in exceptional cases (ibid).

may not be public utilities in the traditional sense, but they should also be responsible in their actions and held accountable.[101] Transparency could also facilitate the acceptance of the decisions or, to the contrary, their rejection and subsequent referrals to the DPO.[102]

10 Is the Procedure Prepared by Google Used in Finland?

Yes, Finnish people utilize the opportunity to request the removal of links from Google's search results. According to news coverage, by the end of June 2014, around 800 requests (concerning almost 3000 web addresses) were made from Finland. This was a lot in relation to population and compared to some other European countries.[103] For example, the national public service broadcaster, Yleisradio (Yle), provided news coverage of the first search results being removed which led to its content (i.e., a link to a web page concerning an episode of a Yle reality show for weddings, called "Satuhäät"; engl. "Dream wedding"), while also including general information on removal requests in Europe.[104] The DPO was also later contacted. By the end of November, a couple of dozen referrals to the DPO had been made. The first orders by the DPO were issued in December 2015; one of these decisions concerning links to information about a crime and a convicted person was appealed by Google (see fn 2). By January 2016, a dozen decisions had been made by the DPO, while referrals kept coming in.[105]

11 Is There Any Upcoming Legal Reform in Finland Whose Purpose Is to Reinforce or Modify the "Right to Be Forgotten"?

The new Data Protection Act (1050/2018) was drafted to supplement the GDPR. The GDPR, including Article 17 on the right to be forgotten, have been directly applicable law in Finland from May 2018.

[101]For codes and computer programs as well as proprietary and public dimensions thereof, see e.g., van Dijk (2010).

[102]See also Ruikka (2016).

[103]Naalisvaara (2014) and Brunila (2014).

[104]Nieminen (2014a).

[105]Rautio (2016a, b), Junttila (2014) and Karhula (2016).

12 What Should Be the Next Step in the Protection of the "Right to Be Forgotten"?

It is unlikely and unrealistic that any legislative changes would occur in the near future since the GDPR was a heavy process. However, guidance by the European Data Protection Board (EDPB)[106] and future CJEU case law in particular is likely to shed light on the interpretation of Article 17 GDPR. The court could provide more detailed guidance on the application of Article 17, especially with regard to the derogations, as well as interpret the provisions in new situations.

With regard to technological development, one possible challenge is blockchain technology which is by nature immutable and time-stamped.[107] Blockchain (BC) technology refers to digital distributed (i.e., decentralized) ledger technology which can be used to keep account of transactions. Recent transactions are recorded as new blocks which in turn form chains following a chronological order (blocks are time-stamped and linked via hashes). There is no centralized intermediary, but every node automatically gets a copy of the BC database. Blocks are validated and the records are indelible, while alterations would require altering and verifying all affected parts via community action.[108] However, blockchains may come with "expirations dates", that is, be fixed-term, while every node need not save all of the history and information.[109]

According to Berberich and Steiner, BC technology challenges implementation of the right to be forgotten, whether based on CJEU praxis or Article 17 GDPR. Firstly, the general problem of identifying a controller able to cater for the data subject's rights applies also to the right to be forgotten; while in a private BC the trusted parties involved would be joint controllers (or regarded as one joint venture), a public BC relying on peer to peer network could mean the status of a controller for every single node (with highly complex situation of joint controllership) or for no one of them. Secondly, erasure in the meaning of the right to be forgotten is practically impossible in the "persistent BC architecture". However, taking into account the operation of BC, personal data could be regarded as "necessary" for the purposes, whereas the functions could represent a specific legal ground for processing (e.g., legitimate interest).[110] Various technological solutions could support data protection by design and enhance privacy, such as inserted "noice" or a combination of on-chain and off-chain storage.[111]

---

[106]See Articles 68–70 GDPR.

[107]Berberich and Steiner (2016), p. 426.

[108]Berberich and Steiner (2016), pp. 422, 426; see also Ministry of Transport and Communications (2017), pp. 3–10; Investopeadia (2017); Wikipedia (2017).

[109]For misconceptions around BC technology, see e.g., Ministry of Transport and Communications (2017), p. 9.

[110]Berberich and Steiner (2016), p. 426; for more on the lack of an intermediary, see ibid, 424.

[111]Berberich and Steiner (2016), pp. 425–426. See e.g. Zyskind et al. (2016).

References

Alén-Savikko A (2015) Pois hakutuloksista, pois mielestä? Lakimies 3–4:410–433

Alén-Savikko A (2016) Finland. In: Ukrov J, Iacino G (eds) Comparative study on investigative journalism. European Centre for Press and Media Freedom, Saarbrücken, pp 160–181. Available via https://ecpmf.eu/files/comparative_study_on_investigative_journalism_emr.pdf

Berberich M, Steiner M (2016) Blockchain technology and the GDPR – how to reconcile privacy and distributed ledgers? Eur Data Protect Law Rev 3:422–426

Brunila M (2014) Googlelle 800 pyyntöä 'unohtua' Suomesta. It-viikko 8.7.2014. http://www.is.fi/digitoday/art-2000001842508.html. Accessed 4 Apr 2017

CMM (2009) Council for Mass Media, Decision on news archives 4069/SL/09, 3.9.2009. http://www.jsn.fi/sisalto/4069-sl-09/?search=4069. Accessed 13 Nov 2018

CMM (2014) Council for Mass Media, Guidelines for journalists and an annex. Guidelines operative from 1 January 2014. http://www.jsn.fi/en/guidelines_for_journalists/. Accessed 7 Apr 2017

CMM (2018) Council for Mass Media, Lausuma verkkosisältöjen poistamisen periaatteista 2018 (Resolution on principles regarding removal of online content), 7.3.2018. https://www.jsn.fi/lausumat/lausuma-verkkosisaltojen-poistamisen-periaatteista-2018/. Accessed 7 Apr 2017

DPO Office (2010) Tietosuojavaltuutetun toimisto: Tietojen poistaminen internetin hakukoneista 15.9.2010. http://www.tietosuoja.fi/material/attachments/tietosuojavaltuutettu/tietosuojavaltuutetuntoimisto/oppaat/6JfqHbbYz/Tietojen_poistaminen_Internetin_hakukoneista.pdf. Accessed 6 Apr 2017

DPO Office (2014a) Euroopan unionin tuomioistuimen päätös Googlen hakutoimintoa koskevassa asiassa, 13.5.2014. http://tietosuoja.fi/fi/index/ajankohtaista/tiedotteet/2014/05/euroopanunionintuomioistuimenpaatosgooglenhakutoimintoakoskevassaasiassa.html. Accessed 7 Apr 2017

DPO Office (2014b) Google 16.5.2014. http://tietosuoja.fi/fi/index/ajankohtaista/tiedotteet/2014/05/google.html. Accessed 7 Apr 2017

DPO Office (2014c) Hakutuloslinkkien poistaminen Googlesta 30.5.2014. http://tietosuoja.fi/fi/index/ajankohtaista/tiedotteet/2014/05/hakutuloslinkkienpoistaminengooglesta.html. Accessed 7 Apr 2017

DPO Office (2017) Tietosuojavaltuutetun toimisto, Hakukonepyyntöasioiden käsittelystä tietosuojavaltuutetun toimistolla, dnro 942/09/2017, 13.4.2017

Government (1993) Government bill on amending the provisions on fundamental rights in the constitutional acts (Hallituksen esitys Eduskunnalle perustuslakien perusoikeussäännösten muuttamisesta) (HE 309/1993 vp)

Government (1998) Government bill for Personal Data Act and certain acts related thereto (Hallituksen esitys eduskunnalle henkilötietolaiksi ja eräiksi siihen liittyviksi laeiksi) (HE 96/1998 vp)

Government (2003) Government bill on amending the Tort Liability Act and certain laws related thereto (Hallituksen esitys eduskunnalle laiksi vahingonkorvauslain muuttamisesta ja eräiksi siihen liittyviksi laeiksi) (HE 167/2003 vp)

Government (2014) Government bill on revising some provisions on appeal in administrative matters (Hallituksen esitys eduskunnalle eräiden hallintoasioiden muutoksenhakusäännösten tarkistamisesta) (HE 230/2014 vp)

Helsinki Administrative Court (2016) Docket no 01135/16/1204, 16/1028/5, 8.12.2016. https://oikeus.fi/hallintooikeudet/helsinginhallinto-oikeus/fi/index/hallintooikeusratkaisut/hallintooikeusratkaisut/1481194726442.html. Accessed 13 Nov 2018

Investopeadia (2017) Blockchain, Explained. http://www.investopedia.com/terms/b/blockchain.asp. Accessed 4 Nov 2017

Junttila J (2014) Kuningaskuluttaja: Googlen unohduspykälä ei suojaa kauppiaita verkkokritiikiltä. Yle 17.12.2014, updated 12.5.2015. http://yle.fi/uutiset/3-7693798. Accessed 4 Apr 2017

Karhula P (2016) "Oikeus tulla unohdetuksi" – Google puhutti. In: Ekholm K, Karhula P, Olkkonen T (eds) Tiellä sananvapauteen. Kansalliskirjasto, Helsinki. Available via https://sananvapauteen.fi/artikkeli/2708. Accessed 4 Apr 2017

Kerkelä L, Peurakoski T (2014) Tietosuojavaltuutettu kiittää Google-päätöstä. EU:n tuomioistuimen mukaan ihmisillä on oikeus tulla "unohdetuksi" Googlen hakutuloksissa. HS 14.5.2014. http://www.hs.fi/kotimaa/art-2000002731032.html. Accessed 30 Apr 2017

Lähdevirta L (2014) Oikeuksia suojaamassa. Vesa Tyni (pictures). Lakimiesuutiset 5:34–37

Mayer-Schönberger V (2009) Delete: the virtue of forgetting in the digital age. Princeton University Press, Princeton

Ministry of Justice (2017) EU:n yleisen tietosuoja-asetuksen täytäntöönpanotyöryhmän (TATTI) mietintö. Oikeusministeriön julkaisu, mietintöjä ja lausuntoja 35/2017, 21.6.2017 (Report of the Working Group for the implementation of the EU General Data Protection Regulation), Ministry of Justice. Available via http://julkaisut.valtioneuvosto.fi/bitstream/handle/10024/80098/OMML_35_2017_EUn_yleinen_tietosuoja.pdf?sequence=1. Accessed 4 Nov 2017

Ministry of Transport and Communications (2017) Lohkoketjuteknologian soveltaminen ja vaikutukset liikenteessä ja viestinnässä. Liikenne- ja viestintäministeriö, julkaisuja 12/2017, 14.9.2017 (Applying blockchain technology and its impacts on transport and communication, Ministry of Transport and Communications). Available via http://julkaisut.valtioneuvosto.fi/bitstream/handle/10024/80667/LVM_12_2017_Lohkoketjuteknologian%20soveltaminen.pdf?sequence=1&isAllowed=y. Accessed 4 Nov 2017

Naalisvaara M (2014) "Suomalaiset haluavat piilottaa tuhansia hakutuloksia Googlesta". Yle 10.7.2014, updated 12.5.2015. http://yle.fi/uutiset/3-7349356. Accessed 4 Apr 2017

Neuvonen R (2014) Yksityisyyden suoja Suomessa. Lakimiesliiton kustannus, Helsinki

Nieminen IM (2014a) Oikeus tulla unohdetuksi: Google poisti Ylen Satuhäät-sarjan sivulle johtavia hakutuloksia. Yle 1.8.2014. http://yle.fi/uutiset/3-7387510. Accessed 30 Apr 2017

Nieminen IM (2014b) Tietosuojavaltuutettu: Google-päätös aiheuttaa vyöryn – "moni tekee tiliä menneisyytensä kanssa". Yle 13.5.2014, updated 12.5.2015. http://yle.fi/uutiset/3-7238879. Accessed 30 Apr 2017

Ollila R (2014a) Henkilötietojen suoja EU:n perusoikeutena. Defensor legis 5:814–824

Ollila R (2014b) Oikeus tulla unohdetuksi 16.5.2014. http://www.jsn.fi/blog/oikeus-tulla-unohdetuksi/. Accessed 4 Apr 2017

Parviainen A (2014) Nöyrtynyt Google avasi linkin tietojen poistamiseksi hakupalvelusta. Yle 30.5.2014, updated 10.1.2016. http://yle.fi/uutiset/3-7270338. Accessed 30 Apr 2017

Pitkänen O, Tiilikka P, Warma E (2013) Henkilötietojen suoja. Talentum, Helsinki

Powles J, Chaparro E (2015) How Google determined our right to be forgotten. theguardian.com 18.2.2015: https://www.theguardian.com/technology/2015/feb/18/the-right-be-forgotten-google-search. Accessed 4 Nov 2017

Rautio (2016a) Google ei suostu poistamaan murhasta tuomitun miehen tietoja internetistä. Yle 4.3.2016. http://yle.fi/uutiset/3-8715424. Accessed 4 Apr 2017

Rautio (2016b) Rautio (2016b) Googlelle ensimmäiset poistomääräykset Suomesta – rikoksesta tuomittujen tietoja halutaan pois netistä. Marjatta Rautio, Yle 10.1.2016, update 19.1.2017. http://yle.fi/uutiset/3-8579862. Accessed 4 Apr 2017

Ruikka T (2016) Mitä tietoja hakukoneen on vaadittaessa unohdettava? Defensor legis 4:693–698

Supreme Administrative Court (2018) ECLI:FI:KHO:2018:112. Available via https://www.kho.fi/fi/index/paatoksia/vuosikirjapaatokset/vuosikirjapaatos/1534308651626.html. Accessed 13 Nov 2018

Supreme Court (2000) KKO:2000:54. Available via http://www.finlex.fi/fi/oikeus/kko/kko/2000/20000054. Accessed 13 Nov 2018

Supreme Court (2005) KKO:2005:137. Available via http://www.finlex.fi/fi/oikeus/kko/kko/2005/20050137. Accessed 13 Nov 2018

Tiilikka P (2007) Sananvapaus ja yksilön suoja. Lehtiartikkelin aiheuttaman kärsimyksen korvaaminen. WSOYpro, Helsinki

Tiilikka P (2008) Journalistin sananvapaus. WSOYpro, Helsinki

van Dijk N (2010) Property, privacy and personhood in a world of ambient intelligence. Ethics Inf Technol 12:57–69. https://doi.org/10.1007/s10676-009-9211-0
Virtanen M (2014) Oikeus tulla unohdetuksi. IPRinfo 3:26–27
Waters R (2014) Google bows to EU privacy ruling. ft.com 30.5.2014. https://www.ft.com/content/b827b658-e708-11e3-88be-00144feabdc0#axzz33BM94Plp. Accessed 4 Nov 2017
Wennäkoski AA (2017) Tietosuojaoikeudellinen vahingonkorvaus murroksessa. In: Korpisaari P (ed) Viestinnän muuttuva sääntely, Viestintäoikeuden vuosikirja 2016. Forum Iuris, Helsinki, pp 68–93
Wikipedia (2017) Blockchain. https://en.wikipedia.org/wiki/Blockchain. Accessed 4 Nov 2017
Working Party (2014) WP 225. Guidelines on the implementation of the Court of Justice of the European Union judgment on "Google Spain and Inc. v. Agencia Española de Protección de Datos (AEDP) and Mario Costeja González" C-131/12, 26.11.2014. Available via http://www.dataprotection.ro/servlet/ViewDocument?id=1080. Accessed 1 May 2017
Zyskind G, Nathan O, Pentland A (2016) Enigma: Decentralized Computation Platform with Guaranteed Privacy. http://livinglab.mit.edu/wp-content/uploads/2016/01/enigma_full.pdf. Accessed 4 Nov 2017

Legislation

Administrative Judicial Procedure Act (586/1996). Unofficial translation by the Ministry of Justice, 17 Oct 2003, amendments followed up to 435/2003. Available via http://www.finlex.fi/fi/laki/kaannokset/1996/en19960586.pdf. Accessed 4 Nov 2017
Code of Judicial Procedure (4/1734). Unofficial translation by the Ministry of Justice, 10 January 2016, amendments followed up to 732/2015. Available via http://www.finlex.fi/fi/laki/kaannokset/1734/en17340004.pdf. Accessed 4 Nov 2017
Constitution of Finland (731/1999). Unofficial translation by the Ministry of Justice, 31 Dec 2011, amendments followed up to 1112/2011. Available via http://www.finlex.fi/fi/laki/kaannokset/1999/en19990731.pdf. Accessed 4 Apr 2017
Penal Code (39/1889). Unofficial translation by the Ministry of Justice, 2 March 2016, amendments followed up to 766/2015. Available via http://www.finlex.fi/fi/laki/kaannokset/1889/en18890039.pdf. Accessed 4 Nov 2017
Personal Data Act (523/1999; PDA). Unofficial translation by the Ministry of Justice, 31 March 2001, amendments followed up to 986/2000. Available via http://www.finlex.fi/en/laki/kaannokset/1999/en19990523.pdf. Accessed 3 Apr 2017
Tort Liability Act (412/1974; TLA). Unofficial translation by the Ministry of Justice, 15 March 2001, amendments followed up to 61/1999. Available via http://www.finlex.fi/fi/laki/kaannokset/1974/en19740412.pdf. Accessed 4 Apr 2017

Germany: The Right to Be Forgotten

Jürgen Kühling

Abstract The article discusses the right to be forgotten in the German jurisdiction. It first states that a special right to be forgotten does not exist in Germany, but arises from general civil law and general data protection law—even before the General Data Protection Regulation (GDPR) came into force. In a second step, the limits of these claims against the service provider as an indirect tortfeasor are shown, who is obliged to review the specific notification of the party affected. In this context, it will be discussed which constitutional requirements—defined by the case-law of the Federal Constitutional Court—the service provider must take into account when balancing different interests. Subsequently, it is shown by which legal remedies the right to be forgotten and possible claims for damages can be enforced. Afterwards, the development of the right to be forgotten is presented, taking into account the reactions of the German courts and commentators, in order to show that in Germany the claiming of service providers as indirect tortfeasors has already been established before the ECJ's decision. The actual effects of the ECJ's decision will then be demonstrated by Google Germany's implementation of the requirements of the right to be forgotten. In the ensuing section, the extent to which claims to deletion arise from the new GDPR will be explained. In the end, the article comes to the conclusion that the core task of the GDPR is to now strike a reasonable balance in the current secondary law while adequately accounting for the basic guidelines from fundamental rights as the competent authorities make use of the consistency procedure.

1 Regulatory Framework

There is no specific right to be forgotten in Germany, if understood in a broader sense as a claim to the deletion of content that infringes on personality rights or understood in a narrower sense as the right to the deletion of links to such content. Under the current legal framework, such claims derive from general civil law or from

---

J. Kühling (✉)
University of Regensburg, Regensburg, Germany
e-mail: Juergen.Kuehling@jura.uni-regensburg.de

F. Werro (ed.), *The Right To Be Forgotten*, Ius Comparatum – Global Studies in Comparative Law 40, https://doi.org/10.1007/978-3-030-33512-0_6

the general right to data protection, as it had been enshrined already in the old German Federal Data Protection Act (Bundesdatenschutzgesetz), which was adopted on the basis of the general Data Protection Directive 95/46/EC. Under this framework, claims to deletion exist against indirect tortfeasors[1] if they transmit content from third parties that infringes personality rights on their websites (such as search engines). A right to deletion is found in the general legal claim to cease and desist analogous to para 1004 § 1 sentence 2 BGB (German Civil Code) read with para 823 § 1 Civil Code. Para 35 § 2 sentence 2 n°1 Federal Data Protection Act read with para 29 § 1 sentence 1 n°2 Federal Data Protection Act also already granted the affected person a claim to deletion. The same holds true for para 823 § 2 Civil Code read together with para 29 § 1 sentence 1 n°2 Federal Data Protection Act. In this respect, one can speak of a claim rooted in general civil law read with general data protection law.

A corresponding claim can be found in Art 17 General Data Protection Regulation (GDPR), which is uniformly applicable in the EU. In the German statutory provision accompanying the GDPR, para 35 Federal Data Protection Act 2018, this claim has only been partially limited, without specific national distinctions (for details, see below, I).

2 The Reach of to the Right to Be Forgotten

Liability of an indirect tortfeasor under an analogous application of para 1004 § 1 sentence 2 BGB (German Civil Code) read together with para 823 § 1 Civil Code depends on specific conditions. The service provider is only liable as an indirect tortfeasor if he infringes his inspection obligations. He has such obligations only if the affected party notifies him explicitly, calling doubt on the legality of a form of expression.[2] It is only then that he is obligated to conduct an in-depth investigation of the legality of the statement.[3]

Moreover, the assessment of the legality of links in any claim to deletion requires the respective affected rights to be balanced against one another. The general right to privacy of the affected party enshrined in art 2 § 1 read together with art 1 § 1 Basic Law (Grundgesetz) must be offset adequately against the colliding public interest in information.[4] Beside the rights of the search engine operator pursuant to art 12 § 1 sentence 1 Basic Law, the website operator's freedom of opinion under art 5 § 1 half-sentence 1 Basic Law and other users' freedom of information under art 5 § 1 half-sentence 2 Basic Law must be taken into account. It is contested whether search engine operators are entitled to protection under art 5 § 1 sentence 1 Basic

[1]BGHZ 209, 139 para 17 clarifies that service providers who merely transmit content generated by third parties can only be indirect tortfeasors.

[2]BGHZ 191, 219 paras 24 ff.

[3]BGHZ 209, 139 paras 39 ff.

[4]BVerfGE 35, 202.

Law or under art 5 § 1 sentence 2 Basic Law themselves or whether they are not protected at all when they merely transmit content.[5] Ultimately, the ECJ seems to have implicitly answered this question in the negative for the entire EU in the case *Google Spain and Google*, though it did not explicitly examine the issue.[6] At any rate, the website operator's freedom of opinion will have significant weight when it is balanced against the personality rights of the affected party.

As far as website operators are concerned, their own as well as third parties' opinions can constitute the content of their statements.[7] Otherwise, the person uttering the statement would be forced to identify himself with the statement, which would amount to a problematic interference with the negative freedom of expression.[8]

The scope of the protection of statements of fact under the German Constitution (Basic Law) is contested, since the wording of art 5 § 1 sentence 1, 1st alternative Basic Law refers only to opinions. The majority of scholars correctly include statements of fact in the protected scope. It is frequently argued that every statement of fact is connected with an evaluation, whether it be through the prioritisation, presentation, or at least through the implicit opinion that the fact is worth stating.[9] While some passages of the Federal Constitutional Court's (Bundesverfassungsgericht) judgments suggest a very broad scope of protection, the case law is inconstant in this respect.[10] At any rate, the Federal Constitutional Court understands the broad notion of the term opinion to include those statements of fact that cannot be separated from the opinion or that serve to form opinions within the scope of protection pursuant to art 5 § 1 sentence 1 alt 1 Basic Law, in order to ensure comprehensive protection of the freedom of expression.[11] Therefore, those facts enjoy indirect protection via the protection of statements of opinion, resulting in broad protection. However, statements of fact that have been proven to be false or that are made in the knowledge that they are untrue fall outside the scope of the protection of the freedom of opinion. This is because false statements of fact

---

[5]So far, there have been no pronouncements from higher or supreme courts in the matter, however, the Higher Regional Court (Oberlandesgericht) Celle, 01.06.2017 – 13 U 178/16 has taken a stand against the protection of search operators; for a corresponding position in academia, see Kutscha and Thomé (2013), p. 51 f; for positions in favour of their protection, see Fechner (2017), p. 50; Trentmann (2017), p. 29; for a cursory presentation with further sources, see Pille (2016), p. 177 ff.

[6]See on this interpretation Buchner and Tinnefeld (2017).

[7]On the following elaborations, see Kühling J. In: Beck'scher Online-Kommentar zum Informations- und Medienrecht Art 5 GG; Schmidt-Jortzig (2009), who correctly rejects the identification requirement sometimes implied in the case law of the Federal Supreme Court as an objectionable quest for the underlying motivation of the statement; see, similarly, Hoffmann-Riem (2001).

[8]Kloepfer (1991), p. 30 f.

[9]Schmidt-Jortzig (2009); Kingreen and Poscher (2016); Grabenwarter (2017); for implicit statements concerning broadcasting, see BVerfGE 12, 205, 260.

[10]Glaeser (1988), pp. 113, 74 ff.

[11]BVerfGE 61, 1, 8; 90, 1, 15; 90, 241, 247; 99, 185, 196 ff.; 114, 339, 352 f.

cannot contribute to the development of opinions and are thus not worthy of protection.[12] The Federal Constitutional Court restricts the latter limitation by affirming that no excessive demands should be made regarding the duty to tell the truth, so as to avoid putting the freedom of opinion at risk.[13] Moreover, the restriction only applies where the statements of fact can be separated from the opinions uttered in the same context without distorting their meaning.[14] Effectively, only deliberate untruths are normally excluded from the scope of protection.

At the centre of the conflict between the various rights, the question thus arises how far the general personality right reaches under due consideration of the meaning of freedom of communication. The Federal Constitutional Court has expressed its views on this multiple times. The point of departure in the Court's reflections is the consideration that the great significance of the freedom of expression requires intensified control by the Constitutional Court, which starts with monitoring the interpretation of statements by the (mostly specialised) courts and also includes the interpretation of legal rules as well as their application.[15] The general right to personality is protected to a different extent depending on whether only the public and social sphere or more sensitive areas such as the private or intimate sphere are concerned.[16]

The court then differentiates between the balancing of statements of fact and evaluations. Since statements of fact create a greater impression of authenticity, stricter standards must apply to them. Effectively, statements of fact therefore enjoy a lower level of protection. If they are deliberate or proven untruths, they fall outside the scope of protection—as mentioned above. With statements of fact that are unintentionally untrue or where there is no proof for their accuracy, respect of the duties of care is decisive. If they have not been sufficiently upheld, the protection of personality rights and the protection of honour prevail over the freedom of communication, otherwise the opposing rights must be balanced against each other. If true factual statements interfere with the most personal sphere of privacy, the protection of honour prevails. If the statement only affects the social sphere, a balancing exercise takes place. This also occurs with the expression of opinions insofar as they affect another person's honour. It is only when they violate human dignity or involve an explicit insult or abusive criticism that the protection of honour again prevails.[17]

These standards for the balancing exercise are also suitable to assess the deletion of links to such webpages as a lesser variation in relation to the deletion of the

---

[12]Most recently BVerfGE 114, 339, 352; see the landmark case BVerfGE 54, 208, 219.

[13]BVerfGE 54, 208, 220; confirmed in BVerfGE 61, 1, 8; 85, 1, 15; 114, 339, 352 f.

[14]Grimm (1995), p. 1699 with references to the case law of the Federal Constitutional Court; with criticism of this "mixing approach" Huster (1996), p. 487 ff.

[15]Grimm (1995), p. 1700 ff.

[16]BVerfGE 80, 367 paras 373 ff.

[17]Grimm (1995), p. 1702 ff., cf also the image on p. 1705; cf also the description by Gosche (2008), p. 63 ff, with further sources.

content on the pages themselves. Incidentally, the duty to delete can also arise from the protection of the personality right for true statements of fact. The public may usually have an especially significant interest in information about true statements of fact, such as when they concern criminals and their offences.[18] This interest in information can, however, be relegated to the back seat to the affected individual's personality rights after a certain amount of time has passed in order to allow the delinquent to be reintegrated into society.[19] The Federal Constitutional Court has, with regard to this specific scenario, stated that the affected individual has the right "to be left alone" after time has passed.[20] Overall, there is no detailed case law on other deletion claims for the protection of personality rights.

3 Legal Remedies Available to Enforce the Right to Be Forgotten

There are no specific enforcement mechanisms or provisions for the "right to be forgotten," not to mention the absence of specific institutions that are entrusted with the enforcement of these claims. Instead, in line with the general substantive provisions, the institutional arrangements from data protection and civil law apply.

From the perspective of data protection law, until the GDPR came into force this meant that the complex and only imperfectly implemented enforcement system of German data protection law applied again, effecting external control, as the provisions in the general Data Protection Directive 95/46/EC were transposed.[21] Enforcement varied depending on whether public authorities at the federal or state (Länder) level or non-state entities were involved. In the main case of external control for the private sector pursuant to para 38 Federal Data Protection Act (Bundesdatenschutzgesetz), the supervisory authorities had informatory, access and inspection rights as well as certain rights to issue orders to undertake and desist from specific actions and can impose fines. Overall, however, the access rights and the sanctions in paras 43 f Federal Data Protection Act were rather weak, causing significant enforcement deficiencies.

This substantially changed once the General Data Protection Regulation (GDPR) entered into force, especially in light of the provisions on fines in arts 82 ff therein.[22] The absolute cap on fines at 20 million euros in art 83 § 5 GDPR is raised to 4% of the company's annual turnover for infringements of the deletion claim in art 17 GDPR, if that happens to be the higher amount. The new law therefore enables theoretical fines of billions of euros against search engine operators such as Google.

[18]BVerfGE 35, 202 paras 230 ff.

[19]BVerfGE 35, 202 paras 233 ff.

[20]BVerfGE 35, 202 paras 233.

[21]On the following, see Kühling et al. (2015), p. 230 ff.

[22]On the following, see Kühling (2017), pp. 1985, 1989.

Fines are exhaustively regulated at the European level and these provisions will replace the previous para 43 Federal Data Protection Act. In this respect, para 41 Federal Data Protection Act 2018 governs only procedural aspects by referring to the provisions on the fining and criminal procedures. The Member States retain a scope for design with regard to criminal penalties for companies, which are now governed by para 42 Federal Data Protection Act 2018. Therefore, the menacing power of the effective implementation of claims for deletion mainly results from the bolstering of the GDPR itself.

From a civil law perspective, the claims against companies have also been enforceable before the civil courts. What legal options remain for the implementation of the adjudicated and legally binding claim does not depend on which substantive provision the claim is based on, but rather on whether the court proclaims an enforceable injunction to take positive action or to desist.[23] If the debtor violates an injunction to desist, a fine or detention can be imposed pursuant to para 890 § 1 Civil Procedure Code (Zivilprozessordnung), which must both be threatened first pursuant to para 890 § 2 Civil Procedure Code. This threat can occur in the judgment itself. For an enforceable injunction to act that the debtor does not obey, a penalty payment or penalty detention can be imposed pursuant to para 888 § 1 Civil Procedure Code. According to para 888 § 2 Civil Procedure Code, there is no need for an advance threat of the penalty measures. It is thus decisive whether the claim to deletion is qualified as an obligation to act[24] or as an obligation to desist.[25] Since the affected party has always already suffered damages and active deletion always forms part of the court's injunction, it is more convincing to find an obligation to act.[26] Therefore, penalty payments or detention can be imposed and no advance threat is needed.

4 Material or Immaterial Damages for Plaintiffs?

A claim to damages for an illegal link could be based on para 823 § 1 Civil Code (BGB), para 823 § 2 read together with a protective law, para 824 or para 826 Civil Code. It is always necessary to find fault on the part of the search engine operator. Moreover, palpable damages, such as lower profits, must be present. The courts only accept immaterial damages in cases of serious infringements of personality rights through gross misconduct and in the presence of the undeniable need for compensation, which cannot be effected through other means.[27] In the situations at hand, claims for damages can derive from different statutory bases. For one, the affected

[23]Bassenge (2017).

[24]LG Heidelberg, 09.12.2014 – 2 O 162/13 para 52.

[25]LG Köln, 28.05.2015 – 28 O 496/14; LG Köln, 7.10.2015 – 28 O 370/14; OLG München, 27.04.2015 – 18 W 591/15; LG Hamburg, 07.11.2014 – 324 O 660/12.

[26]See also in this sense Bassenge (2017) with further sources, on the case of materialised harm.

[27]BVerfGE 34, 269 paras 285 f.; *see also* Wybitul (2016), pp. 253, 254.

party could have a claim to damages because of the illegal link itself. Under current law, however, the search engine operator's liability is excluded prior to his knowledge of the infringement pursuant to para 10 n°1 Telemedia Act (Telemediengesetz).[28]

This legal regime, which lacks strength, will likely also change as the General Data Protection Regulation (GDPR) has entered into force. For instance, the liability standards for service operators' examination and monitoring obligations pursuant to art 17 § 1 GDPR are likely to be raised.[29] A claim to damages could exist when search engine operators do not comply with their examination duties after being informed of the infringement.[30] Art 82 GDPR also ensures a claim to damages when data is processed illegally. This provision now also explicitly covers immaterial harm. Companies must expect an increasing number of complaints concerning data protection infringements, the defence against which is made significantly more difficult by the effective reversal of the burden of proof in art 5 § 2 GDPR and also in art 24 § 1 sentence 1 GDPR.[31]

5 The Implementation of the Right to Be Forgotten

The evolution of the enforcement of a "right to be forgotten" can be divided into three phases: Until the ECJ judgment in the *Google Spain and Google* case, there was no specific mechanism and no specific case law with regard to intermediaries such as search engine operators—notably "Google search." In the current second phase, however, a "right to be forgotten" can be implemented without issue because the general principles on the liability of indirect tortfeasors already exist. It is possible to apply these claims in accordance with the ECJ's *Google* case law without much ado, because there is no differentiated claim to deletion with rigid criteria, but rather a case-by-case balancing exercise of the respective interests. The German courts incorporate the ECJ's considerations on the "right to be forgotten" in this balancing exercise in the current second phase.[32] In Germany, a multitude of

[28]For a different approach, see the District Court (Landgericht) Heidelberg, 09.12.2014 – 2 O 162/13 para 61, which held the search engine operator fully liable pursuant to para 7 § 1 Telemedia Act. However, the outcome did not rest on this, because the required information within the meaning of para 10 n°1 Telemedia Act had occurred.

[29]For details, see below I.

[30]LG Heidelberg, 09.12.2014 – 2 O 162/13 para 60 f.

[31]Cf. Wybitul (2016), pp. 253, 254.

[32]OLG Celle, 01.06.2017 – 13 U 178/16 paras 15 ff; OLG Karlsruhe, 14.12.2016 – 6 U 2/15 para 110; OLG Celle, 06.12.2016 – 13 U 85/16 paras 4 ff; OLG Köln, 31.05.2016 – 15 U 197/15 paras 52 f.; OLG München, 27.04.2015 – 18 W 591/15 para 24; LG Hamburg, 29.01.2016 – 324 O 456/14 paras 39, 54 ff.; LG Hamburg, 10.07.2015 – 324 O 17/15 para 22; LG Köln, 28.05.2015 – 28 O 496/14 para 14; LG Heidelberg, 09.12.2014 – 2 O 162/13 para 45; LG Hamburg, 07.11.2014 – 324 O 660/12 para 105 (cited according to juris).

requests for deletion has been submitted in recent years, according to the mechanisms established in the wake of *Google Spain and Google* (cf H). Nevertheless, there have been lawsuits in some few cases. They were usually carried out between the affected party and the intermediary. When the General Data Protection Regulation entered into force, the situation for effective and harmonised enforcement of the "right to be forgotten" throughout the EU got stricter.

6 Reactions to the ECJ Ruling on Google v González by Courts and Commentators

Most German scholars welcomed the ECJ's judgment, though it was also harshly criticised in parts. While the "right to be forgotten" was partially seen as a milestone[33] and was even denominated as a new European fundamental right,[34] some authors were disappointed that the decision did not deliver rules that could serve as general principles for the balancing exercise.[35] This balancing exercise is still seen as too complex for search engine operators to reliably carry out themselves.[36] On the other hand, the ECJ was lauded for recognizing that large search engine operators such as Google can seriously interfere with personality rights by making content practically accessible.[37]

However, many German scholars criticise that the potential rights of the search engine operators[38] and the website operators were entirely left aside in the ECJ's balancing exercise.[39] It has been prognosticated that the ECJ's case law could in the future make companies such as Google decide to delete links, when in doubt, in order to avoid becoming liable.[40] Independently of this, some scholars welcomed the fact that the ECJ enforced European data protection rights vis-à-vis US search engine operators.[41]

---

[33] Boehme-Neßler (2014), pp. 825, 827; Karg (2014), pp. 359, 361.

[34] Boehme-Neßler (2014), pp. 825, 827.

[35] Karg (2014), pp. 359, 361.

[36] Sörup (2014), pp. 455, 465.

[37] Buchholtz (2015), pp. 570, 574; for differing opinions, see Arning et al. (2014), pp. 447, 450 f.; Rössel (2014), pp. 150, 151.

[38] Holznagel and Hartmann (2016), pp. 228, 231 f.; Trentmann (2017), pp. 26, 29.

[39] Arning et al. (2014), pp. 447, 453; Holznagel and Hartmann (2016), pp. 228, 231 f.; Trentmann (2017), pp. 26, 28.

[40] Rössel (2014), pp. 150, 151.

[41] Kühling (2014), pp. 527, 531 f.

7 The Right to Be Forgotten Prior to the ECJ Ruling

The claims to deletion enshrined in the analogous reading of para 1004 § 1 sentence 2 Civil Code (BGB) read together with para 823 § 1 Civil Code, para 35 § 2 sentence 2 n°1 Federal Data Protection Act (Bundesdatenschutzgesetz) read together with para 29 § 1 sentence 1 n°2 Federal Data Protection Act and para 823 § 2 Civil Code read together with para 29 § 1 sentence 1 n°2 Federal Data Protection Act resemble the "right to be forgotten." It remains unclear if the principles on the liability of indirect tortfeasors are comparable to the ECJ's principles on the liability of service operators.[42] Probably, the ECJ must be understood to be saying that, going beyond the German Federal Court (Bundesgerichtshof), it sees a requirement for Google to search for links that infringe personality rights on its initiative. A claim to desist therefore probably exists both for the ECJ and the German Federal Court only once the search engine operator has been informed of the infringement. This changed in the new legal regime due to art 17 General Data Protection Regulation.[43] Art 17 § 1 General Data Protection Regulation burdens services operators with examination and monitoring obligations, within the bounds of reasonableness.[44] It remains to be seen what degree of scrutiny must be deemed appropriate in this respect. Considering the vast quantity of links, no strict degree of scrutiny seems to be appropriate, however.

The balancing exercise between the affected rights as a means to establish the legality of the link is comparable as well. At most, the ECJ's consideration that even links to legal content can be illegal might possibly be new for the balancing exercise,[45] which may have been limited to the legality of the content until now. This is difficult to assess, however, because there was no case law on the deletion of links prior to the ECJ's *Google* judgment. Comparable rulings, which were adopted before the judgment and which also relate to indirect tortfeasors and liability for third-party content,[46] are the Federal Court's judgments on physician rating portals.[47] Here, the Federal Court also takes account of the legality of the content itself

[42]ECJ, Case C-131/12 *Google Spain and Google* ECLI:EU:C:2014:317, paras 35–38 and 83. In favour of this, see the Higher District Court Munich (Oberlandesgericht München), 27.04.2015 – 18 W 591/15 para 24 and the District Court (Landgericht) Hamburg, 07.11.2014 – 324 O 660/12 para 105 (cited according to juris); for a different approach, see the Higher District Court Cologne (Oberlandesgericht Köln), 31.05.2016 – 15 U 197/15 paras 52 f, which seems to see a qualitative difference.

[43]Buchholtz (2015), pp. 570, 574 criticises this harshly.

[44]Trentmann (2017), pp. 26, 30 f.

[45]Also to this effect Buchholtz (2015), pp. 570, 574; Rössel (2014), pp. 150, 151.

[46]Rössel (2014), pp. 150, 151 draws the inverse conclusion, that the Federal Court was only referring to the illegality of the content on the website, from a comparison with the Federal Court's case law on Google's preview images; however, this comparison seems rather inappropriate because that case law concerned copyright infringements and not infringements of personality rights.

[47]BGHZ 202, 242; BGHZ 209, 139.

because these platforms have a wide reach, a significant threat to personality rights emanates from them[48] and poor reviews can well endanger the affected a person's livelihood.[49]

8 The Mechanisms Established by Google

On the web form on which users can request that Google delete links, Google states that it attempts to reconcile personality rights with the public interest in information through the balancing exercise.[50] At the same time, several categories of content are cited whose links are not deleted, such as information on fraud, serious breaches of professional codes of conduct, criminal sentences or public actions of officials.[51] The exact number as well as the percentage relationship of deletion requests and deleted URLs are also accessible for every individual country and for Europe as a whole.[52] Moreover, examples of requests and the reasons for Google's decisions in response to them can be viewed for every individual country.[53] This manner of proceeding seems sufficiently transparent. Since Google always has to make case-by-case decisions anyway, it seems reasonable that no more detailed information is given on the examination of links. The option of taking legal action against Google's individual decisions sufficiently protects users.

Moreover, Google seems to inform web site operators about the removal of links.[54] However, there does not appear to be information on the search from which the respective search result[55] is removed.[56] This makes it significantly more difficult for the website operator to take legal action because he can hardly show that

[48]BGHZ 202, 242 para 32; BGHZ 209, 139 para 40.

[49]BGHZ 209, 139 para 41.

[50]Cf Google's web form for the request for the deletion of links, available at https://webcache.googleusercontent.com/search?q=cache:3Dz2MvqYkrsJ:https://www.google.com/webmasters/tools/legal-removal-request%3Fcomplaint_type%3Drtbf+&cd=2&hl=de&ct=clnk&gl=de&client=firefox-b (last accessed 25 August 2017).

[51]Cf Google's web form on the request for the deletion of links, available at https://webcache.googleusercontent.com/search?q=cache:3Dz2MvqYkrsJ:https://www.google.com/webmasters/tools/legal-removal-request%3Fcomplaint_type%3Drtbf+&cd=2&hl=de&ct=clnk&gl=de&client=firefox-b (last accessed 25 August 2017).

[52]Cf Google's Transparency Report, available at https://transparencyreport.google.com/eu-privacy/overview (last accessed 28 August 2017).

[53]Cf Google's Transparency Report, available at https://transparencyreport.google.com/eu-privacy/overview (last accessed 28 August 2017).

[54]At least, this was the case for the operator of the website ifun.de, who reported on his experience with deletions by Google, available at https://www.ifun.de/raus-aus-der-google-suche-das-schwierige-recht-auf-vergessen-62511/ (last accessed 28 August 2017).

[55]Google does not delete links to websites entirely, but only for certain search queries.

[56]According to the operator of the website ifun.de, accessible at https://www.ifun.de/raus-aus-der-google-suche-das-schwierige-recht-auf-vergessen-62511/ (last accessed 28 August 2017).

the balancing exercise ought to have had the opposite outcome without knowing which individual is affected by the search. In this respect, the search engine operators' reporting duties should be extended to involve more details on the criteria used in the decision process on deletion.

In the time frame from 29 May 2014 until 21 August 2017, 320158 URLs were examined in Germany in response to requests from affected parties, of which Google removed 47.9%.[57]

9 The Effects of the New General Data Protection Regulation on the Right to Be Forgotten

There is no specific discussion on reform with regard to the "right to be forgotten." This is hardly surprising because the German legislator as well as practitioners are engrossed in adapting to the broad changes to data protection law, which will be determined by EU regulation to an even greater extent in the future. From 25 May 2018, the General Data Protection Regulation (GDPR) took effect in the European Union—and thus, also in Germany. Art 17 GDPR guarantees affected individuals a "right to be forgotten" in the shape of a specific claim to deletion, whose existence as such was the only element guaranteed under current law. This claim is not identical to the claim to deletion the ECJ recognized in its *Google Spain and Google* case law, but partially goes further.[58] Art 17 § 1 lit a GDPR ensures a claim for deletion equivalent to the "right to be forgotten"[59] in the case where the processing of data loses its purpose, art 17 § 1 lit d GDPR ensures this claim for cases of the illegal processing of data within the meaning of art 6 § 1 lit f GDPR and art 17 § 1 lit c GDPR ensures it for cases where data has been processed illegally after the affected individual objects within the meaning of art 21 GDPR.

The processing of data from third parties is only legal within the meaning of art 6 lit f GDPR if there is a legitimate interest in the processing of the data when the respective interests and affected fundamental rights are balanced. This entails a balancing exercise, equivalent to the ECJ's case law on the "right to be forgotten". It is sometimes argued that the legislator did not join the ECJ in its approach rule—exception—balancing exercise.[60] This likely refers to the fact that the ECJ generally ranks data protection over information and communication interests. However, this is not necessarily the design the ECJ must be understood to accord to the "right to be forgotten". The ECJ's balancing exercise may well seem deficient, but it is not at all evident that the ECJ wanted to establish the mechanism rule—exception—balancing

[57]Cf Google's Transparency Report, available at https://transparencyreport.google.com/eu-privacy/overview (last accessed 28 August 2017).

[58]Paal (2016).

[59]Trentmann (2017), pp. 26, 30.

[60]Trentmann (2017), pp. 26, 31.

exercise. This will emerge from the future distinction in the ECJ's jurisprudence, which one may hope will take an equitable approach.

Art 17 § 1 lit c GDPR read together with art 21 § 1 sentence 2 GDPR also requires a balancing exercise. It is, however, designed differently. There, the processor must show compelling reasons that necessitate the further processing of the data.[61] The burden of proof in this proportionality test is therefore reversed.[62] Art 17 § 1 lit c GDPR additionally lays down the requirement of a separate balancing exercise. It seems doubtful in how far this exceeds the balancing exercise in art 21 § 1 sentence 2 GDPR. In this respect, it is decisive if the legal consequence of an objection immediately is the deletion of data. The affected individual has the option of raising an objection. In the balancing exercise, the processor carries the burden of proof. However, the legal consequence of the objection could merely be the obligation to desist from processing the data further, but not deleting it.[63] This is further supported by the existence of art 17 § 1 lit c GDPR, which lays down an obligation to delete, which would otherwise be merely declaratory. Under this provision, however, the usual standard without the shift of the burden of proof applies, which is why the affected party's recourse under art 17 § 1 lit c GDPR is likely no different with regard to the outcome than under art 17 § 1 lit d GDPR. It could also be assumed, however, that the duty to delete is the immediate legal consequence of art 21 § 1 GDPR, depending on how far the objection goes. This is generally the case if the objection is not limited to isolated steps of data processing, but the entirety of the processing.[64] The standard of art 17 § 1 lit c GDPR would thus be equivalent to that of art 21 § 1 GDPR, so that it must be assumed that the legislator merely wanted to clarify that the requirements of art 21 § 1 GDPR must be met by mentioning the balancing exercise again.[65] This would entail a real balancing exercise with a shift of the burden of proof in the affected individual's favour.

Art 17 § 2 GDPR goes beyond the ECJ's case law on the "right to be forgotten", requiring appropriate measures to be taken, and requires that third parties who are processing data must also be informed about the deletion request.[66] Moreover, art 17 § 1 GDPR imposes examination and monitoring obligations on service providers within the bounds of reasonableness.[67] Violations can be sanctioned with

[61]Here, Trentmann (2017), pp. 26, 32 believes data protection must prevail, because the burden of proof for the need to process the data rests with the responsible party and not with the affected individual.

[62]Martini (2016) and Herbst (2017).

[63]This is at least suggested by Martini (2016). According to art 4 n°2 GDPR, the notion of processing is understood broadly and includes both saving and deleting. This has the absurd legal consequence that the processor is prohibited both from processing further and deleting in the case of an objection, so that he has no legal course of action open to him. It can therefore hardly be concluded from the vague wording that a duty to delete already follows from art 21 § 1 GDPR.

[64]Herbst (2017).

[65]*Herbst* (*ibid*) Rn 26.

[66]Buchholtz (2015), pp. 570, 574 criticises this harshly.

[67]Trentmann (2017), pp. 26, 30 f.

"administrative fines up to 20,000,000 EUR, or in the case of an undertaking, up to 4% of the total worldwide annual turnover of the preceding financial year" pursuant to art 83 § 5 lit b GDPR.

With regard to the German legal framework, the flexibility clause for the Member States in art 23 § 1 lit I GDPR must be borne in mind. Para 35 Federal Data Protection Act (Bundesdatenschutzgesetz) 2018 enshrines a limitation on the "right to be forgotten" pursuant to art 17 GDPR. In this vein, para 35 § 1 Federal Data Protection Act 2018 guarantees only the claim to the limitation of the processing of data within the meaning of art 18 GDPR, instead of a claim to deletion, if deleting the data is only possible with disproportionate effort and the affected individual only has a marginal interest in deletion. The Federal Data Protection Act 2018 does not contain an exception on data processing for the purposes of journalism within the meaning of art 85 § 2 GDPR.

10 Next Steps

Since the General Data Protection Regulation (GDPR) took effect in the EU, the design of the right to be forgotten is determined by EU law to an even greater extent than it has already been the case due to the general Data Protection Directive 95/46/EC and the ECJ's *Google* judgment. Incidentally, it has been shown that both the substantive provision on deletion as well as the accompanying procedural rules on effective implementation—notably through sanctions—have been significantly tightened to benefit the protection of personality rights. Therefore, it is now even more important if the entities applying the regulation strike the right balance between the legitimate protection of personality rights of the affected individuals on the one hand and the public interest as well as the freedom of communication and enterprise of website operators as the primary source and search engine operators as intermediaries on the other hand. In light of the ECJ's cryptic case law, this is anything but certain. There is a risk of an excessive emphasis of the protection of personality rights and the inordinate suppression of so-called opposing fundamental rights. Here, the European legal community, as a community for the interpretation of fundamental rights, needs to, i.a., make targeted requests for preliminary rulings from the ECJ[68] in order to participate in creating sensible harmonized EU rules. The more the ECJ charges the interpretation of secondary legislation with interpretations informed by fundamental rights and the primary law, the less of a margin will remain for legislative corrections by the EU legislator. With respect to the division of powers, the ECJ is thus not only moving toward the centre of the debate vertically, vis-à-vis the Member States, but also horizontally in relation to the executive and legislative branches.

[68] On this issue in general, see Kühling and Drechsler (2017), p. 2950.

However, the bulk of the work of fleshing out the rules rests with the national data protection authorities, which ensure the EU-wide approximation of standards together with the European Data Protection Board in accordance with the consistency mechanism in the GDPR (see in particular arts 63 ff). Thus, the European Data Protection Board should specifically flesh out the obligations pursuant to art 17 § 2 GDPR on the basis of art 70 § 1 lit d GDPR by issuing guidelines, recommendations and best practices and thereby specify which measures can be reasonably required of the responsible party that published the content in question, which must be deleted, in order to inform third parties spreading the content of the asserted claims to deletion.[69] This thus involves the case that a website operator who published a certain piece of information must inform other disseminators such as internet search engine operators about the duty to delete (e.g., links). By comparison, the Member States' design will lose significance.

Striking a reasonable balance in the current secondary law while adequately accounting for the basic guidelines from fundamental rights as the competent authorities make use of the consistency procedure are the core task since the GDPR has entered into force in May 2018. Additional legislative reform is not called for, until it emerges if the competent authorities can find an appropriate balance in the interpretation and application of the respective legal framework. In particular, there does not appear to be a need to strengthen the "right to be forgotten" further.

Acknowledgements The author thanks his teaching and research assistant Dr. Anna Kellner for her extensive support in drafting this text and his research assistant Dr. Corinne Ruechardt for the translation.

References

Arning M, Moos F, Schefzig J (2014) Vergiss(,) Europa. CR 447, 450 f., 453

Bassenge P (2017) In: Palandt O (fndr) Bürgerliches Gesetzbuch, 76th edn. C H Beck, § 1004 BGB Rn 53

BGHZ 191, 219 paras 24 ff

BGHZ 202, 242 para 32

BGHZ 209, 139 paras 17, 39, 40, 41

Boehme-Neßler V (2014) Das Recht auf Vergessenwerden -Ein neues Internet-Grundrecht. NVwZ 825, 827

Buchholtz G (2015) Das "Recht auf Vergessenwerden" im Internet – Vorschläge für ein neues Schutzkonzept. ZD 570, 574

Buchner B, Tinnefeld M-T (2017) In: Kühling J, Buchner B (eds) Datenschutz-Grundverordnung, Kommentar. C H Beck, Art 85 DS-GVO para 26

BVerfGE 34, 269; 35, 202; 54, 208; 61, 1; 80, 367; 85, 1; 90, 1; 90, 241; 99, 185; 114, 339

District Court (Landgericht) Hamburg, 07.11.2014 – 324 O 660/12 para 105

District Court (Landgericht) Heidelberg, 09.12.2014 – 2 O 162/13 para 61

[69]See also the reference in Herbst (2017).

ECJ, Case C-131/12 Google Spain and Google ECLI:EU:C:2014:317, paras 35–38 and 83
Fechner F (UTB 2017) Medienrecht, p 50
Glaeser WS (1988) Die Meinungsfreiheit in der Rechtsprechung des Bundesverfassungsgerichts. AöR 113, 74 ff
Gosche A (2008) Das Spannungsverhältnis zwischen Meinungsfreiheit und Ehrenschutz in der fragmentierten Öffentlichkeit. Nomos, 63 ff
Grabenwarter C (2017) In: Maunz T, Dürig G (eds) Grundgesetz, Kommentar, 80th edn. C H Beck, Art 5 GG para 50
Grimm D (1995) Die Meinungsfreiheit in der Rechtsprechung des Bundesverfassungsgerichts. NJW, 1699, 1700 ff, 1702
Herbst T (2017) In: Kühling J, Buchner B (eds) Datenschutz-Grundverordnung. C H Beck, Art 17 DS-GVO Rn 12, 58, Art 21 DS-GVO Rn 22
Higher District Court Cologne (Oberlandesgericht Köln), 31.05.2016 – 15 U 197/15 paras 52 f
Higher District Court Munich (Oberlandesgericht München), 27.04.2015 – 18 W 591/15 para 24
Hoffmann-Riem W (2001) In: Stein E, Denninger E, Hoffmann-Riem W, Schneider H-P (eds) Kommentar zum Grundgesetz für die Bundesrepublik Deutschland. Luchterhand, Art 5 GG paras 29, 32
Holznagel B, Hartmann S (2016) Das 'Recht auf Vergessenwerden'als Reaktion auf ein grenzenloses Internet Entgrenzung der Kommunikation und Gegenbewegung. MMR 228, 231 f
Huster S (1996) Das Verbot der Ausschwitzlüge, die Meinungsfreiheit und das Bundesverfassungsgericht. NJ3rW, 487 ff
Karg M (2014) Anmerkung zu einer Entscheidung des EuGH (Urteil vom 13.05.2014 – C-131/12, ZD 2014, 350), zum Löschungsanspruch gegen Suchmaschinenbetreiber. ZD 359, 361
Kingreen T, Poscher R (2016) Grundrechte, 33rd edn. C F Müller, para 618
Kloepfer M (1991) Produkthinweispflichten bei Tabakwaren als Verfassungsfrage. Duncker & Humblot, p 30 f
Kühling J (2014) Rückkehr des Rechts: Verpflichtung von 'Google & Co.' zum Datenschutz. EuZW 527, 531 f
Kühling J (2017) Neues Bundesdatenschutzgesetz – Anpassungsbedarf bei Unternehmen. NJW 1985, 1989
Kühling J. In: Beck'scher Online-Kommentar zum Informations- und Medienrecht Art 5 GG
Kühling J, Drechsler S (2017) Alles "acte clair?" – Die Vorlage an den EuGH als Chance. NJW 2950
Kühling J, Seidel C, Sivridis A (2015) Datenschutzrecht, 3rd edn. C F Müller, 230 ff
Kutscha M, Thomé S (Nomos 2013) Grundrechtsschutz im Internet, p 51 f
LG Hamburg, 07.11.2014 – 324 O 660/12
LG Hamburg, 10.07.2015 – 324 O 17/15 para 22
LG Hamburg, 29.01.2016 – 324 O 456/14 paras 39, 54 ff
LG Heidelberg, 09.12.2014 – 2 O 162/13 paras 45, 52, 60 f
LG Köln, 28.05.2015 – 28 O 496/14
LG Köln, 7.10.2015 – 28 O 370/14
Martini M (2016) In: Paal B, Pauly D (eds) Datenschutz-Grundverordnung. C H Beck, Art 21 DS-GVO Rn 34 ff
OLG Celle, 01.06.2017 – 13 U 178/16 paras 15 ff
OLG Celle, 06.12.2016 – 13 U 85/16 paras 4 ff
OLG Karlsruhe, 14.12.2016 – 6 U 2/15 para 110
OLG Köln, 31.05.2016 – 15 U 197/15 paras 52 f
OLG München, 27.04.2015 – 18 W 591/15
Paal B (2016) In: Paal B, Pauly D (eds) Datenschutz-Grundverordnung. C H Beck, Art 17 DS-GVO Rn 2
Pille J-U (Nomos 2016) Meinungsmacht sozialer Netzwerke, p 177 ff
Rössel M (2014) Recht auf Vergessenwerden bei Suchmaschinen. ITRB 150, 151

Schmidt-Jortzig E (2009) In: Isensee J, Kirchhof P (eds) Handbuch des Staatsrechts der Bundesrepublik Deutschland, Bd VII, 3rd edn. C F Müller, § 162 Rn 23, para 21
Sörup T (2014) Urteilsbesprechung zu EuGH v. 13.05.2014, C-131/12: Löschungsanspruch gegen Google – "Recht auf Vergessen." MMR 455, 465
Trentmann C (2017) Das "Recht auf Vergessenwerden" bei Suchmaschinentrefferlinks. CR 26, 28–32
Wybitul T (2016) DS-GVO veröffentlicht – Was sind die neuen Anforderungen an die Unternehmen? ZD, 253, 254

The Right to Be Forgotten in Ireland

Patrick O'Callaghan

Abstract This chapter examines the status of the right to be forgotten in Irish law. It pays close attention to data protection law and finds that even before the coming into force of the General Data Protection Regulation (GDPR), a right to be forgotten, rooted in data protection law, was available in Irish law. The chapter also explores whether a right to be forgotten is available beyond data protection law. In doing so, it assesses whether interests in forgetting and/or being forgotten are given expression in other areas of Irish law. The chapter considers the legislation on spent convictions, defamation law and the law of privacy. It finds, however, that data protection law is the most suitable home for a right to be forgotten. The chapter also examines the limits of the right to be forgotten and the remedies available for infringement before commenting on the transparency problem in the context of search engine delisting requests.

1 Introduction

This chapter examines the status of the right to be forgotten in Irish law. When it came into force on 25th Mary 2018, Article 17 of the General Data Protection Regulation (GDPR) introduced a 'right to be forgotten' to Irish law. However, it is possible to argue that this right already existed in Irish data protection law. Indeed, this chapter argues that two versions of the right already existed, the 'full' and the 'narrow' versions.[1] The 'full' version is the right to erasure that was already

This chapter stems from research supported by the Irish Research Council's New Horizons 2016 funding scheme.

[1]This distinction between 'full' and 'narrow' versions of the right is inspired by O'Hara and Shadbolt (2015, p. 178) who draw a distinction between the 'narrow principle' established by *Google Spain* and the 'full term "right to be forgotten" to refer to a more abstract conception.'

P. O'Callaghan (✉)
University College Cork, Cork, Ireland
e-mail: patrick.ocallaghan@ucc.ie

F. Werro (ed.), *The Right To Be Forgotten*, Ius Comparatum – Global Studies in Comparative Law 40, https://doi.org/10.1007/978-3-030-33512-0_7

available under Irish data protection legislation, while the 'narrow' version is the right to be delisted identified by the Court of Justice of the European Union (CJEU) in its *Google Spain* decision ((2014) Case C-131/12).

But an analysis of data protection law is just one aspect of this chapter. The chapter also explores whether a right to be forgotten is available beyond data protection law. In doing so, it assesses whether interests in forgetting and/or being forgotten are given expression in other areas of Irish law. The chapter considers the legislation on spent convictions, defamation law and the law of privacy. It finds, however, that data protection law is the most suitable home for a right to be forgotten.

The chapter will also examine the limits of the right to be forgotten and the remedies available for infringement before commenting on the transparency problem in the context of search engine delisting requests.

2 Interests in Forgetting and Being Forgotten

In recent years, the right to be forgotten has been associated with the privacy challenges of the digital era. However, the idea itself is older still and finds its origins in the civil law doctrine and scholarship of several European legal systems (Brüggemeier et al. 2010; Werro 2009). Indeed, the classic right to be forgotten scenario in the pre-digital age concerned the ex-offender who wished to have a fresh start in life but continued to face media intrusion into his private life (Werro 2009).

If we study closely the language used in several civilian legal systems a 'latent' aspect of the right to be forgotten comes to the fore (de Mars and O'Callaghan 2016). As well as *interests in being forgotten by others*, we also have *interests in forgetting things from our own past* (Koops 2011, pp. 231–232). Consider, for instance, the German *Recht auf Vergessen*. This suggests a 'right to forget' or a right not to be confronted with one's own past. As an aspect of the General Personality Right (*allgemeines Persönlichkeitsrecht*), and therefore rooted in the old Roman doctrine of *iniuria* (Zimmermann 1992, pp. 1050 *et seq*; Hagemann 1998), the harm in question is personal, resulting from being confronted with something from one's past that one would otherwise have forgotten or would rather forget. It is only recently that German scholars have also begun speaking about interests in being forgotten by others (*Recht auf Vergessenwerden*) (Hornung and Hofmann 2013).

The distinction between interests in being forgotten and interests in forgetting takes on added significance against the backdrop of digital technology's capacity for 'perfect memory' (Bellia 2008; Mayer-Schönberger 2009). The ease and extent to which our personal data can be collected, stored and disseminated online makes it necessary to craft 'breathing room' so that episodes from our past can be forgotten by others as well as by ourselves (Cohen 2012).

3 Institutional Forms of Forgetting in Irish Law

It has long been understood that law has an important role in giving expression to and protecting interests in forgetting and being forgotten. To take just one example, throughout history amnesties were sanctioned by States in order to ensure that certain events from the past, usually political crimes, were 'forgotten' and the slate was wiped clean (Lesaffer 2019). The idea has ancient roots. Indeed, the word itself, comes from the Greek ἀμνηστία (*amnestia*) meaning 'forgetfulness'. Official acts such as these are what Ricouer has called 'institutional forms of forgetting' (2004, p. 448). In this part of the chapter, we will examine four areas of Irish law which, at first glance at least, have the potential to give expression to and protect interests in forgetting and being forgotten. These are the law on spent convictions, defamation law, privacy law and data protection law.

3.1 Criminal Justice (Spent Convictions and Certain Disclosures) Act 2016

The classic example of an institutional form of forgetting in modern legal systems is the state-mandated expungement of criminal records. However, it was not until 2016 that Ireland first introduced legislation on spent convictions for adults,[2] making it one of the last member states of the European Union to do so (Kilcommins and O'Donnell 2003). The Criminal Justice (Spent Convictions and Certain Disclosures) Act 2016 allows certain categories of convictions to become spent where the person was convicted seven or more years ago and has served the sentence or otherwise complied with a court order in respect of that conviction (s 5(1)–(2)). Where the conviction is regarded as spent, the person shall not be required to reveal details of the conviction (s 6).[3] This is clearly of particular importance in the employment context given that it is common for employers to enquire about prospective employees' past activities. Most serious offences (including most sexual offences) are excluded (s 4(1) and s 5(2)(c)). In particular, a conviction will not be regarded as spent if the term of imprisonment is more than 12 months (*ibid*). In addition, the Act does not apply to a person who has more than one conviction (s 5(3)).

When compared with similar statutory regimes elsewhere, the Irish legislation is markedly restrictive on what offences may become spent. Contrast, for example, the limit of 12 months in Ireland with the equivalent 48 months in England and Wales (Rehabilitation of Offenders Act (1974), as amended, s 5(2)). Moreover, the so-called 'single conviction rule' has been particularly criticised by commentators.

[2]Spent convictions (in respect of most crimes) were possible in the case of child offenders. See s 258 of the Children Act (2001).

[3]A number of exceptions to this general rule (including, for example, exempted employers) are outlined in 8–11 of the Act.

McIntyre and O'Donnell argue that 'it draws an arbitrary cutoff at one conviction – taking too literally the saying that everyone is entitled to make *one* mistake.' (McIntyre and O'Donnell 2017, p. 33). The authors question whether these two aspects of the Act, the 12 month cutoff and the single conviction rule, meet the requirements of Article 8 ECHR (2017, pp. 38–46). Following an analysis of the recent English court of Appeal decision in *P & Others v Secretary of State for the Home Department* [2017] EWCA Civ 321, the authors argue that the application of 'bright line rules' may have a disproportionate impact on the Article 8 rights of the individual (McIntyre and O'Donnell 2017, pp. 38–46).

Nonetheless, the Irish legislation should be regarded as an important first step in this area. Despite their criticism of the Act, McIntyre and O'Donnell argue that 'its most important contribution is the principle it establishes.' (2017, p. 55). Indeed, the legislation highlights in this particular constellation the importance of interests in forgetting and being forgotten, not just for the ex-offender but also more generally in that society benefits too when ex-offenders are rehabilitated. The Act gives expression to deeply rooted notions about the right to have a 'second chance' in life. As O'Connor J put it in *Nolan v Sunday Newspapers Ltd* [2017] IEHC 367:

> Most reasonable people take a view that convictions given in youthful years, at a time of stupidity and naivety or a decade ago should not engender any particular interest unless those convictions are relevant to a subsequently linked crime. That attitude arises from cultural, religious, historical and societal influences, which is manifested in the recent uncontroversial commencement of the Criminal Justice (Spent Convictions and Certain Disclosures) Act 2016. (at [4])

3.2 *Defamation Law*

In common law systems, what civilian lawyers call 'personality rights' have traditionally been protected by a patchwork of individual torts and equitable causes of action e.g. defamation, privacy, passing off and breach of confidence (Howells and Rott 2010). The most important of these are defamation and privacy. We will start with an analysis of defamation before moving on to privacy law.

At first glance, it may be possible to draw a connection between defamation and interests in forgetting/being forgotten. Indeed, plaintiffs pursing defamation actions have sometimes been accused of seeking to find ways to 'escape their past'.[4] However, it should be emphasised that truth is a complete defence to a defamation action. Where the defendant can show that the statement in question is true 'in all material respects' (Defamation Act 2009, s 1(1)), the plaintiff will not succeed even though the statement may have lowered the plaintiff's reputation in the eyes of the community. This is significant because when an individual has interests in forgetting

[4] In de *Rossa v Independent Newspapers* [1999] 4 IR 432, 470, Hamilton CJ quoted counsel for the Appellant's remarks that the case had 'been brought by [the respondent] in an attempt to escape his past'.

or being forgotten, these interests normally concern true rather than false information from that individual's past.

Of interest, however, is Lord Shaw's speech in *Sutherland v Stopes* [1925] AC 47:

> a statement of fact or of opinion which consists in the raking up of a long-buried past may, without an explanation (and, in cases which are conceivable, even with an explanation), be libellous or slanderous if written or uttered in such circumstances as to suggest that a taint upon character and conduct still subsists, and that the plaintiff is accordingly held up to ridicule, reprobation and contempt. (p 74)

Clearly, context is important. Simply reporting a crime from an individual's past would be protected by the defence of truth. But the defence may not be available if the sting of the statement is that an ex offender, whose crime was in the distant past and has long since been rehabilitated, is still engaging in criminal activity. More difficult is a statement which suggests a 'taint upon character' as Lord Shaw puts it, since under Irish law the defence of honest opinion is potentially available where the opinion in question is based on the fact of a previous conviction (Defamation Act 2009, s 20 (1)–(2)). However, if it can be shown that the statement was actuated by malice (i.e. in the sense that the opinion was not honestly held), then this defence would not be available (*ibid*).

Lord Shaw's reflections are of interest. However, given that, generally speaking, false information falls within the remit of defamation while true information falls within the remit of privacy, privacy law is perhaps a more natural home for a 'right to be forgotten'.

3.3 The Law of Privacy

Traditionally, the common law lacked a general remedy for invasion of privacy but privacy was nonetheless understood as an underlying value in the common law and was given expression through causes of action such as trespass, nuisance and the equitable remedy of breach of confidence (*Wainwright v Home Office* [2004] 2 AC 406, 422 (per Lord Hoffmann)). Following the introduction of the Human Rights Act 1998, English courts expanded the equitable doctrine of breach of confidence in order to give indirect horizontal effect to Article 8 ECHR in cases of misuse of private information (Morgan 2003; Phillipson 2003; Moreham 2005). What emerged was a tort of misuse of private information, under which the plaintiff must show, as a first step, that she had a reasonable expectation of privacy in respect of the misused information in question.[5] In most cases, the competing right of freedom of expression is at stake and courts must therefore engage in a balancing exercise (*In re S*

[5]Lord Nicholls refers to the equitable doctrine as a tort in *Campbell* v. *MGN Ltd* [2004] 2 AC 457, 465: 'The essence of the tort is better encapsulated now as misuse of private information'.

(A Child) [2005] 1 AC 593, 603 (per Lord Steyn); *Campbell v MGN Ltd* [2004] 2 AC 457, at 497 (per Baroness Hale)).

The English courts responded in this way because of a significant gap in the law, one that became glaringly obvious following the introduction of the Human Rights Act. As we will now see, the gaps in the common law have proved less problematic in Ireland because of its constitutional right of privacy and the ways in which plaintiffs can rely on this right in tort actions (McMahon and Binchy 2013, pp. 1403–1446; Quill 2014, pp. 345–355; Carolan and Delany 2008). As McMahon and Binchy explain (2013, p. 1436), unlike the English courts, 'courts in Ireland have been in the happy position that they have not had to resolve the specific question of whether under Irish law, there is a tort of violation of privacy'.

The right of privacy is not expressly mentioned in *Bunreacht na hÉireann* (the Constitution of Ireland) rather it has been recognised as a constitutional right by the courts under the doctrine of unenumerated rights. As the textual basis for this doctrine, it is important to consider the wording of Article 40.3 of the Constitution. It states:

(1) The State guarantees in its laws to respect, and, as far as practicable, by its laws to defend and vindicate the personal rights of the citizen.
(2) The State shall, in particular, by its laws protect as best it may from unjust attack and, in the case of injustice done, vindicate the life, person, good name, and property rights of every citizen.

The Supreme Court has held that the words 'in particular' in Article 40.3.2 indicate that the personal rights of the citizen are not limited to life, person, good name and property rights (*Ryan* v *Attorney General* [1965] IR 294).[6] It follows that there must be other unenumerated personal rights. The right of privacy is considered to be one such right and was recognised in the 1987 case of *Kennedy* v *Ireland* [1987] IR 587 (though note that a right of marital privacy had been recognised by the Supreme Court some 13 years before in *McGee* v *Attorney General* [1974] IR 284). There are two aspects of Irish constitutional law and discourse that are particularly important for the analysis in this chapter.

Firstly, in principle, most fundamental rights can be mobilised as 'constitutional torts' by virtue of the doctrine of horizontal effect of fundamental rights. Where the plaintiff is able to show that the existing common law is 'basically ineffective' in protecting the fundamental right in question, he can take an action for breach of the fundamental right against the defendant (whether a private person or the State) (McMahon and Binchy 2013, pp. 19–50; O'Cinneide 2003).[7] The fact that the

[6]This doctrine remained dormant for several decades until it was reactivated when the Supreme Court identified the unenumerated right to work in *NVH* v *Minister for Justice & Equality and ors* [2017] IESC 35.

[7]The main authorities in this area are *Educational Co Ltd* v *Fitzpatrick (No 2)* [1961] IR 345; *Meskell* v *CIE* [1973] IR 121; *Glover* v *BLN Ltd* [1973] IR 388; *Hanrahan* v *Merck Sharp and Dohme* [1988] ILRM 629; *Herrity* v *Associated Newspapers (Ireland) Ltd* (2009) 1 IR 316; *Sullivan v Boylan Contractors (No 2)* [2013] IEHC 104; *Ogieriakhi v Minister for Justice and*

common law lacks a general remedy for invasion of privacy explains why injured parties have relied on the constitutional right of privacy when seeking to protect this right. However, constitutional torts have been variously described as 'blockbuster' or 'super' torts given that they have the potential to be powerful tools at the disposal of injured parties. For this reason, courts have carefully articulated limiting principles. So, for example, in the case of the right of privacy, the courts have required that the plaintiff's privacy is 'deliberately, consciously and unjustifiably intruded upon' before he can succeed (*Kennedy v Ireland* [1987] IR 587, 593 (per Hamilton P)).[8] As O'Dell (2017b, p. 246) points out, this is a 'particularly strict' test, at least when compared to alternatives such as the 'reasonable expectation of privacy' test associated with the English misuse of private information tort.

The second significant aspect of Irish jurisprudence is that privacy is understood as an evolving concept. Courts do not stick to rigid definitions of what the right entails. This allows for the possibility that a right to be forgotten might be recognised as an aspect of the unenumerated right of privacy should an appropriate case come before the courts. Henchy J's conceptualisation of privacy and its place in the constitutional order as outlined in the case of *Norris v Attorney General* [1994] IR 36 is relevant here. Though this was a dissenting opinion, Henchy J's conceptualisation of privacy in the opinion was later emphasised by Hamilton P in the seminal privacy case of *Kennedy* v *Ireland*. Understood in this way, it is important to pay close attention to Henchy J's observations. He explains that the 'right of privacy inheres in each citizen by virtue of his human personality' and is necessary to secure dignity and freedom (*ibid*, p. 71). The right is actually a 'complex of rights which vary in nature, purpose and range (each necessarily being a facet of the citizen's core of individuality within the constitutional order)' (*ibid*). Some rights in this 'complex' have already been identified by the courts (e.g. the right of marital privacy) but some are:

> yet to be given judicial recognition. It is unnecessary for the purpose of this case to explore them. It is sufficient to say that they would all appear to fall within a secluded area of activity or non-activity which may be claimed as necessary for the expression of an individual personality, for purposes not always necessarily moral or commendable, but meriting recognition in circumstances which do not engender considerations such as State security, public order or morality, or other essential components of the common good. (*ibid*, pp. 71–72)

Henchy J recognised that the constitutional order must ensure that there is sufficient space for the individual personality to express itself. Yet it is impossible to identify in advance all the conditions necessary to ensure such space and so courts may have to give judicial recognition to other rights in the 'complex' of privacy rights as the need arises.

Equality and the Attorney General and An Post (No. 2) [2014] IEHC 582. See also *Clarke v O'Gorman* [2014] 3 I.R. 340, 359 (per O'Donnell J).

[8] See also *Herrity v Associated Newspapers* [2009] 1 IR 316; *K (A Minor) v Independent Star* [2010] IEHC 500.

Taking these two aspects of Irish constitutional law and discourse together, we can identify a general constitutional framework within which a right to be forgotten could emerge. Where a wrongdoer (whether a private party or the State) deliberately, consciously and unjustifiably invades the privacy of another, the injured party can take an action against the wrongdoer. In reaching judgment, a court could identify a right to be forgotten as a right belonging to the complex of privacy rights, falling under the umbrella of the already-established unenumerated right of privacy.

Of course, all of this depends on the right case coming before the courts. To date, the case that perhaps comes closest to a right to be forgotten case in this sense is *Murray v Newsgroup Newspapers et al* [2011] 2 IR 156. In *Murray*, the plaintiff served sentences for serious sexual offences in the United Kingdom and in Ireland. Following his release from prison in 2009, the defendants continued to publish newspaper articles and photographs of the plaintiff. Not only did these articles give details of the plaintiff's previous convictions, they also provided information about his current whereabouts. In the meantime, the plaintiff secured voluntary employment at a hospital in Dublin but stated that he was 'forced to resign' due to the media publicity (*ibid*, p. 164). The plaintiff claimed that he feared for his personal safety and suffered depression and anxiety as a result.

The plaintiff sought an interlocutory injunction against the defendants and based his claim on the right of privacy as protected by the Constitution and the ECHR as well as his rights to life Article 40.3.2 Bunreacht), bodily integrity (Article 40.3 Bunreacht) and inviolability of the dwelling (Article 40.5 Bunreacht). Irvine J in the High Court refused the interlocutory relief, stating that the plaintiff had produced insufficient evidence to demonstrate that his personal rights should outweigh the right of the public to know details of his identity, past crimes and whereabouts (*ibid*, pp. 192–203). On the specific question of the right of privacy, the judge emphasised that the constitutional right of privacy was a qualified right and, likewise, the individual's right to privacy under Article 8 ECHR 'may have to give way to the competing rights of others to exercise their right to freedom of expression under [A] rticle 10' (*ibid*, pp. 193–195).

If the courts were so inclined, they could recognise the privacy interests in cases like *Murray* as interests in forgetting or being forgotten. However, in cases such as these, where the plaintiff has committed serious crimes, this would not necessarily change the outcome of the case since the right to be forgotten is also clearly a qualified right. In effect, then, would recognising a right to be forgotten in a case like this amount merely to a relabelling of the interests at stake? Or would it nonetheless have some sort of meaningful impact? A right to be forgotten emphasises the importance of interests in forgetting and being forgotten not just for the individual but also for society as a whole. In a case involving ex-offenders, this might lead a court, when balancing interests, to pay particular attention to the process of rehabilitation. Rehabilitation is only in some respects a privacy concern. Drawing solely on the language of privacy in this context potentially masks some of the benefits for the individual and society as a whole when an ex-offender is successfully rehabilitated. This chapter submits that the language of forgetting and being forgotten is more likely to highlight such benefits.

Whatever may emerge on this front, perhaps the most obvious place where interests in forgetting and being forgotten are at stake today is online in the face of ever-more expansive data retention and dissemination capabilities. It is to this particular problem that we now turn.

3.4 Data Protection Law

This section will explain how a right to be forgotten, in both full and narrow senses, can be found in Irish data protection law. As a preliminary point, it is important to note that the Data Protection Act 2003 transposed the Data Protection Directive 1995 (DPD) into Irish law and, in doing so, amended the Data Protection Act 1988. When this chapter refers to 'the DP Act 2003' it means the consolidated Acts of 1988 and 2003. This statutory regime was replaced by the GDPR when it became a directly applicable part of Irish law on the 25th May 2018. Alongside the GDPR itself, we should note that the *Oireachtas* (Irish Parliament) passed the Data Protection Act 2018 (DP Act 2018) to give further effect to the GDPR.

3.4.1 A 'Full' Right to Be Forgotten

We might say that a right to be forgotten already existed in Irish law if we understand it as an alternative label for the right to erasure under s 6(1) DP Act 2003, which implemented Article 12(b) DPD. This was replaced by the right to erasure ('right to be forgotten') under Article 17 of the GDPR. We might call this remedy a 'full' right to be forgotten because, in principle at least, it affords the individual the opportunity to secure the *complete erasure* of personal data.

A right to be forgotten, taking the form of a right to erasure in data protection law, is relatively uncontroversial and unproblematic to implement in what we might call the 'classic data protection scenario' i.e. where an individual requests rectification or erasure of data stored in a standalone database, administered by a single data controller. Indeed, it was presumably this sort of scenario the drafters of the DPD had in mind. In securing rectification or erasure of data in this way, the individual was exercising control over her own data, one of the main conceptual justifications for data protection laws. In some respects, the emergence of online service providers (OSPs) has not changed this picture. Where the OSP stores personal data that its users originally provided to the OSP or where the OSP has collected and stored 'back-end data' about its users online activities[9] and this data has not been shared, then we have a situation that is quite similar to the 'classic data protection scenario'.

[9]Keller (2018, p. 292) explains that 'back-end' data are data collected by OSPs when 'tracking their own users' online behaviour. OSPs have plenty of this privately held, "back-end" data – logs tracking users' clicks, profiles used to target advertisements, and more.'

In this context, Article 17 GDPR has strengthened the data subject's ability to remain in control over her personal data since Article 17(1)(b) sets out that a ground for exercising the right to erasure ('right to be forgotten') is the withdrawal of consent.[10] This ground did not expressly feature in the DPD nor did it feature in the Irish DP Act 2003.[11]

However, in other important respects the emergence of OSPs has complicated the picture.[12] Compared to the 'classic data protection scenario', it becomes more difficult to exercise full control over personal data when these data are *published* online whether by the data subject herself or by others (e.g. on social media, discussion boards, blogs or online newspapers).

There are two particular difficulties for the data subject here. *Firstly*, personal information once published online can be disseminated quite easily. This is especially so when information goes 'viral', making it virtually impossible to delete every single trace of this information. In response, the drafters of the GDPR introduced Article 17(2). It is worth setting out the text of this in full:

> Where the controller has made the personal data public and is obliged pursuant to paragraph 1 to erase the personal data, the controller, taking account of available technology and the cost of implementation, shall take reasonable steps, including technical measures, to inform controllers which are processing the personal data that the data subject has requested the erasure by such controllers of any links to, or copy or replication of, those personal data.

Clearly, imposing on obligation on the original controller to ensure that data is deleted from every corner of the Web would be too onerous. Instead, this provision requires the controller to take 'reasonable steps' to inform other controllers processing the data about the erasure request. The provision acknowledges the practical reality ('taking account of the available technology and cost of implementation') that it may not be possible to secure complete erasure of all links to, copies or replications of the personal data.

The *second* difficulty for the data subject concerns personal data that *others* may reveal about him online, without his consent. In certain contexts, it is clear that the data subject will not be able to exercise his right to be forgotten. Article 17(3) states that the right to be forgotten shall not apply 'to the extent that processing is necessary: (a) for exercising the right of freedom of expression and information.' Further Article 85 states that Members States must 'reconcile' the right to the protection of personal data with the right to freedom of expression and information and specifically mentions that this includes 'journalistic purposes and the purposes of academic, artistic or literary purposes'. Section 43 of the DP Act 2018 gives

[10]Article 17(1)(b) should be read alongside Article 7(3) which expressly provides for the right to withdraw consent. Markou (2015, p. 209) argues that this 'innovation' in Article 17 should be welcomes since 'consent-as-permanent-permission' is not consistent with the ability to exercise meaningful control over one's own data.'

[11]A literal interpretation of s 2, 2A, 2B and 6 DP Act 2003 does not afford the data subject a right to erase data on the grounds that he/she has withdrawn consent (assuming that the processing is otherwise lawful).

[12]For a comprehensive analysis of this complicated picture, see Keller (2018).

further effect to Article 85 GDPR. This provision provides an exemption from the operation of the rights of the data subject (including the right to erasure ('right to be forgotten') under Article 17) for data processed for the purpose of exercising the right to freedom of expression and information. Like Article 85 GDPR, s 43(1) of the Irish Act specifically mentions journalistic purposes and purposes of academic, artistic or literary expression. Section 43(1) states that the exemption is available when, 'having regard to the importance of the right to freedom of expression and information in a democratic society, compliance with that provision would be incompatible with such purposes'. So, for example, against the background of Article 40.6.1(i) of the Constitution, which provides for the right to freedom of expression and expressly states that the 'education of public opinion [is] a matter of . . .grave import to the common good', it would seem difficult to exercise a right to be forgotten in respect of personal data, the publication of which is deemed to be in the public interest. Of course, hard cases will arise and the words 'to the extent that' in Article 17(3) GDPR indicate that a balancing of competing rights will be necessary.

More difficult are the presumably much more common situations where personal data about a data subject is revealed by a person not acting in a professional or commercial capacity and there is no public interest in publication of the data nor are there any academic, artistic or literary purposes at stake. Keller (2018, p. 19) provides this useful hypothetical:

> Suppose someone tweets, "Matilda Humperdink served bad fish at her party last night. We all got sick – even Matilda!

The person tweeting this information is revealing personal information about Matilda, including information about her health which is considered a 'special category' of personal data within the meaning of Article 9 GDPR and thus subjected to greater protection. Would Twitter be considered a data controller here? While it is true that the text of GDPR does not expressly state whether OSPs should be regarded as data controllers in respect of user-generated content, Recital 18 of the GDPR states that the original poster of such information would not be regarded as a controller within the meaning of the Regulation.[13] Using the Recital as an interpretative tool, drawing on the definition of a 'controller' within Article 4(7) GDPR[14] and taking into account cases such as *CG v Facebook Ireland Ltd & McCloskey*

[13]Recital 18 states: 'This Regulation does not apply to the processing of personal data by a natural person in the course of a purely personal or household activity and thus with no connection to a professional or commercial activity. Personal or household activities could include correspondence and the holding of addresses, or social networking and online activity undertaken within the context of such activities. However, this Regulation applies to controllers or processors which provide the means for processing personal data for such personal or household activities.'

[14]According to Article 4(7) GDPR: 'controller' means the natural or legal person, public authority, agency or other body which, alone or jointly with others, determines the purposes and means of the processing of personal data; where the purposes and means of such processing are determined by Union or Member State law, the controller or the specific criteria for its nomination may be provided for by Union or Member State law.

[2016] NICA 54,[15] it is arguable that Twitter is a data controller in respect of this user-generated content. In any case, even if there is legal uncertainty on this point, as 'a practical matter', OSPs such as these may nonetheless comply with right to be forgotten requests rather than 'risk being summoned before a [Data Protection Authority].' (Keller 2018, p. 326).

It follows, then, that there is a clear risk of over-removal of content in the face of right to be forgotten requests and this may have a detrimental impact on the right to freedom of expression and information. Drawing on anecdotal evidence and pointing to academic studies, Keller states that 'OSPs receive many inaccurate or bad faith removal requests – and, too often, comply with them.' (Keller 2018, p. 297). This is particularly marked in the area of copyright removals under the US Digital Millennium Copyright Act (DMCA) (*ibid*).

For this reason, Keller (2018, pp. 318–319) argues that it is important that the new European Data Protection Board issues specific guidelines to safeguard against such over-removal[16] and that national legislatures address this issue when fulfilling their obligation to reconcile the right to the protection of personal data with the right to freedom of expression and information under Article 85 GDPR. In this respect, we should turn to consider s 43(5) of the DP Act 2018, which gives further effect to Article 85 GDPR. This provision states: 'In order to take account of the importance of the right to freedom of expression and information in a democratic society that right shall be interpreted in a broad manner.' The explanatory notes in the original Bill state that this provision 'gives expression to the final sentence in recital [153]' and 'is intended to acknowledge activities such as blogging and the expression of views on social media.' While a broad interpretation of freedom of expression in this context is to be welcomed,[17] if we return to the hypothetical tweet about Matilda, would this provision insulate the speaker and Twitter against a right to be forgotten takedown request? Again, let's suppose that Matilda is an ordinary person who is not otherwise in the public limelight. How broad would the interpretation given to freedom of expression need to be to prevent Matilda securing the removal of information relating to her health? The tweet itself may seem trivial enough but there may be all sorts of legitimate personal reasons why Matilda does not want her family, friends or employer know that she was sick (or indeed that she invited the person posting the tweet for dinner!). Certainly, the First Amendment would protect the speaker and Twitter here but, as we will now see, we can be less certain that a court would reach the same conclusion in Ireland.

Where Matilda's complaint is deemed to be an interference with her constitutional right to privacy and there is a potential conflict with the fundamental right to

[15]Here, the Court of Appeal of Northern Ireland held that in respect of information posted by third parties Facebook was a data controller within the meaning of s 5 of the UK Data Protection Act (1998).

[16]This is part of the Board's set of tasks. See Article 70(1)(b)(d) & (e) GDPR.

[17]There is already authority to suggest that a blogger may also be an 'organ of public opinion'. See *Cornec v Morrice & Ors* [2012] IEHC 376.

freedom of expression, the court should attempt to reconcile the rights rather than subordinate one to the other. As recently explained by the Supreme Court in *Gilchrist and Rogers* v *Sunday Newspapers Limited* [2017] 2 IR 284, there is in principle no hierarchy of rights in Irish constitutional law. Where there is a potential conflict between two rights, this conflict 'should be sought to be resolved without the subordination or nullification of one provision'. (*Ibid*). This is because the 'Constitution was intended to function harmoniously' and the 'obligation of a court is to uphold all the provisions of the Constitution.' (*Ibid*, p. 310). According to the European Court of Human Rights (ECtHR), Article 8 and 10 ECHR are of 'equal value' and thus a balancing exercise is necessary when they come into conflict (*Hachette Filipacchi Associés v France* (2009) 49EHRR 23; *Mosley v United Kingdom* (2011) 53 EHRR 30; *Axel Springer AG v Germany* [2012] ECHR 227). Moreover, even though freedom of expression and information is protected under Article 11 of the Charter of Fundamental Rights (CFR), any restrictions on Matilda's right of privacy under Article 7 CFR or her right to the protection of personal data under Article 8 CFR must adhere to Article 52(1) CFR which states:

> Any limitation on the exercise of the rights and freedoms recognised by this Charter must be provided for by law and respect the essence of those rights and freedoms. Subject to the principle of proportionality, limitations may be made only if they are necessary and genuinely meet objectives of general interest recognised by the Union or the need to protect the rights and freedoms of others.[18]

While s 43(5) of the DP Act 2018 encourages a broad interpretation of the right to freedom of expression and information, the jurisprudence of the Irish courts as well as the ECtHR and CJEU clearly set limits on how broadly this right can be interpreted.

What all of this means is that there are open questions about the extent of an OSP's obligations under Article 17 GDPR in respect of user-generated content and that the risk of over-removing content is very real. Keller puts forward an interesting solution, proposing that notice-and-takedown rules based on the E-Commerce Directive be adopted in the GDPR context (Keller 2018). It remains to be seen whether such a solution might eventually be adopted.[19]

A 'full' right to be forgotten is in principle available under Article 17 GDPR. But as emphasised above, this may be difficult to implement once personal data is published online. Firstly, the information may have been disseminated to all corners of the Web and thus it may be difficult or impossible to completely erase

[18]See, for example, Joined Cases C-92/09 and C93/09 *Volker und Markus Schecke GbR v Land Hessen* [2010] ECRI-11063.

[19]It is only a matter of time before these hard questions come before the courts. If there is to be a legislative solution to this conundrum, then this needs to implemented within the context of Article 23 GDPR, which mirrors, to a significant extent, the text of Article 52(1) CFR. Article 23 GDPR states that a national legislature may restrict the scope of the obligations and rights provided for in Article 17 in order to protect, amongst other things, 'the rights and freedoms of others'. But any such measure must be a 'necessary and proportionate' one in a democratic society and must respect 'the essence of the fundamental rights and freedoms'.

it. Secondly, where the right of freedom of expression and information prevails, the individual may prefer that the information is not posted but has little choice in the matter. Nonetheless, all is not necessarily lost for the aggrieved individual who finds herself in one or the other of these situations. Under certain conditions, she may be entitled to a 'narrow' right to be forgotten, which may secure 'good enough' privacy for her (Ohm 2008).

3.4.2 A 'Narrow' Right to Be Forgotten

Search engines were in their infancy when the DPD was drafted—Google, for example, did not become available to the public until 1998. In the meantime, it has become clear that they present particular challenges for privacy protection. As the CJEU explained in *Google Spain* (2014) Case C-131/12:

> [A search engine] enables any internet user to obtain through the list of results a structured overview of the information relating to that individual that can be found on the internet — information which potentially concerns a vast number of aspects of his private life and which, without the search engine, could not have been interconnected or could have been only with great difficulty — and thereby . . . establish a more or less detailed profile of him. (at [80])

A search engine creates a 'collage' of information about an individual.[20] Each piece of information, taken on its own and lawfully reported on the source website, may be unobjectionable, even if this information is very old. But when these pieces of information are patched together into one easily and instantly available collage of information in the search results, we can be less certain that we still have the right balance between freedom of expression and information on the one hand and privacy on the other.

The CJEU found a solution to this problem in *Google Spain* by holding that search engines are data controllers. Thus, as a result of Articles 12(b) and 14(a) of the DPD, individuals can request search engines to rectify or erase search engine results that are inaccurate, incomplete, outdated, or no longer relevant (as set out in the principles of data quality under Article 6 DPD). We might call this solution a 'narrow' version of the right to be forgotten since it may well be (as it was in the *Google Spain* case itself) that the particular data in question remain lawfully accessible on the source website. This version is perhaps better labelled a 'right to be delinked or delisted'.

The right to be delisted survives the introduction of Article 17 GDPR. It is now accepted that a search engine is a data controller. The search engine will thus need to erase links where one of the grounds under Article 17(1) applies and where it cannot rely on any of the exemptions under Article 17(3). Again, Article 17(3)(a) requires a balancing of interests. Perhaps one of the most controversial issues, which the

---

[20] On the idea of collages see of information see Mayer-Schönberger (2009), p. 124.

chapter explores below, is that the search engine itself is charged with this balancing of interests, at least at the initial stage.

The *Google Spain* decision received a mixed reaction in Ireland. Some commentators welcomed it while others were critical. Billy Hawkes, for example, the then Data Protection Commissioner, is reported as saying that the decision was a 'very difficult one' (Slattery 2014). Immediately following the *Google Spain* decision, the Irish Data Protection Commissioner started to receive complaints from data subjects whose delisting requests were rejected by search engines. The Commissioner received 32 such complaints in 2014, 23 complaints in 2015, 26 complaints in 2016 and 21 in 2017 (Annual Reports of the Data Protection Commissioner 2014–2017). According to the available records, the Commissioner rejects the majority of such complaints. In 2015, for example, 16 complaints out of 23 were rejected. But the Commissioner does uphold a minority of complaints. One such example from 2016 concerned what in continental Europe might be regarded as the 'classic' right to be forgotten case constellation: the rehabilitation of ex-offenders (Annual Report of the Data Protection Commissioner 2016, p. 10). The complaint related to the reporting of a conviction for assault causing harm for which the convicted person was sentenced to imprisonment for 6 months suspended for 3 years. Several years after the conviction, however, the search engine in question continued to provide a link to the news story. Though this complaint concerned a crime and the reporting of such information is generally in the public interest, an important factor in this case was that the conviction qualified as a 'spent conviction' within the meaning of the Criminal Justice (Spent Convictions and Certain Disclosures) Act 2016 discussed above. The Data Protection Authority determined 'that the story was no longer relevant on this basis' and the link was removed by the search engine.

To date, only one such decision of the Commissioner relating to the right to be delisted has been appealed to the courts: *Savage* v *Data Protection Commissioner (Respondent) and Google Ireland* [2018] IEHC 122.[21] In this case, the plaintiff stood as a candidate for local elections in Dublin. During his campaign, he produced election material which was examined by members of the discussion board, Reddit.com. The thread title for this discussion was: 'Mark Savage - North county Dublin's homophobic candidate' (*ibid*, [9]). Savage complained about the ensuing search result that appeared on Google once a search for his name was undertaken. This search result was made up of the URL title (worded the same as the thread title of the discussion), the URL link (the web address of the webpage in question) and the snippet text (part of the discussion on the webpage). Google Ireland refused to delist this search result on the basis that, in deciding to run for public office, Savage was a public figure and therefore the public's interest in knowing information relevant to him outweighed any interest he had in privacy (*ibid*, [11]). Savage then complained to the Data Protection Commissioner who rejected his complaint on the basis that

[21]Cf the decision of the High Court of Northern Ireland in *Townsend v Google Inc and Google UK Ltd* [2017] NIQB 81.

'the link remains accurate in that it represents the opinion expressed of [Savage] by a user of the relevant forum.' (*Ibid*, [15]). Following this determination, Savage appealed to the Circuit Court, '[denying] the stance he took [could] accurately define him as a homophobe and that the URL [asserted] this as a fact without any qualification or parenthesis and as a result constitutes inaccurate data…'.[22] The Circuit Court upheld his appeal on the basis that the URL title 'bears the appearance of a verified fact' and thus, in accordance with *Google Spain*, the search result should be edited to make it clear that the statement that appeared in the search results was a statement of opinion.[23] The High Court, however, upheld the Data Protection Commissioner's and Google's appeal against the Circuit Court's decision. Among other reasons, the High Court held that the Circuit Court had not carried out the balancing exercise outlined in *Google Spain* and that *Google Spain* did not envisage an editing process of the sort mooted by the Circuit Court (e.g. inserting quotation marks in the URL Title) (*ibid*, [35–41]).

The case raises a number of interesting questions (e.g. about the boundaries between data protection and defamation law) which cannot be considered within the confines of this chapter. Of pivotal importance in the Circuit Court's reasoning was the premise that the URL title in the Google search result appeared as a statement of fact and could thus be deemed inaccurate within the meaning of data protection law. However, a fundamental principle of defamation law, which is also surely of relevance in this context given that it is a rule of good sense, is that the publication complained of must be read as a whole. In other words, context is key. Understood in this way, even if we were to isolate the Google search result from the discussion page to which it provided a link, the search result ought to be read as a whole. The Circuit Court focused on the URL title only. But the URL link contained the words 'Reddit.com' (a well-known discussion forum) while the snippet text (containing the verbs 'to think' and 'to believe') was suggestive of opinions being expressed. Taken as a whole, then, it should have been clear to the reasonable reader that the search result provided a link to a discussion forum where opinions were expressed.

While Mr. Savage was unsuccessful in his action, the right to be delisted has the potential to be an important remedy. Though this chapter has termed it a 'narrow remedy' because only search engine links are erased, not the content itself, in many ways, the remedy may be more valuable to aggrieved individuals than any of the other options we have considered above. As O'Donnell and McIntyre point out (2017, p. 53), in the context of the discussion on ex-offenders, 'it is likely that ex-offenders will be better protected by the *Google Spain* principle than by legislation specifically addressing spent convictions.' At the same time, we should remain cautious about where this remedy might lead us. Anecdotal evidence in England and

[22]This is how the Circuit Court judge described the appellant's position: *ibid*, para 19.

[23]For the Circuit Court decision see [2016] IECC, Record Number: 2015/02589, [46–51]. Judgment available at: https://www.dataprotection.ie/documents/judgements/Savage_v_DPC_&_Google_Ireland_Circuit_Court_judgment_11.10.16.pdf.

Wales suggests that claimants are pursuing fewer claims in defamation and privacy, seeing the right to be forgotten as a 'quicker' and 'easier' solution (Brock 2015, pp. 58–59).

4 Remedies

We now turn to consider whether damages, both material and immaterial, are available for an infringement of the right to be forgotten. As 'ordinary compensatory damages', both material and immaterial damages are in principle available for invasion of privacy in Irish law (*Conway v INTO* [1991] 2 IR 305, 317). In Ireland, material damages are commonly known as special damages whereas immaterial damages are known as general damages. Where compensatory damages are awarded in privacy cases, they normally take the form of general damages in recognition of the distress caused to the plaintiff by the invasion of privacy (*Herrity* v *Associated Newspapers (Ireland) Ltd* 1 IR 316, 344). Where the conduct of the defendant is particularly 'cavalier' or 'outrageous' aggravated damages may also be awarded (it should be noted that, technically-speaking, aggravated damages are a form of compensatory damages) (*Conway* v *INTO* [1991] 2 IR 305, 317).

There was some uncertainty in Irish law as to whether s 7 DP Act 2003[24] compensated general damages as well as special damages.[25] However, Article 82 (1) of the GDPR should put the answer to the question as to the availability of general damages beyond doubt.[26] The provision states that '[a]ny person who has suffered material or non-material damage as a result of an infringement of this Regulation shall have the right to receive compensation from the controller or processor for the damage suffered.' In a thought-provoking analysis of the wording of Article 82(1), O'Dell (2017a, pp. 111–115) argues that the use of the words 'shall have' are somewhat unambiguous since it suggests that further steps must be taken but the GDPR is silent on what these steps are. O'Dell explains that, if faced with this question, the CJEU is likely to give this provision an expansive interpretation in line with its previous jurisprudence and against the background of Article 47 CFR (*ibid*

[24]s 7 was the Irish legislature's response to Article 23 of the DPD. s 7 read: 'For the purposes of the law of torts and to the extent that that law does not so provide, a person, being a data controller or a data processor, shall, so far as regards the collection by him of personal data or information intended for inclusion in such data or his dealing with such data, owe a duty of care to the data subject concerned...'

[25]*Collins* v *FBD Insurance* [2013] IEHC 137 is a significant Irish authority which interpreted s 7 as compensating special damages only. In his instructive discussion of this case, contrasting *Collins* with the decision of the Court of Appeal of England and Wales in *Google Inc v Vidal-Hall* [2015] EWCA Civ 311 and considering national and EU law, O'Dell (2017a, pp. 128–133) questions the correctness of the decision in *Collins*.

[26]Article 82(1) should be read alongside Article 79 GDPR which provides for the right to an effective remedy against a controller or processor.

pp. 115–121). In the Irish context, likely as a consequence of O'Dell's insights on this point, s 117 of the DP Act, which provides for judicial remedies, sets out in clear terms that such claims are founded on tort and the judicial remedy may take the form of relief (injunction/declaration) and/or compensation for damage suffered by the plaintiff following infringement of his rights.

5 Delisting Requests and Transparency

Finally, we should turn to consider what we might call the 'transparency problem' in how Google reaches decisions on delisting requests. Before we proceed any further, it is necessary to provide some contextual information. It is worth mentioning that as of February 2019, Google had evaluated a total of 2,982,793 URLs following 772,862 requests.[27] Of these URLs, Google had removed links to 44.2%.[28] These may seem like large numbers but two points need to be made. Firstly, the almost three million URLs that Google has reviewed represent a miniscule proportion of the estimated 130 trillion individual webpages that currently exist.[29] The number of URLs removed under right to be forgotten requests should be contrasted with the number removed for apparent copyright infringement. In the 12 months between the 31st May 2016 and the 31st May 2017, 918 million URLs were removed for copyright-related reasons.[30] Secondly, particularly striking is the decrease in applications to Google since 2014.[31] In the 12 month period following the *Google Spain* decision, Google received 260,909 delisting requests. But in the 12 months leading up to January 6th 2019 the total number of requests had dropped to 125,307.

This quantitative analysis is relevant. But it is at the same time true that a qualitative analysis is also necessary given the implications that each successful delisting request may have for freedom of expression and information. It is here where difficulties emerge as relatively little information is available about how Google reaches its decisions. While the Article 29 Working Group and the Google Advisory Council have developed sets of guidelines for implementing *Google Spain*

---

[27] https://www.google.com/transparencyreport/removals/europeprivacy/. Accessed 8th February 2019.

[28] *Ibid.*

[29] This was Google's own estimate set out on their website https://www.google.com/search/howsearchworks/ last accessed on 15th November 2017. The current version of this website (as of 8th February 2019) does not appear to contain this figure but references instead 'hundreds of billions of webpages'.

[30] https://www.google.com/transparencyreport/removals/copyright/ (as accessed on 31st May 2018). Note that the current webpage (as of 8th February 2019) no longer appears to provide a breakdown of the number of removals over particular time periods but states that Google has received requests to remove almost 4 billion URLS.

[31] https://www.google.com/transparencyreport/removals/europeprivacy/ Accessed 8th February 2019.

and Google provides some brief examples of the requests and decisions it makes,[32] Google does not publish detailed reasoning concerning its decisions in the same way a court would do. This transparency problem has been recognised by Google itself. Peter Fleischer, Google's Global Privacy Counsel, is reported as saying: 'Over time, we are building a rich program of jurisprudence on the [RTBF] decision'.... It is a jurisprudence built in the dark.' (Brock 2015, p. 53) The most obvious problem when jurisprudence is 'built in the dark' is that this is at odds with the fundaments of the Rule of Law. As O'Hara and Shadbolt put it (2015, p. 186):

> ...it is essential for accountability, probity, and due process that we know whether requests are being made by private individuals or public figures, and whether the information whose de-indexing is requested has social value or is merely inconvenient for the complainant.

Google and other search engines are profoundly important instruments for exercising the rights to freedom of expression and information. But the way the right to be delisted is currently being implemented means that there has been a privatisation, in effect, of how decisions regarding freedom of expression/information and privacy are made (O'Hara and Shadbolt 2015, p. 186). This is of concern but it is not immediately clear what alternative measures could be implemented, apart from Google publishing more detailed accounts of the requests it receives and the decisions it makes.

6 Conclusions

This chapter has sought to describe the current status of the right to be forgotten in Irish law. The chapter has considered areas of Irish law where interests in forgetting and/or being forgotten could conceivably be at stake. It found that the most suitable home for a right to be forgotten in Irish law is data protection law. Indeed, with the coming into force of the GDPR, the term itself has now formally become a part of Irish law.

This chapter has argued that two versions of the right are potentially available in data protection law, a full and narrow version. In many cases where an individual would prefer to erase personal information online, especially where this information has gone 'viral' or is contained in media reports, only the narrow version of the right (in the sense of a right to be delisted) may be available.

The right to be forgotten is now arguably the best known provision in data protection law. This has resulted from the extensive media coverage surrounding *Google Spain* and the GDPR itself. Even though there may be misunderstandings about the remit of the right (Markou 2015, pp. 215–216), its high-visibility ought to be welcomed as it raises legal consciousness among the general public (Voss and Castets-Renard 2016, p. 307). But this high-visibility may also increase the number

---

[32] *Ibid.*

of inaccurate or bad faith removal requests, as Keller puts it. Moreover, risk-averse OSPs may over-remove content, which would have a significant negative impact on freedom of expression and information. In reflecting on how the right to be forgotten should be developed, perhaps the most pressing concern, then, is finding a solution to the particular conundrum just outlined.

References

Bellia PL (2008) The memory gap in surveillance law. Univ Chicago Law Rev 75:137–179
Brock G (2015) The right to be forgotten. IBTauris, London
Brüggemeier G, Colombi Ciacchi A, O'Callaghan P (eds) (2010) Personality rights in European tort law. Cambridge University Press, Cambridge
Carolan E, Delany H (2008) The right to privacy: a doctrinal and comparative analysis. Thomson Round Hall, Dublin
Cohen JE (2012) Configuring the networked self. Yale University Press, New Haven
Data Protection Commissioner of Ireland (2014–2017) Annual Reports. https://www.dataprotection.ie/docs/Annual-Reports/b/958.htm. Accessed 7 Feb 2019
de Mars S, O'Callaghan P (2016) Privacy and search engines: forgetting or contextualizing? J Law Soc 43(2):257–284
Hagemann M (1998) Iniuria: von den XII-Tafeln bis zur Justinianischen Kodifikation. Böhlau Verlag, Cologne
Hornung G, Hofmann K (2013) Ein "Recht auf Vergessenwerden"? Anspruch und Wirklichkeit eines neuen Datenschutzrechts. JZ 68(4):163–170
Howells G, Rott P (2010) English report. In: Brüggemeier G, Colombi Ciacchi A, O'Callaghan P (eds) Personality rights in European tort law. Cambridge University Press, Cambridge
Keller D (2018) The right tools: Europe's intermediary liability laws and the EU 2016 general data protection regulation. Berkeley Technol Law J 33:287–364
Kilcommins S, O'Donnell I (2003) Wiping the slate clean: rehabilitating offenders and protecting the public. Administration 51(3):73–89
Koops BJ (2011) Forgetting footprints, shunning shadows: a critical analysis of the "right to be forgotten" in big data practice. SCRIPTed 8:229–256
Lesaffer R (2019) Wiping the slate clean... for now: Amnesty in early-modern peace treaties, Oxford Historical Treaties Online. http://opil.ouplaw.com/page/amnesty-peace-treaties. Accessed 7 Feb 2019
Markou C (2015) The "right to be forgotten": ten reasons why it should be forgotten. In: Gutwirth S, Leenes R, de Hert P (eds) Reforming European data protection law. Springer, Dordrecht, pp 203–226
Mayer-Schönberger V (2009) Delete: the virtue of forgetting in the digital age. Princeton University Press, Princeton
McIntyre TJ, O'Donnell I (2017) Criminals, data protection, and the right to a second chance. Irish Jurist 58:27–55
McMahon B, Binchy W (2013) Law of torts, 4th edn. Bloomsbury Professional, London
Moreham N (2005) Privacy in the common law: a doctrinal and theoretical analysis. Law Q Rev 121:628–656
Morgan J (2003) Privacy, confidence and horizontal effect: "hello" trouble. Cambridge Law J 62 (2):444–473
O'Cinneide C (2003) Taking horizontal effect seriously: private law, constitutional rights and the European convention on human rights. Hibernian Law J 4:77–108
O'Dell E (2017a) Compensation for breach of the general data protection regulation. Dublin Univ Law J 40(1):97–164

O'Dell E (2017b) Comparative defamation and privacy law – Irish perspectives. Dublin Univ Law J 40(1):236–249

O'Hara K, Shadbolt N (2015) The right to be forgotten: its potential role in a coherent privacy regime. Eur Data Prot Law Rev 1(3):178–189

Ohm P (2008) Good enough Privacy. University of Chicago Legal Forum. https://chicagounbound.uchicago.edu/uclf/vol2008/iss1/2/. Accessed 7 Feb 2019

Phillipson G (2003) Transforming breach of confidence? Towards a common law right of privacy under the human rights act. Mod Law Rev 66(5):726–758

Quill E (2014) Torts in Ireland, 4th edn. Gill & Macmillan, Dublin

Ricoeur P (2004) Memory, history, forgetting. University of Chicago Press, Chicago

Slattery L (2014) Right to be Forgotten Ruling is Backfiring says Data Watchdog. Irish Times. http://www.irishtimes.com/business/technology/right-to-be-forgotten-ruling-is-backfiring-says-data-watchdog-1.1877421. Accessed 7 Feb 2019

Voss WG, Castets-Renard C (2016) Proposal for an international taxonomy of the various forms of the "right to be forgotten". Colorado Technol Law J 14(2):281

Werro F (2009) The right to inform v. the right to be forgotten: a transatlantic clash. In: Colombi Ciacchi A, Godt C, Rott P, Smith LJ (eds) Liability in the third millennium. Nomos Verlag, Baden-Baden, pp 285–300

Zimmermann R (1992) The law of obligations: Roman foundations of the civilian tradition. CH Beck, Munich

The Right to Be Forgotten in Italy

Virgilio D'Antonio and Oreste Pollicino

Abstract This contribution focuses on the current implementation of the right to be forgotten under the Italian legal order. By answering to twelve questions, the authors provide an overview of the main aspects in this field. In particular, the authors deal with how the right to be forgotten is protected in Italy, its limits as well as the legal remedies available to enforce such right. Once having described such general framework, this contribution analyses the concrete implementation of the right to be forgotten in Italy considering, in particular, the obstacles in its implementation together with the effective use of the Google procedure. Then, the authors share their views regarding the upcoming legal reform of the right to be forgotten from an Italian and European perspective.

This paper is the result of a joint effort of the two authors. However, the answers to the questions 1, 2, 5, 9, 10 are to be attributed to Oreste Pollicino, while, the answers 3, 4, 6, 8, 11 are to be attributed to Virgilio D'Antonio. The answer to the question 12 is the result of the joint effort of the two authors.

V. D'Antonio (✉)
University of Salerno, Fisciano, Italy
e-mail: vdantonio@unisa.it

O. Pollicino
Bocconi University, Milano, Italy
e-mail: oreste.pollicino@unibocconi.it

F. Werro (ed.), *The Right To Be Forgotten*, Ius Comparatum – Global Studies in Comparative Law 40, https://doi.org/10.1007/978-3-030-33512-0_8

1 How Is the Right to Be Forgotten Protected Under Your Law? Does Your Law Specifically Grant a Right to Be Forgotten or Does This Right Derive from a More General Framework?

As many other Member States, Italy has not yet adopted a specific regulation on the enforcement of the *right to be forgotten*, and accordingly it relies upon the Court of Justice case law and national courts' rulings as well as the decisions issued by the Italian Data Protection Authority (DPA).

In Italy, the *right to be forgotten* finds its roots in the judicial decisions related to the right to privacy, and in its relationship with the protection of personal identity, which results the only legal support for national courts and the authority to recognise the *right to be forgotten*. In particular, in 1985, the Italian Supreme Court recognised that the right to personal identity constitutes an interest which deserves protection in order avoid that the personal heritage would be prejudiced or altered, finding the basis of this legal protection in Art. 2 of the Italian Constitution.[1]

For this reason, the core of the right to be forgotten consists in ensuring that the personal identity of each individual is not compromised over the time. The relationship with the right to privacy is particularly evident. In fact, the persons concerned should have the right to have their data processed in such a way as to avoid any damage to their personal identity. For example, the information that has been lawfully published online, about the involvement of the data subject in a criminal proceeding, should be corrected or updated in case of his or her acquittal. Furthermore, it is not related to the processing of data, or defamation, because, in this case, an illegal conduct occurs for which the legal system provides other procedures to which each individual can have recourse.

Hence, the main issue related to the *right to be forgotten* regards the competing right of society to know information that, although the relevant public interest at the time of publication has expired, continues to represent a subject in a specific manner over the time. In this case, the *right to privacy* should be balanced with the freedom of expression: enforcing the *right to be forgotten* requires affecting the freedom of information and vice versa.

For this reason, the main criterion to take into account, in order to recognise the right to be forgotten, consists in the verification of the public interest, which justifies a specific data processing, regardless of the data subject's consent. Only through the balancing of two rights introduced above, it is possible to recognise the right to be forgotten, protecting also the other fundamental rights at stake.

In general, the regulation of the right to be forgotten evolves in parallel with the diffusion of mass media (i.e., newspapers, TV, Web): like the changes that mark society and technology, it is characterised by its dynamic and multiform nature.

[1]Italian Supreme Court, case No. 3769/1985.

Traditionally the *right to be forgotten* has been defined by the most attentive scholars and, then, by the jurisprudence, as having a dual nature, strictly related to the regulation of personality rights.

On one hand, in fact, it is understood as the right of an individual to prevent information concerning their private lives, previously diffused in a legal manner, from re-emerging in the public sphere. The individual's request to be forgotten, hence, becomes an expression of the right to privacy,[2] aimed at protecting the privacy of news that, even though in the past was the subject of a legal act of communication to the public—motivated by its importance and the interest the news held for all those involved—no longer has any reason to be recalled publicly due to the amount of time that has passed.

In this sense, the right to be forgotten is understood in relation to a static vision of the communication activity, as, for instance, with printed newspapers: they are published in a certain moment and the opportunity for a possible re-publication must always be evaluated under the principles of truth, social utility, moderation of expression, and whether the news can still be regarded as current.[3]

The reprinting of information belonging to an individual's past, on the other hand, could have negative consequences in terms of protection of personal identity, as it could influence a correct and current social image of one's own person.[4]

With the entry into force of the EU legislation for the protection of personal data (Directive 95/46/EC) the so-called *right to be forgotten* can be found under Art. 7, par. 3 of Italian Legislative Decree no. 196/2003 (so-called Privacy Code), under which the individual concerned can ask an ordinary judge or the Authority responsible for the protection of personal data: (a) the update, the rectification or, where relevant, the integration of his or her data; (b) the cancellation, the transformation into an anonymous form, or the blocking of the data processed in violation of the law, including data that are not required to be stored in relation to the purposes for which the data were initially collected and processed; and (c) the attestation that the actions referred to in points (a) and (b) have been communicated to those who received or diffused the data, except in cases where this is either impossible or involves the use of means which are clearly disproportionate with respect to the right protected.

Art. 15 of the Privacy Code obliges any person who, treating personal data, causes material or immaterial damages to third parties, to compensate them for the damage suffered.

From an analysis of the case-law prior to the judgment of the EU Court of Justice given in the case *Google Spain*, it is possible to ascertain that both Italian case law and the Data Protection Authority have applied these rules, but with results that have not always been uniform.

---

[2]Ferri (1990), p. 801; Auletta (1983), p. 127.

[3]Civil Court of Rome, 15.5.1995; Italian Supreme Court, case No. 3679/1998.

[4]Italian Supreme Court, case No. 3769/1985.

On one hand, the Italian Data Protection Authority had to tackle this issue frequently: in some cases, it has ordered the administrator of a source page (in this case, the online historical archive of a daily newspaper) to perform all technical updates necessary to prevent the applicant's personal details from being viewed directly through the use of "external" search engines. In other cases, it has requested the contextualization and updating of the news related to an individual's past information.[5]

A greater sensitivity to the different nature and the particular effects of the visualization of research results can be seen from another decision. In that case, the autonomy of search engine processing was explicitly recognized, even though in the general context of a rejection of the complaint due to the inapplicability of the legal discipline given the lack of the requirement of establishment in Italy.[6]

The jurisprudence of legitimacy, on the other hand, has highlighted the neutrality of the search engine—understood as a mere intermediary that is limited to making third-party sites accessible, without playing any "active" role—by requiring the managers of an online historical archive to update personal data: to prevail in this case, according to the Supreme Court, is the data processing of the source site, which disseminates its information to the public even though it is "filtered" through the search engine activity.[7]

Both the Authority and the Supreme Court have identified the managers of the source site as those who have the task of intervening in cases where the complaint of an individual is upheld, but they take a different view regarding the role of the search engine.

In contrast, in the *Google Spain* case, the Court of Justice unequivocally affirmed the existence (and the operational autonomy) of the data processing done by the source page and the one carried out by the search engine, pointing out that the purpose underlying the indexation of a web page, for example, may not coincide with the original journalistic aim, with the consequence that the exemptions outlined in Art. 9 of Directive 95/46/EC are not valid.

Following the transposition of the European Court of Justice ruling into Italian law, the jurisprudential activity guaranteed the application of the "right to be forgotten", without deviating from the EU model.

In one of the first hypotheses concerning a quite recent publication of a news story (e.g. 2 years) related to an important judicial inquiry, public interest in having access to information indexed by the search engine has prevailed, claiming also that the search engine manager was not responsible for any damages resulting from the inaccuracy or falsehood of the information published on third-party websites.[8]

[5]Italian Privacy Authority, 8.4.2009, No. 1617673; 11-19.12.08, No. 1583152; 24.1.2013, No. 2286820.

[6]Italian Privacy Authority, 18.1.2006, No. 1242501.

[7]Italian Supreme Court, case No. 5525/2012.

[8]Civil Court of Rome, 3.12.2015, no. 23771.

2 What Are the Limits to the Right to Be Forgotten Under Your Law?

The limits to *the right to be forgotten* are based on the decisions of national courts and the Italian DPA and should be read in the light of the case-law of the Court of Justice, in particular the *Google Spain v AEPD and Mario Costeja González*[9] and *Camera di Commercio, Industria, Artigianato e Agricoltura di Lecce v Salvatore Manni* cases.[10]

The first decision of the DPA related to the right to be forgotten was adopted in 2004. The case regarded the request of an entrepreneur to deindex news published on the website of the Italian Competition Authority (AGCM), about a sanction imposed on the company he owned many years before. He complained that, notwithstanding the passing of time, the search engine researches based on its name showed as result the link to the website of the Italian Competition Authority which included the news above-mentioned. The Italian DPA ordered the delisting of the content at issue, however without imposing the removal of the relevant web page. In other words, although the website could continue to make the online content at issue publicly available, the news had not to be subject to indexing by search engines. In this way, for the first time, the Italian DPA has balanced two different interests: on one hand, it ensures that administrative measures are duly made public on the institutional website, and, on the other hand, the right to be forgotten of the subject which was harmed as a consequence of the indexing of the relevant news.

In another case, the Italian DPA dealt with the issue of cache copies of web pages. In that case, although this technique is implemented with the aim to allow users to access search results more efficiently, cache copies do not make available updated web pages and, for this reason, even dated news can be shown as recent results by the search engine. The case at issue concerned news related to a criminal proceeding where the person under investigation was finally acquitted. In this case, the DPA recognised that the memorisation of cache copies constitutes a processing of personal data and, for this reason, this kind of processing is subject to the Privacy Code. In other words, even before the *Google Spain* decision of the Court of Justice, the Italian DPA had recognised the role of search engines as data controllers. Although in this specific case the search engine was found to have no *locus standi* because of the territorial scope of application of the Italian Data Protection Law, this dismissal decision has, nevertheless, contributed to shape the boundaries of the right to be forgotten creating a right to be represented online through accurate information.

In another case, the Italian DPA balanced the exercise of freedom of expression, which namely consisted in the storage of the articles in the online archive for informative and historical purposes, and the right to privacy. In particular, although the Authority had recognized the possibility to remove news which was indexed by

---

[9]C-398/15.

[10]C-131/12.

search engines in cases the relevant articles were no longer of public interest, there was also an interest to storage that news for historical purposes, as pointed out by the Italian DPA in different occasions.[11] In another case, which still constitutes a landmark decision as far as online archives are concerned, the DPA has balanced the right to know with the right to be forgotten recognising, on one hand, that the news should remain available to the public, but requiring, on the other hand, that the related content was not indexed by search engines, in order to protect the data subject.[12]

A more recent decision of the Italian Supreme Court is particularly significant.[13] The Court held that, according to Arts. 7 and 10 of the Privacy Code, personal data should be kept up to date over the time in order to represent the individual according to his or her current personal and moral identity. For this reason, individuals have the right to ask the modification or cancellation of their data. In this case, although the information related to the arrest still had public relevance inasmuch as related to a political figure, the Court found a violation of the right to an accurate and updated identity under Art. 10 of the Privacy Code. Accordingly, the Supreme Court held that, in this case, the data subject has the right to ask for a "contextualization" of his or her personal data.

However, in 2014, a decision of the Court of Appeal of Milan[14] has contributed to clarify some aspects in the field of online archives. In particular, the inclusion of an article in the online archive, although enhances the chances to access the content, does not create a new publication, but is comparable to the physical access to the article. For this reason, it seems that the only obligation which applies to online newspapers consists in the updating (i.e. contextualization) of the news of the online archive in order to represent the evolution of the relevant events. The only relevance of the online publication refers to the amount of damages in case the article is found to be defamatory. In fact, the online publication increases the prejudice suffered by the injured party. Moreover, differently from the previous rulings, the publisher and the owner of the archive are not generally obliged to update the information by inserting a link that allows users to be informed about the new events, but must act only upon specific request of the person concerned. According to the Court of Appeal, this obligation is based on Art. 7 of the Privacy Code which attributes to the data subject the right to ask the data processor for the updating of his or her data. In other words, the Court of Appeal has enforced the obligation to keep data updated based on Art. 7, but it has specified that there is not a general obligation for the publisher and the owners of the archive to update the news without a request of the data subject.

The decisions mentioned above establish the main limits to the right to be forgotten. In particular, the legislative gap clearly entails that the limits of the right

[11] *Ex multis*, Italian Privacy Authority, 8.04.2009, No. 1617673; 25.06.2009, No. 1635966.

[12] Italian Privacy Authority, 11.12.2008, No. 1582866.

[13] Italian Supreme Court, case No. 5525/2012.

[14] Milan Court of Appeal, January 2014.

to be forgotten are mainly the result of the national courts and DPA decisions balancing, on one hand, the right to privacy and, on the other hand, different rights such as freedom of expression.

Hence, it is possible to identify the main limits of the *right to be forgotten*. First of all, the enforcement of the right to be forgotten is strictly related to the public relevance of the subject involved. When it comes to non-public figures, the *right to be forgotten* is more likely to be enforced because there is no actual public interest of the society to know behind the indexing of online content. On the contrary, the *right to be forgotten* is more difficult to apply if a public figure is concerned.

Another crucial aspect is the time factor. In this case, the conflicting interest at issue against the right of privacy refers to the existence of historical archives which could justify that some information is kept available to the public. However, it is necessary to point out that the Court of Appeal of Milan has specified that, although it is fundamental to contextualise and update information, this burden cannot be generally imposed on newspapers, but only as a consequence of the request of the subject which aims at obtaining the removal of the content.

3 What Are, in Your Law, the Legal Remedies Available to Enforce the Right to Be Forgotten?

The legal remedies available to enforce the *right to be forgotten* in Italian Law are: (a) injunctive relief; (b) damages; (c) cancellation; (d) deindexing; and (e) updating (with regard to digital archives).

Within traditional civil rights protection, the right to be forgotten—understood as the right to the removal of information regarding one's own person which is no longer considered to be current, updated or in conformity with the principles of appropriateness or the need for processing—can be enforced through recourse to an injunctive action, and the judicial remedy for compensation damages, nominally provided for by the Italian Civil Code in the context of the protection of the right to the name, the pseudonym or image (Arts. 7, 9 and 10 Civil Code), but more generally extended to the so-called protection of personal rights and personal identity under the clause referred to in Art. 2 of the Italian Constitution in conjunction with Art. 700 of the Civil Procedure Code concerning emergency injunction measures.

The other remedies of cancellation, deindexing and updating are provided by Art. 7, par. 3, lett. a) and b) of the Privacy Code, which give to the data subject the right to obtain the update, rectification or, where interested therein, integration of the data and also the right to obtain erasure, anonymization or blocking of data that have been processed unlawfully, including data whose retention is unnecessary for the purposes for which they have been collected or subsequently processed.

4 As a Follow-up to the Previous Question, Does Your Law Allow the Plaintiff to Receive Material or Immaterial Damages? If Yes, Is Such Remedy Realistic in Practice?

Italian Law allows the plaintiff to receive material and immaterial damages. In practice, the first Italian case law on the right to be forgotten and cancellation seems to indicate a substantial softening of the objective or semi-objective civil liability regime envisaged in the case of the misuse of personal data.

The rejection (or confirmation) of the decision adopted towards the provider by the court or the Data Protection Authority has not in fact led to an order to pay for material or non-material damages, but only the full payment of the legal expenses incurred for the lawsuit.[15]

This trend appears to be confirmed in the new wording of the norms regarding compensation responsibilities hosted under the EU General Regulation on Personal Data Protection No. 679/2016, where there is a move towards a more analytical provision of the obligations and duties of behaviour of data owners and those responsible for data processing.

According to Art. 82 of the Regulation 679/2016, anyone who has suffered material or non-material damage, related to the violation of the norms contained in the regulation, has the right to obtain compensation from the holder or the processor of the data. The letter of the law indicates a greater propensity to mitigate the civil liability regime for misuse of personal data, most likely leading it towards parameters of guilt.

First, unlike the current discipline in Italy, the passively legitimated subjects are the sole holder and the processor and not "anyone" who has caused the damage. Furthermore, according to Art. 23 of the Directive 95/46, the regulation exempts the holder and the processor where they have proved that the harmful event is by no means imputable to them, in the sense that they have processed the data in accordance with the said legal regulation, or, in the case of the data processor, in compliance with the tasks specifically assigned to them by the law or by the owner. In other words, the rule of conduct is divided between the various figures and modulated with respect to the fulfilment of the precise obligations set out in the regulation, abstractly softening the "*probatio diabolica*" of the extraneousness of the cause of the damage to its own sphere of risk—in Italian Law understood as mere proof of fortuitousness, force majeure or an equivalent—as can be inferred from the reading of the combined provision referred to in the second and third paragraphs of Art. 82.

[15]Italian Privacy Authority, 31.3.2016, no. 4988654, in which the Authority rejected the request for de-indexation brought by a former terrorist, dividing the legal costs between the parties, a provision which was subsequently annulled by the Civil court of Milano, 12.10.2016.

5 In General, How Do You Assess the Implementation of the Right to Be Forgotten in Your Law? Is It Effective? Is It Used in Practice? Are There Particular Obstacles in the Implementation of This Right?

The implementation of the right to be forgotten resulting from judicial or administrative orders is effective, but these procedures require users to devote time and resources due to the procedural rules that they must comply with.

However, the implementation of the right to be forgotten is the result not only of the decision already mentioned, but it is based on the interpretation given by the Court of Justice ruling in the *Google Spain* case which recognized the right of individuals to ask search engines, such as Google, to delist some results. The Court of Justice ruled that search engines have to assess individuals' requests for delisting, taking into consideration whether there is a public interest which justifies the search engines to continue to display the contents at issue.

For this reason, the implementation of the right to be forgotten is strengthened by the measures made available by private actors, which will probably be enhanced once the General Data Protection Regulation will become applicable.

6 How Did Courts and Commentators in Your Country Welcome the ECJ Ruling on *Google v González*?

The majority of Courts and scholars welcomed the decision of the European Court of Justice on the right to be forgotten, reflecting its innovative or evolutionary profile. At the same time, commentators have raised some doubts about the economic sustainability of private complaint procedures for deindexation and the associated risk of delegating public powers to private individuals over whom weighs the rule of objective or semi-subjective liability for the misuse of personal data.[16] In fact, according to the principles established by the CJEU, the first "judgment" on the merits of the instance of cancellation is remitted to the search engine, called to the difficult task of identifying the right balance between the fundamental rights of the data subject and the legitimate interest of Internet users to have access to information.

[16]See for example, Resta and Zeno-Zencovich (2015); see also Palmieri and Pardolesi (2014), p. 295; Giannone Codiglione (2014), p. 1054. For the case law see questions no. 3 and 4.

7 For Those Who Are from a Country That Is Not Part of the European Union, Did Your Courts Follow the ECJ Ruling on the Right to Be Forgotten? Is It Likely Do That They Will Follow It?

Question not relevant in the perspective of the Italian report.

8 Did Your Law Already Grant a Similar Right to Be Forgotten Than the One Stated in the ECJ Ruling?

Yes, in addition to the case law mentioned above, some Courts of Appeal have recognized a right to deindexation and cancellation with respect to the protection of personality rights on search engines and to the visualisation of results which, due to the automated completion algorithms used by the data manager, automatically list the terms most looked for by users in relation to the personal data of the person searched for, with the risk of damaging their honour and reputation.[17]

9 To Implement the ECJ Ruling, Google Has Created a Form in Which Anyone Interested Can Submit a Request to Have Information About Him of Herself Be Delisted. Based on This Request, Google Will Weigh Between the Private Interest of Petitioner and the Public Interest to Be Informed. Google Does Not Disclose the Ways in Which It Deals with Requests. In Particular, Google Does Fully Not Disclose, the Category of Requests That Are Excluded or Accepted, the Proportion of Requests and Successful De-listings and, Among Others, the Reason for the Denial of Delisting. Do You Think That Google Should Be More Transparent About the Ways It Uses to Implement the Right to Be Forgotten?

Transparency is one of the main issues in the procedures for the implementation of the right to be forgotten. Differently from national courts and the Italian DPA, private actors are not obliged to disclose any kind of information explaining the

[17]Civil Court of Pinerolo, 23.7.2012, Civil Court of Milano, 23.5.2013, *contra* Civil Court of Milan, 25.3.2013 and 24.3.2011.

modalities according to which the right to be forgotten is balanced with other conflicting interests. On one hand, this lack of transparency increases the efficiency of the implementation mechanisms provided by private actors due to the lower level of workload. On the other hand, individuals do not have access to the reasons for the denial of their requests. However, such proceedings do not hinder the possibility for individuals to rely on national courts or the Italian DPA in order to obtain a delisting order.

For this reason, due to the lack of any legislative framework, the private model of enforcement of the right to be forgotten offered by Google is only a first step for the data subject and such procedure should not be weighted by specific obligations related to transparency, considering both the necessity to ensure the freedom to conduct business and the possibility for any individual to rely on other legal remedies offered by the judicial authority and the DPA.

10 Is the Procedure Prepared by Google Used in Your Country?

According to the data provided by Google online, the Google procedure for the implementation of the right to be forgotten has been used in many occasions. Since 2014, the Italian privacy requests for search removals submitted to Google have been more than 45,000 involving more than 145,000 URLs. Only 33% have been removed (about 41,000), while the others have been rejected (about 83,400).[18]

11 Is There Any Upcoming Legal Reform in Your Country Whose Purpose Is to Reinforce or Modify the Right to Be Forgotten?

No, but in the future Art. 7, par. 2, lett. b) of the Privacy Code may be completed and integrated with the provision on the "right to cancellation (right to be forgotten)" under the new European Regulation on data protection (Art. 17). In detail, Art. 7, par. 2, lett. b) of the Privacy Code already guarantees full protection of the right to be forgotten, but it may be extended to the balance with other fundamental rights and with regard to the duty of the controller to inform controllers which are processing the personal data that the data subject has requested the erasure by such controllers of any links to, or copy or replication of, those personal data (Art. 17, par. 2 GDPR).

[18]Data updated at 27 April 2017.

12 In Your Opinion, What Should Be the Next Step in the Protection of the Right to Be Forgotten? Do You Think That One Must Go Further and Strengthen the Right to Be Forgotten? Do You Think That the European Union Should Modify or Adapt Its Legislation on the Right to Be Forgotten?

The right to be forgotten is a key-remedy in order to enforce fundamental rights especially in the digital age. Art. 17 of the new Regulation 679/2016 already offers a wider formulation of the right to be forgotten, and spells out its application and procedural limits in detail. Moreover, the general applicability of the Regulation will ensure a uniform application of the right to be forgotten across the EU.

Among the most interesting aspects from a forward-looking perspective of the right to be forgotten, it is worth to mention the exercise of "quasi" public powers by online platforms consisting in balancing conflicting interests in order to satisfy the request to be forgotten submitted by data subjects, the review of civil liability as an implicit "lever" acting on the correct application of the normative criteria, as well as the risk of generating chilling effects with a progressive increase in out-of-court requests for deindexing that do not deserve protection.

A first step to deal with this framework would consist in avoiding the introduction of specific national rules which oblige online platforms, and in particular search engines, to comply with additional obligations related to the management of the data subjects' requests aimed at obtaining the delisting. Such additional obligations would make particularly difficult for online platforms to exercise their core activities. Moreover, in the European framework, this would be particularly important in order to reduce the fragmentation of the application of the right to be forgotten, considering also that the rules introduced by Regulation 679/2016 will apply uniformly across the EU.

References

Auletta T (1983) Diritto alla riservatezza e droit à l'oubli. In: Alpa G et al (eds) L'informazione e i diritti della persona. Jovene, Napoli, p 127

Ferri G (1990) Diritto all'informazione e diritto all'oblio. Riv. dir. civ. 1:801

Giannone Codiglione G (2014) Nuova giur. civ. comm. 11:1054

Palmieri A, Pardolesi R (2014) Foro it. IV:295

Resta G, Zeno-Zencovich V (eds) (2015) Il diritto all'oblio su Internet dopo la sentenza Google Spain. RomaTre Press, Rome, with contributions by Frosini TE, Pollicino O, Finocchiaro G, Caggiano G, Piroddi P, Sartor G and Viola De Azevedo Cunha M, Mantelero A, Sica S and D'Antonio V, Comella C, Riccio GM, Flor R, Pizzetti F

Cases and Regulations

Civil Court of Milan, 25.3.2013 and 24.3.2011
Civil Court of Milano, 23.5.2013
Civil Court of Pinerolo, 23.7.2012
Civil Court of Rome, 15.5.1995
Civil Court of Rome, 3.12.2015, no. 23771
Italian Privacy Authority, 8.4.2009, No. 1617673; 11-19.12.08, No. 1583152; 24.1.2013, No. 2286820; 18.1.2006, No. 1242501
Italian Supreme Court, case No. 3679/1998
Italian Supreme Court, case No. 3769/1985
Italian Supreme Court, case No. 5525/2012
Milan Court of Appeal, January 2014

The Right to Be Forgotten in Romania: Before and After the ECJ Judgment in Google V. González

Simona Şandru

Abstract The right to privacy and the right to personal data protection are two fundamental rights enshrined in the European Union's treaties and Charter, providing individuals with proper tools of control over their private life. The "right to be forgotten" may be considered one of these tools. After being expressly consecrated by the European Court of Justice in a 2014 ruling which involved a search engine on the Internet, this right seems to have its own path, apart from the existing legal regime of the data subjects' rights, as judicial and administrative practice shows it (in Romania, as well). As of the 25th of May 2018, a new European Union regulation directly and uniformly applies in all the Member States and specific legal provisions on the right to be forgotten have come into force. Romania is taking part in all these reforms, so the legal transplant of the right to be forgotten was smoothly put in place with all its relevant guarantees, besides the domestic civil legal protection of the personal rights which is under the judiciary's scrutiny.

1 Introduction

One of the most common phrases related to the right to privacy, when it comes to its definition, is the "right to be let alone", as the "parents" of this right, Louis D. Brandeis and Samuel D. Warren, inspired by the European legal environment (where the private life was legally protected against intrusive press, in France and Germany, especially—Whitman 2004)[1] have stated in their famous paper (Warren and Brandeis 1890). However, since 1890 the way people communicate their ideas has been subject to a lot of changes, so an ordinary printed newspaper article still may cause harm to an individual by disclosing intimate facts, but an electronic version of the same article, displayed and indexed on the Internet could instigate

[1] "In fact, it is best to think of the Warren and Brandeis tort not as a great American innovation, but as an unsuccessful continental transplant." Whitman (2004), p. 1204.

S. Şandru (✉)
Law School, University of Bucharest, Bucharest, Romania

F. Werro (ed.), *The Right To Be Forgotten*, Ius Comparatum – Global Studies in Comparative Law 40, https://doi.org/10.1007/978-3-030-33512-0_9

much more damage to a person, given the possibility to be revealed to an indefinite number of readers (Internet users) and to be constantly or at random disseminated and replicated, with no control. Therefore the newly-invented "right to be forgotten" has two main functions: to reinstate the control of the individual over his/her personal related information and to ensure his/her right to be let alone, to vanish from the sight of the entire world. Nevertheless, the scope of this right is mainly contextualised to the cyberspace, having in view the risks therein above-mentioned.

The "right to be forgotten" is a modern right, born in the legal space of the European Union (EU), even if there were some previous attempts made by the French Government, in 2010, when they tried to raise public awareness of "le droit à l'oubli numérique" by two charters[2] destined to advertising companies, websites and search engines. The real revolution actually came in 2014 when the European Court of Justice (ECJ) rendered a judgment in *Google Spain* case[3] and the "right to be forgotten" was officially and de facto recognised. Although the case involved a search engine (not even European), the arguments brought by the European judges may be as well used *a fortiori* in other similar circumstances of revealing personal data on the Internet. As the subsequent case law revealed, the scope of the "right to be forgotten" has two components: the right to have the personal data deleted (or anonymised) from a website and the right to obtain the de-indexation of the information from the search engines. The ECJ judgment pre-announced the legislative reform on data protection in EU[4] which entailed, among other many novelties, the introduction of the "right to be forgotten", as an autonomous right of the data subject[5]; so, as of the 25th of May 2018, every person subject to the new General Data Protection Regulation (GDPR) might make use of the right to have his/her data erased, according to the conditions and limitations provided therein.[6]

---

[2]"Charte du droit à l'oubli numérique dans la publicité ciblée"; "Charte du droit à l'oubli numérique dans les sites collaboratifs et moteurs de recherche" (Google and Facebook did not sign this charter). See: http://archives.gouvernement.fr/fillon_version2/gouvernement/charte-du-droit-a-l-oubli-numerique-mieux-proteger-les-donnees-personnelles-des-interna.html (Accessed 19 April 2017).

[3]Case C-131/12, Google Spain SL and Google Inc. v Agencia Española de Protección de Datos (AEPD) and Mario Costeja González, ECLI:EU:C:2014:317.

[4]The Regulation (EU) 2016/679 of the European Parliament and of the Council of 27 April 2016 on the protection of natural persons with regard to the processing of personal data and on the free movement of such data, and repealing Directive 95/46/EC (General Data Protection Regulation) (Text with EEA relevance), OJ L 119, 4.5.2016, pp. 1–88, is equally and mandatory applicable in all Member States of the EU as of the 25th of May 2018.

[5]The "data subject" is the term used for the individual whose personal information is being processed.

[6]Article 17

Right to erasure ('right to be forgotten')

1. The data subject shall have the right to obtain from the controller the erasure of personal data concerning him or her without undue delay and the controller shall have the obligation to erase personal data without undue delay where one of the following grounds applies:

So, the "right to be forgotten" is nothing else but an ancient right (right to be let alone), "reloaded" for the modern times and equipped with proper tools, established by law and jurisprudence, under the umbrella of the right to personal data protection, in order to ensure, under certain circumstances, the individual's control over his/her own private life, especially on the Internet.

In Romania, the "right to be forgotten" was subject to a similar process as in most EU countries, from the non(legal)existence to a jurisprudential recognition and a future/current clear legal status, in accordance with the new GDPR. Our paper will show first the main legal aspects of this right not expressly written yet, as reflected in two categories of legal rules, enshrined in the Civil Code and the legislation on data protection. Then, we shall focus on the way the jurisprudence and the legal doctrine have assessed the claims of the individuals and the legal regime based on the ECJ judgment in *Google Spain* case. As a Member State of the EU, Romania will also benefit of the new legal framework provided by the GDPR which expressly gives legal value to the right to be forgotten.

(a) the personal data are no longer necessary in relation to the purposes for which they were collected or otherwise processed;
(b) the data subject withdraws consent on which the processing is based according to point (a) of Article 6(1), or point (a) of Article 9(2), and where there is no other legal ground for the processing;
(c) the data subject objects to the processing pursuant to Article 21(1) and there are no overriding legitimate grounds for the processing, or the data subject objects to the processing pursuant to Article 21(2);
(d) the personal data have been unlawfully processed;
(e) the personal data have to be erased for compliance with a legal obligation in Union or Member State law to which the controller is subject;
(f) the personal data have been collected in relation to the offer of information society services referred to in Article 8(1).

2. Where the controller has made the personal data public and is obliged pursuant to paragraph 1 to erase the personal data, the controller, taking account of available technology and the cost of implementation, shall take reasonable steps, including technical measures, to inform controllers which are processing the personal data that the data subject has requested the erasure by such controllers of any links to, or copy or replication of, those personal data.
3. Paragraphs 1 and 2 shall not apply to the extent that processing is necessary:

(a) for exercising the right of freedom of expression and information;
(b) for compliance with a legal obligation which requires processing by Union or Member State law to which the controller is subject or for the performance of a task carried out in the public interest or in the exercise of official authority vested in the controller;
(c) for reasons of public interest in the area of public health in accordance with points (h) and (i) of Article 9(2) as well as Article 9(3);
(d) for archiving purposes in the public interest, scientific or historical research purposes or statistical purposes in accordance with Article 89(1) in so far as the right referred to in paragraph 1 is likely to render impossible or seriously impair the achievement of the objectives of that processing; or
(e) for the establishment, exercise or defence of legal claims.

2 The Romanian Legal Framework: Is There a "Right to Be Forgotten"?

2.1 Main Provisions

Although the "right to be forgotten" is closely linked to the online environment, a natural need of any individual to be let alone and "forgotten" is also available in the offline life, especially in relation to facts from someone's past, that the persons wishes to remain or become unknown. The ECHR *Rotaru* case[7] reveals the right to have deleted from any archives (of the former political *Securitate*, in this case) the personal information which is no longer needed, even inaccurate, and sometimes painful for the memory. So this right, which had no chance of existence some time ago, could be very well used nowadays by Romanian citizens as a tool for defending their privacy.

From the outset it must be mentioned that there was no "right to be forgotten" regulated *expressis verbis* in the current legal framework of Romania (until GDPR). Having as premise the following (possible) definition of the "right to be forgotten" as *the right to request some measures to be applied in order to have the personal information wrongfully used removed*, there are two major situations when this right could be extracted from the existing legal provisions.

First, in accordance with the existing legislation on data protection before GDPR,[8] any data subject might request a data controller[9] to delete data which were not legally processed, notably because they were incomplete or inaccurate or to stop the processing of his/her personal data, based on legitimate grounds. The rights regulated by the Romanian law which were called "to intervention"[10] and "to

[7]European Court of Human Rights, Case of *Rotaru v Romania* (application no. 28341/95), judgment of 29 March 2000.

[8]Law 677/2001 on the Protection of Individuals with Regard to the Processing of Personal Data and the Free Movement of Such Data, OJ 790/12.12.2001 (for the purpose of this paper, the English version of the Law 677/2001 is the one available on http://www.dataprotection.ro/index.jsp?page=legislatie_primara&lang=en. Accessed 20 April 2017).

[9]Art. 3 of Law 677/2001: "e) data controller: - any natural or legal person, including public authorities, institutions and their legal bodies, that establishes the means and purpose of the personal data processing; if the purpose and means of the personal data processing is set out or based on a legal provision, the data controller shall be the natural or legal person assigned as data controller by that specific legal provision".

[10]Art. 14: The Right of Intervention upon the Data

(1) Every data subject has the right to obtain from the data controller, upon request, and free of any charge: a) as the case may be, rectification, updating, blocking or deletion of data whose processing does not comply with the provisions of the present law, notably of incomplete or inaccurate data; b) as the case may be, transforming into anonymous data the data whose processing does not comply with the provisions of the present law; c) notification to a third party to whom the data were disclosed, of any operation performed according to letters a) or b), unless such notification does not prove to be impossible or if it does not involve a disproportionate effort towards the legitimate interest that might thus be violated.

object"[11] reflected similar (equivalent) provisions regarding the rights of access (Art. 12 (a)) and to object (Art. 14 para. 1 (b)) as they were set up by the EU directive[12] which was transposed by Law 677/2001. Having said that, it is worth mentioning that the Romanian legislation on personal data protection is relatively new, and it is actually the result of the obligations taken by the country in order to comply with the *acquis communautaire* before joining the EU in 2007. Law 677/2001 was the primary legislative act transposing the EU directive.[13] Its provisions were quite similar to the European text, with a few specific norms as regards the scope[14] and exceptions to the rights.[15] Moreover, a data protection authority was set up (the

---

(2) In order to exert the right stated in paragraph (1), the data subject shall fill in a written, dated and signed petition. The petitioner may state his/her wish to be informed at a specific address, which may also be an electronic mail address, or through a mail service that ensures confidential receipt of the information.
(3) The data controller has the obligation to communicate the measures taken, based on the provisions of paragraph (1), as well as, as the case may be, the name of a third party to whom the data concerning the data subject were disclosed, within 15 days from the date of the petition's receiving, whilst complying with the petitioner's possible option, according to paragraph (2).

[11]Art. 15: The Right to Object

(1) The data subject has the right to object at any moment, based on justified and legitimate reasons linked to his particular situation, to a processing of data regarding him/her, unless there are contrary specific legal provisions. In case of justified opposition, the processing may no longer concern the respective data.
(2) The data subject has the right to object at any moment, free of charge and without any justification, to the processing of the data concerning his/her person for overt marketing purposes on behalf of the controller or of a third party, or to be disclosed to a third party for such a purpose.
(3) In order to exercise the rights stated under paragraphs (1) and (2), the data subject shall fill in and submit to the data controller a written, dated and signed petition. The petitioner may specify if he/she wishes to be informed at a specific address, which may also be an electronic mail address, or through a mail service that ensures confidentiality.
(4) The data controller has the obligation to inform the data subject of the measures taken, based on the provisions of paragraph (1) or (2), as well as, as the case may be, the name of the third party to whom the data concerning the data subject were disclosed, within 15 days of the date of the petition's arrival, in compliance with the petitioner's option, according to paragraph (3).

[12]Directive 95/46/EC of the European Parliament and of the Council of 24 October 1995 on the protection of individuals with regard to the processing of personal data and on the free movement of such data, OJ L 281, 23.11.1995, pp. 31–50.

[13]As of the 25th of May 2018, the Law 677/2001 was repealed by Law 129/2018 (OJ 503/ 19.06.2018), that ensures the implementation of GDPR in Romania. The legal framework was completed by Law 190/2018 (OJ 651/26.07.2018), as regards the domestic implementation of a few GDPR provisions which were left in the margin of appreciation of the Member States.

[14]Law 677/2001 had a restricted scope for the processing operations related to criminal law and public order activities and excludes from its application the processing operations related to national defense and national security activities (Art. 2 para. (5) and (7)).

[15]Law 677/2001 restricted the exercise of the rights to information, to access, to intervention and to object if their enforcement could affect the efficiency of the action or the legal objective followed by

National Supervisory Authority for Personal Data Processing—DPA) by Law 102/2005,[16] and it has an independent and autonomous status[17] and is endowed with all the powers of intervention and investigation[18] as required by the EU directive (Art. 28). As a conclusion, the legislation on data protection is the result

---

a public authority engaged in criminal law or public order activities (Art. 16). However, the restriction was limited only to the period necessary to achieve this goal, so afterwards, measures had to be taken in order to comply with the requests of the data subject.

[16]Law 102/2005 regarding the setting up, organisation and functioning of the National Supervisory Authority for Personal Data Processing, OJ 391/9.05.2005.

[17]The management of this authority (President and Vice-President) has to be politically independent, their functions being incompatible with any other public or private functions, except for the academic ones. The Romanian Senate is vested with the power of appointing and revoking the management. The yearly report of the authority is also submitted to the Senate. The authority's staff is directly employed and the authority has its own budget, stipulated as a distinct part of the State budget. No one can give instructions to the authority or subject it to an imperative mandate or mandate of representation. All these legal stipulations ensure institutional, functional and financial independence of the authority.

[18]Art. 21 para. (3) of the Law 677/2001:

> The supervisory authority shall monitor and control with regard to their legitimacy, all personal data processing, subject to this law. In order to achieve this purpose, the supervisory authority exerts the following attributions: a) issues the standard notification forms and its own registers; b) receives and analyses the notifications concerning the processing of personal data and informs the data controller on the results of the preliminary control; c) authorizes personal data processing in the situations set out by law; d) may dispose, if it notices the infringement of the provisions of the present law, temporarily suspending the data processing or ending processing operations, the partial or total deletion of processed data and may notify the criminal prosecution bodies or may file complaints to a court of law; d1) informs the natural or legal persons that work in this field, directly or through their associative bodies on the need to comply with the obligations and to carry out the procedures set out by this law; e) keeps and makes publicly accessible the personal data processing register; f) receives and solves petitions, notices or requests from natural persons and communicates their resolution, or, as the case may be, the measures which have been taken; g) performs investigations –ex officio, or upon requests or notifications; h) is consulted when legislative drafts regarding the individual's rights and freedoms are being developed, concerning personal data processing; i) may draft proposals on the initiation of legislative drafts or amendments to legislative acts already enforced, in the fields linked to the processing of personal data; j) collaborates with the public authorities and bodies of the public administration, centralizes and analyzes their yearly activity reports on the protection of individuals with regard to the processing of personal data, issues recommendations and assents on any matter linked to the protection of fundamental rights and freedoms regarding the processing of personal data, on request of any natural person, including the public authorities and bodies of public administration; these recommendations and assents must mention the reasons on which they are based and a copy must be transmitted to the Ministry of Justice; when the recommendation or assent is requested by the law, it must be published in the Official Journal of Romania, Part I; k) co-operates with similar foreign authorities in order to ensure common assistance, as well as with foreign residents for the purpose of guaranteeing the fundamental rights and freedoms that may be affected through personal data processing; l) fulfills other attributions set out by law; m) the manner in which the National Supervisory Authority for Personal Data Processing is organized and functions is set out by law.

of a legal "transplant" of the EU law into the Romanian legal order, having as a main model the EU directive on data protection, to which a few particular features were added, just like in the case of most other EU Member States.

The above mentioned provisions of the Law 677/2001 providing for the right to intervention and the right to object may be considered as a strong legal basis for the exercise of the un-written "right to be forgotten", having in mind that the equivalent legal provisions of the EU directive were also taken into account by the ECJ in the *Google Spain* case,[19] in order to interpret the existence of a "right to be forgotten" in the context of the processing made by indexation of Internet web pages, carried out by a search engine on the basis of a person's name.

If we consider the legal provisions of the Law 677/2001 and the arguments of the ECJ judgment,[20] the following requirements had to be met in order to have a valid exercise of the "right to be forgotten", in the context of a processing on the Internet by an operator of a search engine:

- the data subject may request the controller to delete his/her personal data which have been illegally processed, especially, if they are incomplete or inaccurate or

[19]"On those grounds, the Court (Grand Chamber) hereby rules:

(...)

3. Article 12(b) and subparagraph (a) of the first paragraph of Article 14 of Directive 95/46 are to be interpreted as meaning that, in order to comply with the rights laid down in those provisions and in so far as the conditions laid down by those provisions are in fact satisfied, the operator of a search engine is obliged to remove from the list of results displayed following a search made on the basis of a person's name links to web pages, published by third parties and containing information relating to that person, also in a case where that name or information is not erased beforehand or simultaneously from those web pages, and even, as the case may be, when its publication in itself on those pages is lawful.

4. Article 12(b) and subparagraph (a) of the first paragraph of Article 14 of Directive 95/46 are to be interpreted as meaning that, when appraising the conditions for the application of those provisions, it should inter alia be examined whether the data subject has a right that the information in question relating to him personally should, at this point in time, no longer be linked to his name by a list of results displayed following a search made on the basis of his name, without it being necessary in order to find such a right that the inclusion of the information in question in that list causes prejudice to the data subject. As the data subject may, in the light of his fundamental rights under Articles 7 and 8 of the Charter, request that the information in question no longer be made available to the general public on account of its inclusion in such a list of results, those rights override, as a rule, not only the economic interest of the operator of the search engine but also the interest of the general public in having access to that information upon a search relating to the data subject's name. However, that would not be the case if it appeared, for particular reasons, such as the role played by the data subject in public life, that the interference with his fundamental rights is justified by the preponderant interest of the general public in having, on account of its inclusion in the list of results, access to the information in question."

[20]As regards the way the European data protection authorities, assembled in the Working Party Article 29, interpret and apply the ECJ ruling, see "Guidelines on the implementation of the Court of Justice of the European Union judgment on "Google Spain and Inc v. Agencia Española de Protección de Datos (AEPD) and Mario Costeja González" C-131/12" (14/EN WP 225, adopted on 26.11.2014, available at http://ec.europa.eu/justice/data-protection/article-29/documentation/opinion-recommendation/files/2014/wp225_en.pdf. Accessed 20 April 2017).

to stop processing his/her personal data for legitimate grounds; at this point, the data subject must justify that all these conditions are satisfied (for instance, his/her personal data were processed without consent, or they are stored for a longer period than necessary, or they are no longer accurate, or that by revealing some sensitive information, such as a disease, the data subject could suffer a serious prejudice); anyhow, the data subject is not obliged to prove that the information revealed by the search is causing him/her damage;
- the processing is determined by the search based on the person's name and the links reveal information about that person;
- the legality of the first disclosure on the original websites does not constitute a deterrent factor in complying with the request (however, this condition has to be checked on a case-by-case basis);
- the data subject may approach firstly the search engine and not the owner (publisher) of the website or websites where the information is originally presented;
- after analysing the interests at stake, the operator of the search engine could validly refuse such a request only if the interest of the general public in having access to specific information prevails over the personal interest of the data subject to have his/her rights to privacy and data protection defended (for instance, if the data subject plays an important role in public life and that information is linked to that activity).

From a procedural point of view, the Romanian law imposed that all requests are made in written form and the data controller replies in 15 days.

These conditions apply for the second aspect of the right to be forgotten, which is linked to the de-indexation (or "de-listing") from the results displayed on the Internet by searching a person's name.

As for the first component of the right, the deletion of personal data is the traditional core object of the rights to intervention and to object, irrespective of the context of the processing (on the Internet or otherwise, by automated or manual operations). Therefore, the "right to be forgotten" may be as well exercised in other circumstances of processing, by applying *mutatis mutandis* the conditions emerged from the ECJ ruling.

Otherwise, the new GDPR, when expressly providing the "right to be forgotten", does not restrict its scope to a specific kind of processing or sector of activity. However, the preamble (para. 65) makes reference to the particular relevance of this right for the situations where the consent for processing some personal data was given when the data subject was a child and he/she was not fully aware of the risks involved and he/she later wants to have such personal data removed, especially on the Internet. Also, the preamble puts emphasis on the need to strengthen this right in the online environment (para. 66), by extending it to the correlative obligation of the controller to take reasonable steps in order to inform third parties about the request of the data subject to have any links, copies or replications of the published personal data erased. So, by way of a systematic interpretation of the preamble and the provisions themselves, the "right to be forgotten" could be acknowledged as a special tool for obtaining the deletion of personal data in particular circumstances related to the *age of the data subject* when the information was first revealed and to the *means of processing*, the disclosure on the Internet being seen as a more risky operation than others.

Secondly, a (still un-written) "right to be forgotten" might be extracted from the Romanian Civil Code[21] provisions on personality rights. According to the civil norms, fundamental values like life, dignity, physical and psychological integrity, privacy and "processing personal data"[22] are considered as personality rights and enjoy full legal protection. Among other infringements of the privacy, non-exhaustively listed by the Art. 74 of the Civil Code,[23] a few are highly relevant for our paper: broadcasting news or other written or audio-video materials on private life, without the consent of the interested person, dissemination of materials containing images or personal data about health condition or medical treatments without the consent of the person or his/her relatives, when deceased, deceitful use of the name, image, voice or resemblance to another person, dissemination or using the correspondence, manuscripts or other personal documents, including data on domicile, residence and phone numbers of a person or family members, without consent. In these cases, the prejudiced person may request a court to impose measures aimed at stopping the illegal behaviour, including by publishing the judgment in the respective case (Art. 253 of the Civil Code). One of these measures (the Civil Code includes only a general norm on that subject) could consist in the deletion of the personal information or other materials containing private details from any type of the used storage medium.

The remedies provided by the Civil Code are available against any (natural or legal) person who violated a personal right (in private law relations, mostly), and not only against a data controller that was strictly defined by the scope of the Law 677/2001 (as being a natural or legal person, pertaining to the private or public sector, that decides the purpose and the means of a specific data processing). Therefore, the two legal regimes are quite distinct from one another.

---

[21]The new Romanian Civil Code entered into force on the 1st October 2011. Before that date the Decree 31/1954 on natural and legal persons contained some provisions concerning the personality rights, such as the right to a name or the right to reputation. See more in Şandru (2016), pp. 303–309.

[22]As regards this right, the Civil Code invokes the special law on data protection, as the governing law.

[23]For instance, the judge admitted that the recording without consent of an incident between two neighbours, in the private yard of one of them is a privacy infringement, and awarded to the plaintiff compensation for the moral damages—civil judgment 601/2017 of City Court of Aiud, available at rolii.ro. Accessed 3 November 2017. Other cases are cited below (see footnotes 33 and 34).

2.2 Limitations

As a general rule, the Romanian Constitution (Art. 53) sets forth that no restriction can be imposed to a fundamental right (as the right to privacy), except when the following conditions are satisfied: the restriction on the exercise of the rights has to be provided by law; the restriction has to be necessary in a democratic society and only for a few limited reasons (the defence of national security, of public order, health, or morals, of the citizens' rights and freedoms; conducting a criminal investigation; preventing the consequences of a natural disaster, or an extremely serious catastrophe); the measure shall be proportional to the situation that has caused it, applied without discrimination, and without infringing on the existence of such right or freedom. These rules have to be read in conjunction with the ones derived from the European Convention for the Protection of Human Rights and Fundamental Freedoms and the relevant jurisprudence of the European Court of Human Rights, especially when it comes to the right to privacy, protected under Art. 8 of the Convention. Romania ratified this international instrument on human rights which has (at least) equal value with the other domestic norms governing the protection of fundamental rights (Art. 11 and 20 of the Romanian Constitution), where the principle of *lex mitior* is applicable (Muraru and Tănăsescu 2008).[24]

As regards the "right to be forgotten", its limits, besides the general ones derived from the Constitutional provisions, could be categorized accordingly: they are the same as for any other subjective right which were set up by the Law 677/2001, or, as the case may be, as for the other personality rights on dignity and privacy, reflected by the Civil Code. In the first category, there are the exceptions already mentioned which involve the restriction of the rights in the related law enforcement activities (criminal law, in particular). As for the second category, the Civil Code (Art. 75) refers to general limitations implied by the legal norms, either domestic ones, or international covenants on human rights that allow restrictions on the personality rights, whenever other rights or liberties are exercised in good faith. The right to free speech is one of the limits expressly provided by the Civil Code, as applicable whenever the object of the judicial action an interested person might fill in concerns an infringement produced in this context (information published by the press). Also, the new legislation implementing the GDPR in Romania provides derogations from the data subjects' rights (including the right to be forgotten), in order to reconcile the right to the protection of personal data with the right to freedom of expression and information, including processing for journalistic purposes and the purposes of academic, artistic or literary expression. The derogations (Art. 7 of Law 190/2018) are strictly applicable whenever the processing is related to personal data that were manifestly made public by the data subject or are linked to their public status or public events in which they were involved.

[24]For an elaboration, see Muraru and Tănăsescu (2008), p. 173.

3 Means of Protection of the "Right to Be Forgotten"

In terms of personal data protection legislation, Law 677/2001 provided both administrative and judicial means of protection, by way of the control carried out by the Romanian data protection agency (DPA) and by the ordinary courts.

As regards the "right to be forgotten", any data subject may turn to the DPA in order to have his/her right protected, by filling out a complaint. Art. 25 of the Law 677/2001 imposed some prior procedural steps: not to go first to a court of law for the same subject matter and against the same data controller and to previously complain about that issue to the data controller itself (15 days ahead). These conditions underline the importance of preserving the independent and impartial act of justice, without any interference, if a court action was already introduced and assure, on the other hand, a possible amicable settlement of the case. The DPA may have recourse to all the administrative means of action the law provides, by carrying out an investigation which might result in applying administrative sanctions (warning or fine)[25] and ordering by decision to stop the illegal processing and to delete the personal data unlawfully processed. During the procedure of solving complaints, the DPA may also order the suspension of the processing of personal data. A court action may be introduced either by the DPA, or by the data subject, in both cases the action being exempt from stamp duty. The Romanian DPA's powers are currently consolidated and adjusted to the ones established by the GDPR for all the EU Member States' DPAs, therefore corrective measures could also be imposed in case of a breach. According to the yearly reports made public on its website, the Romanian DPA solved complaints concerning the deletion of excessive or outdated data published on the Internet before[26] and after the ECJ ruling in *Google* case.[27]

[25]For violating any rights (including the rights to intervention and to object) a fine might be applied, between 1000 lei and 25,000 lei (1 Euro is around 4.50 lei, and 1 USD is around 4.10 lei), according to Art. 32 of the Law 677/2001. The GDPR raised the maximum amount of a fine up to 20,000,000 EUR, or in the case of an undertaking, up to 4% of the total worldwide annual turnover of the preceding financial year, whichever is higher, in case of infringement of the provisions on the data subjects' rights (Art. 83 para. 5).

[26]For instance, according to the 2013 report, the Romanian DPA found that a public institution disregarded the applicable legal provisions, disclosing personal data on the Internet consisting of national identity number and home address or the address of the properties listed in the declarations of assets and interests of some people who have held public office. After the investigation, the data were made anonymous. Other cases involved the omission to delete personal details associated with an account available on a selection of workforce related website (in 2007), the continuous disclosure on the Internet of the national identity number associated with fiscal data (in 2008 and 2011), the excessive publication of personal data on the official websites of the courts of law and of a prosecutor's office, in 2012 (a few yearly reports are available in English at the http://www.dataprotection.ro/index.jsp?page=Annual%20reports&lang=en. Accessed 1 May 2017).

[27]For instance, there was a case involving the refusal of a company to delete personal data associated with judicial information republished from the official website and indexed on the Internet, where the court of law upheld the position of the Romanian DPA (the 2014 DPA's report of activity, pp. 44–48, available in Romanian at http://www.dataprotection.ro/index.jsp?page=Rapoarte%20anuale&lang=ro. Accessed 1 May 2017).

Starting with 2014, the DPA included in its yearly reports a special section with regard to the complaints on the "right to be forgotten", in relation to the indexation on the Internet of a person's name. According to the information provided therein, most cases refer to the Google's refusal to delete the results published on the Internet by its search engine, as requested, on the grounds of the allegedly public interest over the information's content; after the intervention of the DPA, most cases were solved in favor of the complainant.[28]

As regards the judicial remedies for defending the "right to be forgotten", the person concerned may have recourse either on the basis of the provisions of the Law 677/2001 (Art. 18),[29] now replaced by the GDPR (Art. 79 and 82) and Law 102/2005 (Art. 14), or on the common grounds of the Civil Code. In each case, the party concerned may require not only restoration of his/her right (by deletion of illegally used personal details, for instance), but also material and moral damages, with the limits imposed by the freedom of speech, already mentioned (Art. 253 and 255 of the Civil Code).

Even if it is a case prior to the ECJ ruling in *Google Spain* case, it is relevant to mention here the efficiency of this sort of civil action,[30] related to the effect of the dissemination of personal data on the Internet: a plaintiff requested from a public administration body (and received 10,000 EUR) civil damages for the moral harm caused by mistakenly releasing the personal details on the Internet, including health related data, associated with his status of a socially assisted person.[31] This could be a good example of the possible remedies available for the individuals whose personal data are still made available on the Internet, despite their "right to be forgotten" request.

[28] See p. 74 of the 2015 DPA report of activity. The cited cases concern the disclosure of personal data on the Internet, associated with judicial cases, republished by private websites or images and defamatory information indexed from private blogs.

[29] "Art. 18: The Right to Refer to a Court of Law

(1) Without prejudice to the possibility of addressing the supervisory authority, the data subject has the right to address to a court of law in defense of any rights, guaranteed by the present law, that have been infringed. (2) Any person that has suffered a prejudice as a consequence of unlawful processing of personal data may address a competent court of law in order to obtain compensation for the prejudice suffered. (3) The competent court of law is the one whose territorial jurisdiction covers the complainant's domicile. The complaint addressed to the court of law is exempt from stamp tax."

[30] At the time of the ruling, the plaintiff's personal data were no longer available on the Internet.

[31] For other comments on this judgment, see Zanfir (2012).

4 A Glance Over the Romanian Case Law and Legal Doctrine After the Google Case

In general, the "right to be forgotten", even if it is not yet expressly provided by the applicable law, is effectively recognized by the existing case law. Moreover, individuals are aware of having this right, after the *Google Spain* case, and they exercise it by requesting this search engine to have their data deleted and then complaining to the Romanian Data Protection Authority, if their requests are not positively solved. Perhaps a greater harmonization of the jurisprudence is likely to be achieved by taking rather into account the safeguards provided by the legislation on data protection (former Law 677/2001 and GDPR), which directly relate to this right, than the general provisions based on the Civil Code, which seems to be the current tendency.

As already said, judicial review is one of the important means to effectively and efficiently protect the right to privacy and data protection in Romania. However, an exhaustive overview of the case law on this matter would be an impossible mission to accomplish. After the ECJ ruling in *Google Spain* case, the number of the cases brought before the courts is expected to grow since the legislative framework is strongly backed-up by the arguments to be drawn out of the decision, which plays the role of a "guiding light" for the EU Member States whenever they apply the EU law.

Further to the analysis of the judicial practice two types of litigation might be outlined: requests for deleting information from Internet and actions brought by the search engines.

The scrutiny undertaken for the purpose of this paper[32] shows a relatively small number of cases reported by the courts, which involved possible infringements of the "right to be forgotten", since 2014. Relevant cases concern the request for deleting from the Internet some personal photographs related to a medical treatment and claiming moral damages,[33] the request for deleting information, photos and

[32]For the purpose of this paper, in 2017 a few questions were addressed to all the appeal courts of Romania (except for the military one) and to the High Court of Cassation and Justice. The answers received show that the vast majority of the courts have no records of judgments regarding the deletion of personal information from the Internet or delisting of personal data by the search engines on the Internet, neither based on Law 677/2001, nor on other legal grounds. In two cases, the courts also indicated they received ordinary requests for deleting information from their public sites, related to the judicial files where the petitioners were involved. Besides this "public access exercise", we have also consulted the publicly available sources of judicial information, such as: portal.just.ro, www.scj.ro, www.rolii.ro, www.jurisprudenta.org. Judicial information published on the official sites of the Romanian courts is not indexed on the Internet.

[33]Civil judgment no. 192/2015 of the Cluj County Court (according to the answers received during the public access exercise). The case was against a physician and the company owning the website where images of the plaintiff were disclosed for advertising purposes, over the administered medical treatment ("before" and "after" a facial plastic surgery), without the plaintiff's consent. The court found that a breach of the plaintiff's rights to image and privacy occurred and ordered the deletion of those images from the Internet, publication of the judgment on the said website and compensation for moral damages.

defamatory articles from Facebook and claiming moral damages,[34] request for deleting personal details associated with judicial files from the official electronic database of the Romanian courts.[35] All the cited jurisprudence reflects judgments ruled in favor of the claimants and it was based on the Civil Code provisions; no judgment was grounded on Law 677/2001. There is no available information that foreign citizens (outside the EU) have filed any action related to this subject matter to Romanian courts. However, there is no legal impediment in this respect, since the rights protected by Law 677/2001/GDPR and Art. 26 of the Romanian Constitution may be exercised by any individual, regardless of their citizenship. Also, the Civil Code recognizes a similar status of the foreign citizens as for the Romanian citizens, as regards their civil rights and liberties (Art. 27).

The second category of judicial files concerns the cases that Google LLC brought against the Romanian Data Protection Authority in order to get the annulment of the administrative acts issued by the latter. Between 2015 and 2017, all of the three cases were ruled in favor of the Romanian DPA.[36] One of these cases[37] was related to the refusal of Google LLC to delete a number of links concerning pictures, defamatory and untrue information about a professor who complained to the Romanian DPA. The Bucharest Court of Appeal rejected the Google action, taking into account the unfounded nature of the information about the complainant, who was no longer a public person.[38]

[34]Civil judgment no. 97/2017 of the Vișeu de Sus City Court (according to the answers received during the public access exercise). The case was against a number of individuals who published on Facebook photos of the plaintiff and several negative comments with a defamatory character, related to her personal profile, sexual habits, etc. The court found that a breach of the plaintiff's right to image occurred and ordered the defendants to delete those images and comments from Facebook, to refrain from having again this kind of conduct and to pay compensation for moral damages.

[35]Civil judgment no. 14/CA/2015 of the Constanța Court of Appeal (according to the answers received during the public access exercise). The case was against this court and the Ministry of Justice.

[36]According to the information on portal.just.ro (Accessed 3 November 2017). In the latter case, Google refused to delete the link associated with a blog where defamatory information was published in relation to the personal activity of the individual who complained to the Romanian Data Protection Authority. The Google action was rejected as inadmissible by the Bucharest Court of Appeal (extract from this decision—civil judgment no. 3283/2017—is available on the www.rolii.ro).

[37]According to a press release on 22.04.2016 available at http://www.dataprotection.ro/?page=drept_de_interventie_fata_de_google&lang=ro (Accessed 12 May 2017).

[38]The court upheld that by publishing on the Internet, the personal data become accessible to an indefinite number of persons, so the data subject has no knowledge about the entities downloading the personal information and the way they are subsequently used; the potential impact of the means of communication is highly significant, thus publishing news online has a faster and stronger effect than in printed press, as regards the dissemination and further use of the information by various entities. As a result, the court stated that the data subject's interest to obtain deletion of personal data revealed on the Internet prevails over the economic interest of the controller, as the ECJ also ruled in the case C-131/12.

Romanian legal doctrine has showed a constant interest towards the "right to be forgotten" after the ECJ ruling in *Google Spain* case, in the context of analyzing the provisions of the existing legal framework on data subjects' rights[39] and of the new EU regulation (GDPR)[40] or in a specific examination of the arguments derived from ECJ ruling, from a neutral evaluation of its content to a critical one. The "right to be forgotten", as it was admitted by the ECJ, was also seen as a possible tool against discrimination or harassing.[41] Some authors discussed this new right in the context of making an (unexpected) comparison with the right to pay tribute to the memory of a deceased person. It is not the oblivion *per se* which was consecrated by the ECJ's jurisprudence, but a right whose exercise creates the premises of the natural oblivion that is an inherent phenomenon of the human being (Şchiopu 2017). In other words, the "right to be forgotten" (or "le droit à l'oubli") is a new guarantee for the protection of privacy, having particular connotations for the virtual space. The arguments upheld by the Court were considered pertinent to justify the prevailing interest in protecting the right to privacy over the economic interest of a search engine controller: this one allows Internet users to combine various information about a person's private life and help to create a (more or less) complete profile of that individual, with a potential effect of aggravated interference (Jugastru 2017). However, the ECJ ruling also received some critical comments, as regards the lack of clarity on deciding the territorial scope of the applicability of the "right to be forgotten", namely, if a search engine like Google is compelled to delete data related only to European websites, or also from outside the EU (such as ".com"—Costescu 2016).[42]

5 How Is Google Dealing with the "Right to Be Forgotten"

The ECJ ruling set up new standards of interpreting the European rules on data protection as regards the territorial scope of Directive 95/46/EC and the limits of the right of having personal data deleted, imposing the same obligations on European and non-European controllers, whenever they address their processing activities to European individuals; also, for the first time, ECJ considered a search engine as being a data controller. In the specific case, the ECJ ruling of 13 May 2014 targeted primarily the global search engine of Google LLC which had to promptly implement mechanisms of dealing with the increasing number of requests of delisting results containing personal data from the Internet. For this purpose, a web form was put in

---

[39]For an elaboration, see Şandru (2016), p. 217; Zanfir (2015), pp. 145–165.

[40]More comments in Şandru (2016), p. 179 et passim.

[41]For an elaboration on this subject, see Opre and Şandru (2015), pp. 270–280.

[42]"We appreciate that this cannot be considered a sufficient mean to satisfactorily guarantee the rights of data subjects according to the ruling and in order for the de-listing to be effective, it should apply on all relevant domains, including .com."—p. 70.

place in all the EU Member States languages.[43] Unfortunately, Google understands ECJ ruling in a restrictive way, so only requests related to European websites are taken into consideration when they receive a request for deletion. In each case, Google has to put in balance the right of the petitioner to have control over his/her personal data and the right of the public to be informed, as the ECJ stated. In order to admit a request, the results must be inadequate, irrelevant, excessive or outdated; cases where the public interest is at stake will not be solved positively.[44] This "balancing test" puts a great burden on Google's shoulders, as it is only a private organization, so eventually, a local data protection authority[45] or a national court of law will decide if a request was properly handled.

On its official website (google.com), Google LLC publish a regularly updated transparency report on "European privacy requests for search removals"[46] with statistics per country and examples of cases and URLs they processed (positive and negative ones). According to this report, between 29 May 2014 and 20 May 2017 Google received 725,372 requests and evaluated 2,046,180 URLs for removal (43.1% URLs were removed).

As for Romania, Google received 10,231 requests for the removal of 43,214 URLs (a total of 30.6% URLs were removed, which is below the European medium score mentioned above). This report contains no practical example from Romania. However, it seems that only a small number of people who have received no positive answer from Google have turned to Romanian DPA, compared to the total number of complaints received by this authority in 2014 (940) and 2015 (1074),[47] irrespective of their subject matter. This is a general trend in all EU countries, showing that not all the requests submitted to Google are really justified.

Besides the current content of the information Google has made public, a list of concrete criteria for de-listing should be available in order to make their procedures more transparent. Examining the examples Google has provided, a pattern for the denial of requests is linked to the assessment of the (potential) public interest in the facts a petitioner was involved into, considering their gravity or their age.[48] Both these criteria may be subjectively and inconsistently handled by Google, thus

---

[43] Available at https://www.google.com/webmasters/tools/legal-removal-request?complaint_type=rtbf&visit_id=0-636309063139388868-4096106632&hl=ro&rd=1 (in Romanian). Accessed 20 May 2017.

[44] Google offers examples such as: financial scams, malpraxis, criminal conviction, public behaviour as an official public figure.

[45] Otherwise, Google recommends in this web form and in their standard answers to contact the local data protection agency, in case a petitioner is not satisfied with their solution.

[46] The report is available at https://www.google.com/transparencyreport/removals/europeprivacy/?hl=en. Accessed 20 May 2017.

[47] According to the Romanian DPA reports (available at http://www.dataprotection.ro/?page=Rapoarte%20anuale&lang=ro).

[48] For instance, Google provides an example where it refused to remove recent articles about a high ranking public official, referring to a decades-old criminal conviction (request from a Hungarian individual).

supporting the previously expressed doubt as to the qualification of this company to make an a posteriori assessment of the interests at stake, since it has not carried out an a priori evaluation of the potential impact on privacy or the likely public interest in having access to personal information when it first decided to automatically collect and combine information from the indexed sites on the Internet.

6 Future of the "Right to Be Forgotten"

Once the EU GDPR is being put in place (as of the 25th of May 2018) in all EU Member States, including Romania, the "right to be forgotten" will benefit from a distinct and stronger legal regime, with clear prerogatives for data subjects and obligations of data controllers, expressly providing the reasons justifying a request and the exceptions from this right (Art. 17). The current ambiguity (related to the jurisdiction over European/non-European websites) will hopefully disappear, since the regulation changes completely the criteria of establishing the territorial scope of European data protection rules: GDPR applies to any establishment of a controller or a processor in the Union (regardless whether the processing takes place in the Union or not) and to all data controllers who process data in relation to the offering of goods or services (irrespective of whether a payment of the data subject is required) to data subjects in the Union, or to the monitoring of their behaviour as far as their behaviour takes place within the Union (Art. 3). Although the legal provisions of the Art. 17 are generally applicable (except for the second paragraph which also refers to the due diligence obligation to inform third parties to delete some links where personal data are re-published), the preamble of the regulation underlines the particular relevance of this right especially where the data subject has given his/her consent as a child and is not fully aware of the risks involved by the processing, and later wants to remove such personal data, especially on the Internet (paragraphs (65) and (66)). Therefore, the "right to be forgotten" will remain a right which particularly concerns the processing of personal data in the online environment, where the potential risks of harming privacy by enduring replication of data are higher than in the offline activities, as ECJ also ruled in *Google Spain* case.

The ECJ judgment of 2014 was a milestone for data protection on the Internet and marked a big step forward on the way of conceptualizing and legally admitting the "right to be forgotten", ahead of the regulators. The judges' arguments must not be interpreted narrowly, so it should be clear that the scope of this right is not limited to national or European domain names associated with a site, and the criteria for searching a person on the Internet could not be limited to his/her name, as other pieces of information could also be linked to his/her identification and possible profiling. A restricted interpretation would eventually lead to an incomplete protection of this fundamental right to privacy. The new EU regulation bestows upon individuals enhanced prerogatives of control over their personal data, whenever they are in the European Union. However, as the Internet is probably the most relevant proof of the globalisation, we think that the "right to be forgotten" has to become a universal right, so an international legal instrument could provide equivalent means for its protection all over the world, regardless of the citizenship of the interested persons, irrespective of the nationality of the data controllers or the territorial scope of their processing activities. This objective might be achievable by the adoption of a legal instrument to such effect by an international human rights organisation, or by the accession of many other countries or international organisations outside the Council of Europe, to the European Convention on Data Protection[49] which was also recently amended in 2018.

References

Costescu ND (2016) Google Spain decision – an analysis of the right to be forgotten - a regression from past interpretations of ECJ. Analele Universităţii din Bucureşti 1:64–71

Jugastru C (2017) Tradiţie şi inovaţie în materia protecţiei datelor cu caracter personal (Tradition and innovation in data protection field). Universul Juridic 2:74–84

Muraru I, Tănăsescu ES (2008) In: Muraru I, Tănăsescu ES (eds) Constituţia României. Comentariu pe articole (Romanian Constitution. Comments by articles). C.H. Beck, Bucharest, pp 169–175

Opre AG, Şandru S (2015) Dreptul de a fi uitat pe Internet, mijloc de combatere a discriminării (The right to be forgotten on the Internet, a tool for combating discrimination). In: Tomescu M (ed) Exercitarea dreptului la nediscriminare şi egalitate de şanse în societatea contemporană; Lucrările celei de a IX-a Conferinţe a nediscriminării şi egalităţii de şanse – NEDES 2015 (The right to non-discrimination and equal opportunities in contemporary society; the 9th NEDES Conference 2015). ProUniversitaria, Bucharest, pp 270–280

Şandru S (2016) Protecţia datelor personale şi viaţa privată (Protection of personal data and private life). Hamangiu, Bucharest

Şchiopu S-D (2017) Dreptul la ştergerea datelor şi dreptul la aducerea ultimului omagiu: a fi uitat sau a fi ţinut minte? (The right to delete data and the right to pay a last tribute: to be forgotten or to be remembered?). Universul Juridic 2:85–93

Warren SD, Brandeis LD (1890) The right to privacy. Harv Law Rev 4:193–220

Whitman JQ (2004) The two western cultures of privacy: dignity versus liberty. Yale Law J 113

Zanfir G (2012) Protecţia datelor cu caracter personal. Categorii speciale de date. Despăgubiri morale (Jud. sect. 1 Bucureşti, sentinţa din 16 martie 2009, irevocabilă) (Protection of personal data. Special categories of data. Moral damages. Court of District 1, Bucharest, judgment of 16 March 2009, final). Pandectele Române 2:143–155

Zanfir G (2015) Protecţia datelor personale. Drepturile persoanei vizate (Protection of personal data. The rights of data subjects). C.H. Beck, Bucharest

[49]More information about the modernisation process of the Convention no. 108/1981 for the Protection of Individuals with Regard to the Processing of Personal Data (started in 2011) is available at https://www.coe.int/en/web/data-protection/convention108/modernised (Accessed 30 October 2018). Article 27 of the new Convention regulates on accession by non-member States and international organisations to the Convention.

The Right to Be Forgotten in the UK: A Fragile Balance?

Sabine Jacques and Felix Hempel

Abstract This chapter comprehensively illustrates the recent status of the right to be forgotten in the UK and unveils the significance of the changes caused by recent developments. Particularly, the latest reforms in both domestic and international law have had a drastic impact on the application of the right to be forgotten. With the General Data Protection Regulation (GDPR) scheduled to have direct effect in all EU member states, the Government introduced the Data Protection Act (DPA) 2018 in order to retain the regulation post-Brexit. Significantly, the GDPR emphasises the need for a statutory right to be forgotten in Article 17, which goes beyond what was guaranteed under the old legal framework in the UK. In addition, the crucial judgment of *NT1 and NT2 v Google LLC* handed down by the High Court in 2018 established and clarified under which circumstances a person can successfully 'erase' unwanted information from the digital landscape under UK law. It further contains novel and significant conclusions as to how UK courts should balance out the different interests involved in a right to be forgotten case. By drawing upon these developments in both legislation and case law, this chapters provides a unique overview of how the right to be forgotten has been conceptualised over time and what issues have already been raised under the new legal framework. Also, it offers an insight into the rationales underpinning the right to be forgotten from a UK perspective and explores whether further protection would be desirable.

S. Jacques (✉) · F. Hempel
University of East Anglia Law School, Norwich, England
e-mail: sabine.jacques@uea.ac.uk; f.hempel@uea.ac.uk

F. Werro (ed.), *The Right To Be Forgotten*, Ius Comparatum – Global Studies in Comparative Law 40, https://doi.org/10.1007/978-3-030-33512-0_10

1 Introduction

Until 2018,[1] the right to be forgotten (aka 'right to erasure'[2]) did not exist as such in UK law but derived from a more general framework for a right of privacy and data protection laws. However, recent developments in both domestic and international law have led to a drastic change in the national legal landscape. With the UK on course to leave the European Union yet the General Data Protection Regulation (GDPR) scheduled to have direct effect in all member states beforehand, the Government introduced the Data Protection Act (DPA) 2018. The aim of this act is to implement derogations in the GDPR into national law during the pre-withdrawal period and then retain the regulation in domestic law post-Brexit.[3] Most importantly for the purposes of this chapter, the GDPR emphasises the need for a statutory right to be forgotten in Article 17. Further, the crucial judgment of *NT1 and NT2 v Google LLC* handed down by the High Court in 2018 established and clarified under which circumstances a person can successfully 'erase' unwanted information from the digital landscape under UK law.[4] This decision marks the first delisting order made by a UK court compelling Google to remove links where the search engine operator had rejected a request.[5] It is seen as perhaps the most high-profile consideration of this right in a common law jurisdiction following the ruling in *Google Spain*.[6] As explored below, this judgment provides crucial guidance on when search listings should be delisted and how UK courts should balance out the different fundamental rights and interests at in a case involving the right to be forgotten. To provide an overview of these issues and developments, this chapter aims to illustrate the trajectory of the present status of UK and (still) relevant EU law relating to the right to be forgotten, and particularly, attempts to unveil the significance of the changes caused by the GDPR.

To do so, this chapter is divided into four sections. After briefly setting out the legal basis for the right to be forgotten in the UK (Sect. 2), it highlights the remedies available under privacy law in general (Sect. 2.1), and more specifically, the recently updated data protection legislation (Sect. 2.2). This chapter then outlines the limitations to the right to be forgotten in the UK (Sect. 2.3). Subsequently, it examines the relevant UK case law, showing how the right to be forgotten has been conceptualised over time and how the jurisprudence has changed in response to *Google Spain* (Sect.

---

[1]An earlier version of this chapter has partly been included in the conference report of the 'Congress of the International Society of Comparative Law' held in Fukuoka in 2018. The authors would like to thank Professor Franz Werro, and Ms Claudia Hasbun for comments on an earlier draft. All errors remain the authors' responsibility.

[2]This is the preferred name in the UK.

[3]For further detail see Mc Cullagh (2018).

[4]See *NT1 and NT2 v Google and The Information Commissioner* (2018) EWHC 799 (QB).

[5]For further detail see Sect. 3.3 below.

[6]See Case C-131/12 *Google Spain SL & another v Agencia Espanola de Proteccion de Datos (AEPD) and another* (2014) ECLI:EU:C:2014:317. For further detail see Sect. 3 below.

3). Lastly, the final part of this chapter summarises the findings and ventures an evaluation of whether further protection would be desirable (Sect. 4).

2 The Legal Basis for the Right to Be Forgotten Under UK Law

In the UK, the right to be forgotten primarily derives from privacy rights. From a legal standpoint, privacy itself is based in international human rights law and protected by numerous areas of UK law, mainly tort and data protection laws but also, enshrined in constitutional law.[7] Therefore, the right to be forgotten under UK law flows from a more general framework for a right to privacy.

Historically, the Data Protection Act (DPA) 1984 (repealed in 2000) was the first piece of legislation relevant to data protection in the UK. This enacted the Council of Europe's Convention for the protection of individuals with regard to automatic processing of personal data 1981 (Council of Europe 1981). Later, the DPA 1998 implemented the EU Data Protection Directive 95/46/EC[8] in domestic law. However, this Directive (and its national counterpart) had been criticised for being obsolete, given that it predates the digital era.[9] More recently, the GDPR has replaced this Directive. Significantly, article 17 of the GDPR provides a statutory right to be forgotten, which will be further discussed below.[10] It inter alia enables data subjects to request that data controllers erase links to data which data subjects see is as detrimental to them (House of Lords European Union Committee 2014, p. 15).

Although privacy laws and data protection laws are distinct, they are also complementary insofar as data protection laws provide procedural rights through which substantial rights (privacy rights) can be enforced (Ghezzi et al. 2014, p. 67). In order to provide a comprehensive overview, the following subsection first outlines how privacy rights are protected under the UK law regime and how this impacts the right to be forgotten. Second, this part of the chapter then examines the right to be forgotten under the GDPR and the updated UK Data Protection Act. The final subsection sets out the limitations to the right to be forgotten in the UK. In this context, it should be noted that an examination of the relevant case law is conducted separately in Sect. 3.

[7]Privacy laws: general tort of common law (libel, breach of confidence) but also harassment laws (Protection for Harassment Act 1997) and the misuse of private information tort action. For further detail see: Ghezzi et al. (2014), p. 35.

[8]See Directive 95/46/EC (1995).

[9]See for example Ghezzi et al. (2014), p. 17.

[10]See Sect 2.2 below.

2.1 The Framework for the Protection of Privacy Under UK Law

Historically, the UK has been criticised for not having a general tort remedy for invasion of privacy.[11] Yet, the right to privacy is a long-standing value underpinning common law.[12] Traditionally linked to the invasion of the physical private sphere, the right to privacy was seen as an extension of the protection of one's property.[13] Later in the nineteenth century, the right to privacy acquired a new youth when celebrities and other public figures sought the protection of their private life in the public forum.[14]

The introduction of the Human Rights Act 1998 (HRA) sought to implement the European Convention on Human Rights (ECHR) in national law. This instrument allowed English courts to expand the equitable doctrine of breach of confidence in relation to the misuse of information by the media,[15] which later evolved in a tort of misuse of private information (Moreham 2005). Besides, it is possible to obtain damages in compensation for moral prejudice. For example, in cases of misuse of

---

[11]See for example *Wainwright v Home Office* (2004) 2 AC 406. In *Wainwright v Home Office*, the House of Lords was required to declare whether an action for the invasion of privacy was available under UK law for the first time. This case dealt with the strip-search of a mother and son during a prison visit in 1997 in breach of Prison rules. There is no overall remedy for the invasion of one's privacy. To the contrary, the UK follows a piecemeal approach to the protection of privacy. See also Aplin (2007).

[12]In 1974, the Third Royal Commission on the Press was established to '...inquire into the factors affecting the maintenance of the independence, diversity and editorial standards of newspapers and periodicals and the public freedom of choice of newspapers and periodicals, nationally, regionally and locally.'; In 1989, Sir David Calcutt's 'Inquiry into Privacy and Related Matters' was established. For further detail see Select Committee on Communications (2015), p. 53; *Wainwright v. Home Office* (2004) 2 AC 406, para 422; *The Law Society and others v Kordowski* (2011) EWHC 3182 (QB).

[13]There was no overall remedy for the invasion of one's privacy. See also *Semayne's Case* (1604) 5 Coke Reports 91a.

[14]See *Tolley v Fry* (1931) AC 333; *Kaye v Robertson* (1991) FSR 62. *Tolley v. Fry* deals with the publication of adverts for chocolate bars of a golfer without his consent whilst *Kaye v Robertson* deals with the trespassing of a celebrity's hospital room by a journalist and photographer. Whilst having to rely on other areas of the law such as the law of trespass and tort of malicious falsehood, the judges criticised the lack of tort of privacy sending a direct message to the Parliament at the time.

[15]See Lord Nicholls in *Campbell v MGN Ltd* (2004) 2 AC 457, in relation to an article published by the 'Mirror' on Naomi Campbell's drug addiction. The article included photographs of Naomi Campbell as she was leaving a Narcotics Anonymous meeting, together with additional information on her treatment. Here, the Court held that the confidential information and claimant's right to article 8 of the ECHR prevailed over article 10. *Campbell v MGN Ltd* (2004) 2 AC 457, para 14 details: 'The essence of the tort is better encapsulated now as misuse of private information'; reliance on the human rights framework also allowed courts to expand damages of distress in situations absent of pecuniary loss. In *Vidal-Hall v Google Inc* (2014) EWHC 13 individuals sued Google for the distress and anxiety caused due to the tracking and collation of browser activity the used for advertising purposes (see para 98-99). This decision was upheld on appeal in *Vidal-Hall v Google Inc* (2015) EWCA Civ 311.

private information, damages are awarded to claimants in repair for moral harm (e.g., hurt of feelings or distress).[16] However, instead of developing a general tort remedy available to all, English courts erred on the side of caution and limited this doctrine to public figures.

Nevertheless, there are few exceptional successful rulings where the claimant was not a public figure. In *Applause Store Productions v Raphael*,[17] the High Court dealt with a case of Internet libel against a Facebook group and profile pages that contained private information in relation to the claimant's credit standing, sexuality, religious beliefs and political views. The fact that at least some of the comments were defamatory was not disputed.[18] The court found that the allegations were serious and could be detrimental to the claimant's business, despite the relatively limited reach of the publication.[19]

Under defamation laws, the court can award a monetary sum in damages,[20] and grant an injunction preventing future publication (Defamation Act 2013, s 13).[21] In those decisions, the terms of the settlement agreement are negotiated between the parties and range from an apology, retraction or correction of the published statement to the payment of a sum in damages, payment of the resulting legal fees and undertakings not to repeat the allegations complained of. Yet, most of the defamation cases are generally settled out of court.

Whereas privacy laws developed in the analogue world have adapted to extend into the digital world, data protection laws deal predominantly with digitalisation. As such, data protection laws have evolved separately from privacy laws (Walden 2011, p. 585).

---

[16]See *Applause Store Productions v Raphael* (2008) EWHC 1781 (QB). This case dealt with a Facebook group and profile containing defamatory materials in relation to the financial status of a business and the misuse of private information regarding the sexuality of the owner of said business; *McKennitt v Ash* [2006] EMLR 178, para 162. Here, the claimants were a Canadian folk musician and her recording companies against a former friend and personal assistant of the musician. This latter published a book revealing some of the musician's personal and private life. The Court noted that each excerpt from the book had to be examined in turn to appreciate whether the threshold test of reasonable expectation of privacy was met. Eventually, the Court held that the musician was entitled to damages for hurt feelings and distress (see para 162); *Campbell v MGN Ltd* (2004) 2 AC 457; *Campbell v MGN Ltd* [(2002) EWHC 499 (QB); *Vidal-Hall v Google Inc* (2014) EWHC 13.

[17]See *Applause Store Productions v Raphael* EWHC 1781 (QB).

[18]For further detail see paras 38 and 69 of the judgment.

[19]For further detail see paras 69 and 82 of the judgment.

[20]Several forms of damages are available such as compensatory damages and exemplary damages, see *John v MGN Ltd* (1996) 2 All ER 35.

[21]It is worth noting that certain parts of the Defamation Act 2013 only extend to England & Wales and therefore do not apply to Scotland and Northern Ireland. For an overview, see UK Government (2013).

2.2 *The Recently Revised Data Protection Legislation in the EU and Its Impact on the UK Data Protection Law*

In addition to providing an overview of the right to be forgotten under the GDPR and its implementation into domestic law, this section briefly outlines the historical development of data protection legislation in the UK. This is significant not only to understand the status quo but also sets the scene for the examination of the relevant case law in Sect. 3 below.

As noted above,[22] the shape of UK data protection laws has been heavily influenced by the UK's regional and international obligations. Prior to being superseded by the GDPR, article 12(b) of Directive 95/46/EC provided a right for the 'data subject to obtain from the controller as appropriate the rectification, erasure or blocking of data the processing of which does not comply with the provisions of this Directive, in particular because of the incomplete or inaccurate nature of the data.' This provision should be read in conjunction with articles 6 and 7 of said Directive, which set out the principles in relation to data quality and lawfulness of data processing. In order to respect the proportionality principles, the Directive also provided other remedies than the removal of the data. As such, a more adequate remedy may be to anonymise the information for the data subject not to be associated with the publication anymore.

The UK implementation was slightly different from the text of the Directive. For example, article 12(b) of the Directive had been transposed in section 14(1) DPA 1998 as:

> If a court is satisfied on the application of a data subject that personal data of which the applicant is the subject are inaccurate, the court may order the data controller to rectify, block, erase or *destroy* those data and any other personal data in respect of which he is the data controller and which contain an expression of opinion which *appears* to the court to be based on the inaccurate data (the emphasis is the rapporteur's own).

In other words, under the old regime data subjects had a right to apply for an order to rectify, block, erase or destroy inaccurate information about them, as well as any other personal data which contains an opinion based on the inaccurate information. In addition, a court was also able to order the data controller to notify any third party to whom inaccurate data has been disclosed of the rectification, blocking, erasure or destruction. Nevertheless, the DPA 1998 granted UK courts the power to refuse a right to erasure (section 14(2) DPA 1998) and request that the data be supplemented by true facts where the data controller (e.g. the publisher) acted honestly in publishing the information.

Having come into effect on 25 May 2018, the GDPR introduced an update to those rules and practices. One of the first differences to the old legal framework is the fact that the EU chose to harmonise the new data protection laws by means of regulation rather than directive as done before. This is significant, as a regulation is a

---

[22]See the Introduction.

legal instrument that has immediate binding effect in all Member States, without any further action being required by those states. Although regulations can (and the GDPR does) include provisions that allow Member States to create additional rules, such as exceptions to requirements,[23] and, where permitted, clarifications on the GDPR's applicability, the main cut and thrust of the provisions of the legal instrument is identical across the EU (Carey 2018).

Most importantly for the purposes of this chapter, Article 17 of the GDPR contains both a 'right to erasure' as well as a 'right to be forgotten'. More specifically, Article 17(1) notes that personal data must be erased immediately where the data are no longer needed for their original processing purpose, or the data subject has withdrawn his consent and there is no other legal ground for processing. The same applies to situations where the data subject has objected, and there are no overriding legitimate grounds for the processing or erasure is required to fulfil a statutory obligation under the EU law or the right of the Member States. In addition, data must naturally be erased if the processing itself was against the law in the first place. Thus, the controller is therefore on the one hand automatically subject to statutory erasure obligations, and must, on the other hand, comply with the data subject's right to erasure. This provision should be read in conjunction with articles 5 and 6 of said regulation, which set out the principles in relation to data quality and lawfulness of data processing.

In addition to the right to erasure, article 17(2) contains a right to be forgotten. As a result, if the controller has made the personal data public and if one of the above reasons for erasure exists, he or she must take reasonable measures,[24] 'to inform controllers which are processing the personal data that the data subject has requested the erasure by such controllers of any links to, or copy or replication of, those personal data'. Considering these requirements set out in the GDPR, this right can be exercised by data subjects even if the processing is not causing the data subject any damage or distress.[25] Also, article 19 contains a duty for the controller to inform any recipient of whom personal data have been disclosed of any erasure carried out pursuant to article 17 unless this proves impossible or involves disproportionate efforts. Further detail is contained in recital 66 of the GDPR. However, like the right to erasure under article 17(1), the right to be forgotten is not absolute and therefore subject to exceptions detailed in article 17(3). Those exceptions are covered separately in Sect. 2.3 below.

In addition to those rights, one should note article 16, which contains a 'right to rectification'. This remedy grants the data subject the 'right to obtain from the controller without undue delay the rectification of inaccurate personal data concerning him or her' and 'to have incomplete personal data completed, including by means of providing a supplementary statement.'

---

[23]See Sect. 2.3 below.

[24]Considering the specific circumstances.

[25]See Sect. 2.3 below.

Significantly, this new legal framework goes beyond the 'right of erasure' that was guaranteed under the old regime in the UK. This is because instead of having to apply for a court order with the aim of erasing inaccurate personal data, individuals now have a statutory right to have personal data erased as long as they meet the requirements and none of the exceptions applies (UK Government 2018a). In other words, the new legal framework grants individuals a qualified right to have personal data relating to them deindexed from search engines (Garstka and Erdos 2017, p. 127). This lowers the bar for successfully making a delisting request and therefore strengthens the position of the person seeking to protect their privacy and personal data. However, it is yet to be determined what steps, if any, third party controllers are required to take once they have been notified about the erasure of the data pursuant to an erasure request (Lloyd-Jones and Carey 2018, p. 146). This uncertainty primarily stems from the fact that article 19 does not determine what recipients are expected to take when notified of an erasure (Lloyd-Jones and Carey 2018, p. 146).

For comprehensiveness, one must also note the DPA 2018. This act has repealed and replaced the DPA 1998 and primarily aims to implement derogations in the GDPR into national law. Thus, the new statute contains several exemptions and clarifications to the GDPR,[26] which are relevant to the purposes of this chapter and therefore further discussed in Sect. 2.3 below. Significantly, the power to pass an act that contains exceptions and clarifications to the GDPR stems from the regulation itself. Article 23 of the GDPR holds that under certain circumstances, 'Union or Member State law to which the data controller or processor is subject may restrict by way of a legislative measure the scope of the obligations and rights provided for in Articles 12 to 22 [...] in so far as its provisions correspond to the rights and obligations provided for in Articles 12 to 22 [...].'[27]

The second main purpose of the DPA 2018 is to retain the GDPR in domestic law post-Brexit. As a recap, a referendum took place on 24 June 2016 after which the UK decided to leave the European Union (EU). Having triggered article 50 of the TFEU on 29 March 2017, the UK is due to exit the EU on 29 March 2019.[28] However, the data protection law of the regulation will still apply to the UK, as section 22(1) of the DPA 2018 emphasises that: 'the GDPR applies to the processing of personal data [...] as if its Articles were part of an Act extending to' the UK. In addition, section 3 of the European Union (Withdrawal) Act 2018 highlights that any 'direct EU legislation [...] operative immediately before the exit day' like the GDPR will be incorporated into domestic law. However, the Government has made it clear that after the cut-off date, the Court of Justice of the European Union (CJEU) will cease to have any UK jurisdiction. Therefore, while the CJEU is likely to retain some influence in a post-Brexit UK, not least in the name of maintaining stability, the UK Supreme court will then be free to depart from the CJEU's rulings. For comprehensiveness, one should note that a new Digital Charter, described by the Government

[26]See in particular UK DPA 2018, schedules 2 and 6.

[27]See Sect. 2.3 for further detail.

[28]At the time of writing.

as a 'rolling programme of work to agree norms and rules for the online world and put them into practice', is also envisaged (UK Government 2018b).

2.3 *Limitations to the Exercise of the Right to Be Forgotten in the UK*

It is crucial to reiterate that despite the recent developments in both the domestic and the international legal landscape, the right to be forgotten is not an absolute right in the UK. Instead, the exercise of this right is subject to certain requirements and limitations, which this section aims to outline.

Under privacy laws, the tort of misuse of private information, enabling claimant (s) to sue the defendant(s) for disclosure of information, which claimant (s) reasonably expected to remain private, is generally limited to celebrities and public figures.[29] The general limitations of these tort actions also apply.

In defamation (libel) actions, there is a limitation period of 1-year from the publication date.[30] Further, the publication must be defamatory, meaning that it must bring others to think less of the person targeted by the publication. There is no possibility to get an injunction to restrain an individual from future publications unless there is evidence that the libel action would succeed.[31]

Similarly, an action brought under the Protection for Harassment Act 1997 suffers from several limitations. First, the period to bring an action is a 6-year limitation period. Second, the arduous limitations described in Sect. 2.3 above (e.g. limitation period and high chances of success at trial) linked to injunctions to restrain a defendant from publishing a piece under defamation laws are not applicable under the harassment laws.[32] Third, there might be difficulties to demonstrate that an intermediary, such as Google, represents a common law publisher (Hurst 2015,

[29]See Sect. 2.1 above.

[30]See for example the case of *Reed Elsevier v Bewry* (2014) EWCA Civ 1411*; Steedman v BBC* (2001) EWCA Civ 1534, [2001] All ER (D) 316 (Oct). The facts concerned the claimant's application to disapply the limitation period in his proceedings for libel under section 32A of the Limitation Act 1980. The claimant was a local authority approved foster carer and the first defendant is the owner of the LexisNexis website. The proceedings where brought in relation to the words used in a case note about a judicial review published on the LexisNexis website. This case note was prepared prior to the judgement and was made available to the website's subscribers. Whilst the High Court of England granted the claimant's application to disapply the limitation period, the Court of Appeal eventually overturned this decision in 2014. In sum, it is very hard to extend a statutory limitation.

[31]See for example *Bonnard v Perryman* CA (1891) 2 Ch 269 reaffirmed in *Greene v Associated* (2004) EWCA Civ 1462. In the latter, Mrs Greene failed in her attempt to obtain and injunction preventing a publisher to disseminate a story based on emails she had sent to a conman. Applying the *Bonnard* libel rule whereby a libel injunction will be granted only if the claimant shows that chances of success at trial are high.

[32]See in particular Protection for Harassment Act 1997, s 3(3)a(a).

p. 189). Finally, pecuniary compensation could be possible where the publication is truthful. Yet, in this particular situation, there might be practical barriers in establishing that the defendant's actions were 'unreasonable'.[33]

As a consequence of adopting the GDPR, there is no longer a right to erasure for inaccurate and incomplete data like it was set out under the now replaced EU Data Protection Directive 95/46/EC. Instead, following article 16 of the GDPR, individuals are entitled to have personal data *rectified* if data is found to be inaccurate or incomplete. In contrast, Directive 95/46/EC set out several options for how to deal with incomplete or inaccurate data in article 12, as it recognised a right to 'rectification, erasure or blocking of data the processing of which does not comply [the Directive] [...] in particular because of the incomplete or inaccurate nature of the data'. Thus, the GDPR seems to establish a presumption for inaccurate or incomplete data to be rectified rather than erased. However, this presumption might still be overturned if one of the grounds set out in article 17 of the GDPR applies.[34] In essence, one might argue that this amendment can be seen as an attempt to lower the potential 'chilling effect' of the right to forgotten and strengthen freedom of expression. If inaccurate or incomplete data is rectified instead of being erased, it remains accessible to the public. Leaving information within the public domain rather than erasing it serves a society's interest in the pluralism of information and freedom of expression in general.

However, it is yet to be determined whether this change will have any negative impact in the UK as the new data protection legislation significantly extends the data subject's right and clarifies uncertainties that were present under the old legal framework. These uncertainties existed primarily because the now replaced UK DPA 1998, which implemented Directive 95/46/EC into domestic law, deviated from the 'right to rectification' set out in article 12 under said Directive. For example, section 14(1) of the DPA 1998 required data subjects to apply for a court order in order to rectify, block, erase or destroy inaccurate data. This was despite the fact that the right to rectification under the Directive had a mandatory character and did not require courts to get involved. Thus, critics argued that this discretion granted to UK courts (instead of the mandatory character enshrined in the Directive) was not compliant with EU law (Boulanger 2010). The UK provision was further limited insofar as data subject rights could not be triggered by incomplete data under the Directive. However, despite these limitations, the UK DPA 1998 appeared to have also broadened the scope of the Directive. First, it included the word 'destroy' where the Directive was solely focused on rectification, erasure or blocking without expanding to destruction. Second, under the UK provision, the inaccuracy of the data did not have to be established.[35] The appearance of the inaccurate suffices. Nevertheless, where the controller honestly believed the information to be accurate, UK courts were able to require an appending statement with the rectified information

[33]See in particular Protection for Harassment Act 1997, s 1(3)(c).

[34]See Sect. 2.2 above.

[35]See for example *The Law Society and Ors v Kordowski* (2011) EWHC 3185 (QB), para 134.

instead awarding data erasure. Finally, where article 6(1)(e) provided that data subjects shall not be identifiable for longer than necessary, Principle 5 stated that personal data shall not be 'kept for longer than necessary'. Scholars have argued that this perhaps meant that the DPA 1998 did not appear to grant an anonymisation remedy but seems to support the right to be forgotten (O'Callaghan and de Mars 2016, p. 46). As established in Sect. 2 above, these uncertainties regarding the relationship between domestic and international law no longer exist under the updated legal framework.

As opposed to privacy or personality rights based actions, which ought to be invoked within a certain time period, the right to be forgotten and the right to erasure under the GDPR can be invoked at all times. It should further be noted that the updated data protection legislation has removed some of the major hurdles for an individual to erase or rectify personal data. Under the old legal framework, there was a burden of proof on the person applying for a court order to demonstrate that there are compelling legitimate grounds to stop the data processing (including the disclosure). As the claimants needed to prove that they suffer from *all* the forms of processing, the data controller enjoyed a wide discretionary power. For example, the data controller was able to decide that the request is allowed in relation to the disclosure and not the storage, which might be seen as less detrimental to one's privacy. Therefore, the data subject would have had to bring a new action if the data infringes the privacy rights of the data subject again in the future (Ghezzi et al. 2014, p. 95). The deletion of information may have not always been the preferred outcome of a court action. If section 7 of Part II of Schedule 1 DPA 1998 is respected, the court could have had decided that the request for erasure of the data is not the best remedy. Hence, in a scenario in which the information had been legitimately obtained and the data controller had taken reasonable steps to ensure the accuracy of the information, the DPA 1998 stated that the court may order the inaccurate information to be supplemented with true facts. As shown above, the GDPR has streamlined the process significantly, as it now up to the data subject whether he or she prefers rectification or erasure of data. Further, individuals are now able to request erasure or rectification from the controller directly rather than having to obtain court order.

In addition, the right to erasure under the DPA 1998 was limited to processing that caused unwarranted and substantial damage or distress. Significantly, this threshold is not present under the GDPR. However, if the processing does cause damage or distress, this is likely to make the case for erasure stronger.

It is further worth noting that the so-called 'domestic purposes exception', which was expressly mentioned in the DPA 1998, has not been replicated in the DPA 2018. However, this is only because personal data processed in the course of a purely personal or household activity, with no connection to professional or commercial activity, is outside the GDPR's scope anyway.[36] Thus, if one only uses personal data

[36]As you know, this exception was already established under Directive 95/46/EC and had been interpreted in Case C-101/10 *Bodil Lindqvist v Sweden* (2011) ECLI:EU:C:2011:462, para

for such things as writing to friends and family or taking pictures for your one's enjoyment, one's is not subject to the GDPR (Information Commissioner's Office 2018). Therefore, it was not necessary to expressly include this exemption.

One of the most controversially discussed topics of data protection legislation, whether there should be an exemption for journalists, had also been revisited during the drafting process of the UK DPA 2018. Under the old legal framework, section 32 of the DPA 1998 provided an exemption for journalistic purposes meaning the 'processing... undertaken with a view to the publication by any person of any journalistic, literary or artistic material'. Significantly, this exemption has been reproduced and its scope widened under the DPA 2018. In contrast to section 32 DPA 1998, Schedule 2, part 5, para 26(3) of the DPA 2018 is wider as it stipulates that the disapplication of certain GDPR provisions for journalists will apply 'to the processing of personal data carried out for the special purposes, whether or not the data are being processed for a second or ancillary purpose.' It is crucial to mention that although the DPA 2018 contains new criminal data offences, explicit journalism public interest defences have also been introduced (Mc Cullagh 2018). When forming a belief that publication is in the public interest, a data controller must have regard to relevant codes of practice, namely the BBC Editorial Guidelines, the Ofcom Broadcasting Code and the Editors' Code of Practice (UK DPA 2018, schedule 2, part 5, para 26(5)).

Guidance on the interpretation of the exception for journalistic purposes under the old legal framework can be found in *Steinmetz v Global Witness* (2014) EWHC 1186 (Ch).[37] The judgment underlines that media companies wishing to uphold a publication will need that prove that there is a reasonable belief that the information is in the public interest at the time when the request is received (rather than at the time of publication).[38] In certain circumstances, this exemption can be expanded to citizen

47, where the CJEU held that this exception relates 'only to activities which are carried out in the course of private or family life of individuals, which is clearly not the case with the processing of personal data consisting in publication on the internet so that those data are made accessible to an indefinite number of people.' See also Ghezzi et al. (2014), p. 71.

[37]This case deals with a claim brought against the NG Global Witness for publication of corruption allegation against a mining conglomerate (BSGR) in Guinea. Steinmetz is one of the four individuals targeted by the report and founder of BSGR. He and three other individuals introduced a claim under section 7 DPA to obtain their personal data held by NG Global witness. Failing to answer these requests, Steinmetz and al. issued new proceedings seeking disclosure, deletion of personal data and damages. In turn, NG Global Witness argued the application of the exemption for journalistic purposes enshrined in section 32 of the DPA 1998. The ICO decided that campaigning non-governmental organisation (NGO) can rely on the 'journalism' exemption even if it is not a professional journalistic organisation. In essence, the scope of this exemption extends to anyone engaged in public interest reporting.

[38]See also Information Commissioner's Office (2014).

journalists such as bloggers,[39] and non-media organisations.[40] It is yet to be determined whether this jurisprudence will also be applied to the new data protection legislation.

Finally, the rights guaranteed under article 17 of the GDPR do not apply to the extent that processing is necessary:

- to exercise the right of freedom of expression and information;
- to comply with a legal obligation for the performance of a public interest task or exercise of official authority.
- for public health purposes in the public interest;
- for archiving purposes in the public interest, scientific research historical research or statistical purposes; or
- for the exercise or defence of legal claims.

The right to erasure and the right to be forgotten deriving from this regulation are further limited by the DPA 2018. In particular, the rights do not apply to personal data processed for the purpose of preventing or detecting crime, apprehending or prosecuting offenders or assessing or collecting taxes, to the extent that compliance would prejudice any of these purposes (Lloyd-Jones and Carey 2018, p. 146).

In essence, cases in relation to the right to be forgotten require balancing of two fundamental rights: freedom of expression (article 10 ECHR) and privacy right (article 8 ECHR). Thus, it will sometimes be necessary for controllers to carry out a balancing between, on the one hand, the data subject's interest in erasure and, on the other, competing legitimate interests of the controllers and others (Lloyd-Jones and Carey 2018, p. 146). In order to resolve possible conflicts between such rights of equal weight, member states are granted a certain margin of appreciation. This margin varies depending on the information shared. For example, McGoldrick (2013) argues that this margin of appreciation is likely to be greater in relation to archives of past events than news in relation to current events. In other words, some limitations to the application of the right to be forgotten might result from the balance struck between privacy/identity and freedom of expression (including freedom of information) in order to respect the proportionality principle. As further discussed below in Sect. 3.3, the case of *NT1 and NT2 v Google LLC*, which draws on both the judgment in *Google Spain* as well as the EU Article 29 Working Party's 'Guidelines, provides assistance on how to balance the competing rights and interests at stake.

In addition to these exemptions stated expressly in the GDPR, one must also note the ongoing case of *Google v CNIL*.[41] It concerns the territorial scope of the right to

[39]See for example *The Law Society and others v Kordowski* (2011) EWHC 3182 (QB). Mr Justice Tugendhat granted an injunction ordering the closing down of a website vilifying solicitors and law firms. As the journalism exemption enshrined in section 32 DPA relates to communication of information in the public interest and that today, any individual can engage in journalism, this exemption is not limited to traditional journalists.

[40]See in particular *The Law Society and Ors v Kordowski* (2011) EWHC 3185 (QB), para 134.

[41]See Case C-507/17 *Google v CNIL* OJ 2017/C-347/30.

be forgotten as established in *Google Spain* on the basis of the relevant provisions in Directive 95/46/EC. Although a decision has not been handed down at the time of writing, the Advocate General (AG) of the CJEU has recently published an opinion setting out his view on the most pressing issues of the case.

The background to this case is a disagreement between the French data protection authority (CNIL) and Google. In 2015, the CNIL served formal notice on the search engine provider requesting that 'when acceding to a request from a natural person for the removal of links to web pages from the list of results displayed following a search performed on the basis of that person's name, it must apply that removal to all of its search engine's domain name extensions'. In other words, the data protection authority is of the opinion that in order for the right to be forgotten to be effective, it requires erasure from all domains worldwide. Ultimately, the reason for serving this notice was the fact that Google only proceeded to delist results in relation to EU domains, such as Google.de or Google.fr, but not domains outside of the EU, such as Google.com (CJEU (Press release) (2019). However, Google refused to comply with that formal notice, arguing that there was no legal basis for extending this right globally.[42] After the CNIL imposed a penalty of €100,000 on Google for failing to comply with the formal notice, the search engine operator challenged the regulator's decision before the French administrative court, which then referred this matter to the CJEU for a preliminary ruling. Most importantly for the purposes of this section, one of the referred questions asks the CJEU to define the territorial scope of EU data protection law. These issues arose because the provisions of EU law applicable to the present case do not expressly govern the issue of the territorial scope of de-referencing. In January 2019, the AG published his view on the case.[43] For the sake of brevity, this chapter only touches on the issues that are most important to comprehend the limitations of the right to be forgotten.

Highlighting the significance of the balancing exercise that must be undertaken in cases concerning the right to be forgotten,[44] the AG proposed that Google should not be required to carry out the requested dereferencing on all the domain names of its search engine worldwide (AG 2019, para 79(1)). This conclusion is primarily based on the argument that if worldwide dereferencing were permitted, the EU authorities would not be able to define and determine a right to receive information, let alone balance it against the other fundamental rights to data protection and to privacy (AG 2019, para 60). Further, the AG argued that broadening the scope as suggested by the CNIL would run the risk of having a 'chilling effect' on the freedom of information. This is because the AG fears that if an authority within the EU could order a worldwide application of the right to be forgotten, this would prevent citizens in 'third states' from accessing information (AG 2019, para 61). If those 'third states' would then decide to limit EU citizen's access to information equally, this would have a negative effect on freedom of expression worldwide (AG 2019, para 61). Yet,

[42]For further detail see Finck (2018a).

[43]It should be noted that The Advocate General's Opinion is not binding on the Court of Justice.

[44]See Sect. 3 above.

it is crucial to highlight that the AG did not rule out the possibility that, in certain situations, Google may be required to take de-referencing actions at the worldwide level.[45] Nevertheless, he takes the view that the situation at issue in the present case does not justify this (AG 2019, para 62).

However, the AG also emphasised that once a 'right to de-referencing' within the EU has been established, the search engine operator must take every measure available to it to ensure full and effective de-referencing within the EU (AG 2019, para 79(2)). This includes the use of the 'geo-blocking' technique, in respect of an IP address deemed to be located in one of the Member States, irrespective of the domain name used by the internet user who performs the search (AG 2019, para 79(2)).

Although this case is concerned with the old data protection framework, the findings of this decision might also be relevant for the interpretation of the GDPR. This is because even though the GDPR contains provisions regarding its territorial scope in article 3, it remains somewhat unclear if, and if so, to what extent, organisations outside the EU/EEA are subject to the rules of the regulation.[46] Thus, depending on its exact findings, this case might be helpful for interpreting the scope of the right to be forgotten under the GDPR.

3 The UK Jurisprudence and the Impact of *Google Spain*

Historically, the UK judiciary has been rather reluctant to recognise a right to be forgotten, preferring to advance freedom of expression over the removal of content. However, this stance seems to be evolving slowly in response to *Google Spain* and is likely to continue doing so given the provisions in the GDPR. Recently, the High Court of England and Wales handed down its judgment in the case of *NT1 and NT2*, which marks the most high-profile consideration of the right to be forgotten in a common law jurisdiction following the decision in *Google Spain* (Costello 2018). This is because it is the first time that an English court directly applied the seminal ruling of the CJEU.

In order to set the scene for a comprehensive overview of cases relating to the practical application of the right to be forgotten in the UK, this section first provides a brief summary of the ruling in *Google Spain*. This includes an insight into how academic commentators have welcomed this decision. Subsequently, it examines the relevant UK jurisprudence, showing how the right to be forgotten has been conceptualised in previous case law and how this has changed over time.

[45]However, he did not specify which situations might trigger a need for a 'worldwide' right to be forgotten.

[46]For further detail see Welfare and Carey (2018), p. 6; European Data Protection Board (2018); Finck (2018b), pp. 27–28.

3.1 Reception of Google Spain *in the UK*

For the sake of brevity, the facts off this well-known case are not repeated here. In summary:

- The territorial scope of the Directive extends to search engine based outside the EU;
- Google was held to be a data controller under the DP Directive;
- The CJEU confirmed that the Directive's requirement that personal data is not retained for longer than it is justified equates to 'a right to be forgotten'.

This judgment was of utmost significance for the UK, primarily because domestic law at the time of the ruling did not have a similar right to be forgotten like the one stated in the ECJ decision. As discussed in Sect. 2, the right to be forgotten did not exist as such in UK law but derived from a more general framework for a right of privacy and data protection laws. Focusing on the DPA 1998 (transposing the data protection Directive), the provisions were stronger in places,[47] but weaker elsewhere, given their discretionary nature.

Nevertheless, traces of comparable forms of protection were to be found in English law even before the adoption of the GDPR.[48] There were parallels in analogue contexts, including insolvency laws, defamation laws, the retention and communication of information by the police relating to offenders,[49] and the Rehabilitation of Offenders Act 1974. The latter recognises that rehabilitation goals are advanced if, in certain instances at least, a defendant is not required to disclose any previous convictions.[50] O'Callaghan and de Mars (2016, pp. 42–56) also highlighted that the information which journalists may publish concerning the private details of ex-offenders was curtailed.[51] Consequently, prior to the *Google Spain* decision, privacy laws and data protection laws framing the 'right to be forgotten' were, and remain, distinct concepts in UK law, although overlapping in places (Walden 2011, p. 584).

Despite some sceptical reactions to the Commission's proposal to strengthen the right to be forgotten and the controversial interpretation by the CJEU in *Google Spain*, the goals underpinning the judgment were generally supported by UK

---

[47]See Sect. 2 in relation to destruction of data

[48]Insolvency laws, defamation laws, the retention and communication of information in relation to offenders by the police, and the Rehabilitation of Offenders Act 1974.

[49]See for example *R (Wood) v Commissioner of Police for the Metropolis* (2010) 1 WLR 123; *S and Marper v United Kingdom* (2008) ECHR 1581; *R (Ellis) v The Chief Constable of Essex Police* (2003) EWHC 1321 (Admin).

[50]For an application by courts, see for example: *R v Secretary of State for the Home Department* (2014) UKSC 35; *MM v UK* (2012) ECHR 1906.

[51]O'Callaghan and de Mars (2016) specifically refer to *X (a Woman Formerly known as Mary Bell) and Y v News group Newspapers Ltd, MGN Ltd* (2003) EWHC 1101; *Venables v News Group Newspapers Ltd and Others* (2001) EWHC 32 (QB)*; Thompson v News Group Newspapers Ltd and Others* (2001) Fam 430.

commentators. Primarily, there was an agreement that subjects should have greater control over their data and its processing by data controllers.[52] Yet, the Head of Policy of the ICO seemed to have mixed feelings. While agreeing with the principles set out in the decision, he has expressed concerns surrounding the practical difficulties that might arise.[53] The Government, and some commentators have been even more negative, emphasising not only these practical difficulties but also disagreeing with the CJEU on matters of principle.[54] One of the recurring criticisms of the ruling related to the appropriate balance between freedom of expression and the right to privacy. More specifically, some commentators contest that instead of mediating between these two fundamental rights, the CJEU had prioritised the right to privacy over freedom of expression. This is perceived as having a detrimental effect on the role of the media in a democratic society (Peers 2014).

Further, the majority of scholars have welcomed the *Google Spain* because it demonstrates that UK courts are, albeit cautiously, expanding the tort of misuse of private information over time.[55] As already mentioned above,[56] very few cases relate to private individuals. That being said, the Information Commissioner's Officer (ICO) has been active. In its annual report for 2016, the ICO stated that it had received 16,388 data protection complaints (Information Commissioner's Office 2016). During the same year, the ICO issued an enforcement notice to Google to remove nine search results (Information Commissioner's Office 2016). Initially, Google agreed to remove the search results from the European versions of its search engine but the ICO required the search results to be removed from all the versions of the search engine accessible from the UK territory. Having appealed this decision, Google eventually complied. Furthermore, the ICO has issued three preliminary enforcement notices for delisting. Overall, during 2015–2016, nearly 400 people sought the ICO's help to remove results under the right to be forgotten. The ICO requested to remove search results in about a third of these complaints. Around 125 complaints related to past criminal convictions. Here, the ICO agreed with the removal of the results, except in those cases relating to recent or serious convictions.

3.2 *Cases in the Aftermath of* Google Spain

Since the *Google Spain* decision, UK courts seem more receptive to privacy rights concerns. In *Hegglin v Persons Unknown* (2014) EWHC 2808 (QB), the Queen's Bench Division of the High Court dealt with a claim for an injunction under section 14 of the DPA 1998 relating to highly defamatory content posted anonymously on

[52]For a list of reactions in the aftermath of the Google Spain case, see Powles and Larsen (2015).

[53]See European Union Select Committee (2014, question 21).

[54]See European Union Select Committee (2014, question 34).

[55]See for example *Vidal-Hall v Google Inc* (2015) EWCA Civ 311.

[56]See Sect. 2.3.

various websites. While this case was ultimately settled out of court, the initial proceedings show the possible wide-reach of the right to be forgotten (at least its interpretation by UK courts), arguably beyond that envisaged in *Google Spain*. There seems little doubt that the judge would have found Google liable if the case had gone to trial.

Later in 2015, *Mosley v Google Inc*.[57] provided another indication of the courts' willingness to countenance the erasing of information. This case dealt with the making available, through Google's search engine, of sadomasochist sex tapes featuring the claimant. In the wake of *Google Spain,* Google was confirmed to fall within the definition of a data controller for the purposes of DPA 1998. This was even though Google argued that it was shielded from liability for content shared by third parties as a result of the 'safe harbour' provisions enshrined in the E-Commerce Directive. This case identifies an important point relating to the hierarchy of norms. As the judgment notes, either the DP Directive is self-contained, or the E-Commerce and the DP Directive must be interpreted in harmoniously. Although stating a personal preference for the latter, Mitting J. concluded that the claimant seemed to have: 'a viable claim which raises questions of general public interest, which ought to proceed to trial' (para 54).[58] It is further noteworthy that, since *Google Spain*, UK courts have entertained claims for damages under the DPA 1998 for distress where no pecuniary loss has been suffered.[59]

On balance, the criticisms tend to focus on the potential 'chilling effect' the judgment might have on the exercise of freedom of expression in particular cases.[60] This scepticism has to be put into its wider context as illustrated by O'Callaghan and de Mars (2016). Post-Leveson,[61] there is a sense that freedom of expression in the UK is under threat, and a greater commitment is required to preserve freedom of expression and freedom to receive information in preference to privacy rights (O'Callaghan and de Mars 2016, p. 59).[62] Scholars in this camp also concede that any possible chilling effect on freedom of expression derived from a right to be forgotten should not be exaggerated (O'Callaghan and de Mars 2016, p. 50). For example, they point out that the right to be forgotten does not prevent any individual from exercising their right to freedom of expression. The right to be forgotten simply recognises the possibility for some information to be removed when its retention is not justified.

[57]See *Mosley v Google Inc* (2015) EWHC 59 (QB).

[58]For further detail see Hurst (2015), p. 193.

[59]See *Vidal-Hall v Google Inc* (2015) EWCA Civ 311.

[60]This relates to the idea that there should be limits as to how much we should allow rewriting history.

[61]The Leveson inquiry is a judicial public inquiry examining the culture, practices and ethics of the British press following phone hacking scandal by the *News International* in 2007, chaired by Lord Justice Leveson. Most of the hearings took place in 2011–2012.

[62]See also de Baets (2016), p. 59.

In addition, a few cases under the DPA 1998 relating to the issues surrounding the right to be forgotten have reached UK courts. Yet, those few decisions available demonstrate that the courts are able to provide remedies in circumstances in which privacy actions would not be possible or would prove unsuccessful. Most of these tend to deal with clear-cut cases of unlawful data processing rather than borderline cases providing an opportunity for courts to explore the boundaries of the DPA 1998. For example, *The Law Society and Ors v Kordowski*,[63] dealt with a website which vilified hundreds of solicitors and their law firms. Here, the court departed from the position of the ICO and held that the DPA requires the ICO to consider how third parties use their right to freedom of expression. In the words of Justice Tugendhat:

> I do not find it possible to reconcile the views of the law expressed in the Commissioner's letter with authoritative statements of the law. The DPA does envisage that the Information Commissioner should consider what it is acceptable for one individual to say about another, because the First Data Protection Principle requires that data should be processed lawfully... As Patten J made clear in Murray, where the DPA applies, if processing is unlawful by reason of it breaching the general law of confidentiality (and thus any other general law) there will be a contravention of the First Data Protection Principle within the meaning of s.40 (1), and a breach of s.4(4) of the DPA...The fact that a claimant may have claims under common law torts, or under HRA s.6, does not preclude there being a claim under, or other means of enforcement of, the DPA. (*The Law Society and Ors v Kordowski, para 100)*

As noted above, the right to be forgotten under English law derives from different areas of law and there is a certain overlap between libel, harassment, breach of confidence and data protection in the digital world. Therefore, despite following a different regime, these actions can be seen as complementary to some extent to address the harm caused by the publication of the information.

3.3 NT1 and NT2 v Google LLC*: A Balance in Jeopardy?*

However, none of these decisions directly applied the findings made in *Google Spain*. Instead, the judgment in the case of *NT1 v NT2* handed down by the High Court in early 2018 marks the first delisting order made by a UK court compelling Google to remove links where it had rejected a request.[64] It is seen as perhaps the most high-profile consideration of this right in a common law jurisdiction following the decision in *Google Spain*.[65] Although this ruling was made before the introduction of the GDPR and the DPA 2018, i.e. under the old legal framework, it contains significant conclusions as to how UK courts should balance out the different interests involved in a right to be forgotten case. As the High Court was concerned with a

[63]See *The Law Society and Ors v Kordowski* (2011) EWHC 3185 (QB).

[64]See *NT1 and NT2 v Google LLC* (2018) EWHC 799 (QB).

[65]For further detail see Sect. 3.1 above.

balancing exercise similar to what is required under the GDPR,[66] this judgment is therefore likely to be relevant for the interpretation and application of the principles contained in article 17 of the regulation.

In essence, the case concerns two separate claims against Google, which were brought by two businessmen (anonymised as NT1 and NT2) who were previously convicted of criminal offences. The two trials took place sequentially in February and March of 2018, both before Mr. Justice Warby and with the claimants sharing the same legal team, which led to the joint judgment this section is referring to (Wilson 2018). Both claimants requested the removal by the defendant, Google, of several search results concerning their previous convictions on the basis that the results conveyed inaccurate, out of date and irrelevant information, failed to attach sufficient public interest and/or otherwise constituted an illegitimate interference with their right to be forgotten as established in *Google Spain*.[67] Thus, in accordance with *Google Spain,* the claimants each sought delisting orders under section 14 of the DPA 1998, which permitted erasure of personal data that is inaccurate, and compensation under section 13 of the DPA 1998 which gave effect to the Directive 95/46/EC (Costello 2018, p. 269). Additionally, both NT1 and NT2 demanded compensation for the tort of misuse of private information as a result of Google's conduct in continuing to return search results in the period following their complaints (Costello 2018, p. 269).

The court identified the main issues of the case to be first, whether the records in question needed correcting; second, 'whether the data protection or privacy rights of the claimants extend to having shameful episodes in their personal history eliminated from Google Search'; and third, whether damages should be paid (*NT1 and NT2 v Google LLC,* para 9).

In its assessment of these issues, the court started by ruling that NT1 did not provide sufficient evidence that the publications he complained of were inaccurate (*NT1 and NT2 v Google LLC,* paras 92–94). Thus, there was no duty to erase those statements under the DPA 1998. In contrast, the judge found that a national newspaper article that NT2 complained of contained inaccuracies in that it gave 'a misleading portrayal of the claimant's criminality', and therefore issued a delisting order in respect of that link (FitzPatrick et al. 2018, p. 936).

Subsequently, the High Court examined whether Google's processing complied with the requirements of *Google Spain* or if it unjustifiably interfered with the claimant's privacy rights. This included an evaluation of each of the 13 criteria published in the EU Article 29 Working Party's 'Guidelines on the Implementation of [Google Spain]'.[68] Noting that 'neither privacy nor freedom of expression has [...]

[66]For further detail see Sect. 2.3 below.

[67]For the full facts of the case, see *NT1 and NT2 v Google LLC* (2018) EWHC 799 (QB) paras 5–12.

[68]The 'Article 29 Working Party' was an advisory body made up of a representative from the data protection authority of each EU Member State, the European Data Protection Supervisor and the European Commission. On 25 May 2018, it has been replaced by the European Data Protection Board (EDPB). For further information see Article 29 Data Protection Working Party (2014).

precedence over the other', (*NT1 and NT2 v Google LLC,* para 132), the judge highlighted that the Working Party Guidelines identified the overall purpose of its criteria as assessing whether the information is relevant according to the interest of the general public (*NT1 and NT2 v Google LLC,* para 142). Thus, in addition to assessing whether there is evidence showing that the availability of the search results is causing prejudice to the claimant's privacy rights,[69] the judge was also required to balance the claimant's rights against the public interest. Significantly, the Guidelines note that whether the claimant was still engaged in the same professional activity and was a public individual would be particularly relevant in such an assessment (Costello 2018, p. 274).

Based on this legal framework, the court decided to dismiss NT1's delisting claim because he had failed to satisfy the criteria established in *Google Spain*. Further, the judge rejected NT1's claim for misuse of private information. These decisions were primarily based on the argument that the claimant was a 'public figure' with a role in 'public life (*NT1 and NT2 v Google LLC,* para 138). In fact, the court had adopted the very broad definition of 'public figure' put forward by the EU Working Party. This definition includes anyone with a degree of media exposure, members of the regulated professions, and business figures (Wilson 2018). Thus, search results concerning NT1's previous convictions and his punishment could not be considered of a private nature but was regarded as a matter of public interest, specifically a business crime, its prosecution and punishment (*NT1 and NT2 v Google LLC,* para 140). This finding was underpinned by the argument that because NT1 continued to conduct business activities including lending money to businesses and individuals, and the information in question was still relevant to a (limited) number of people with a legitimate interest (FitzPatrick et al. 2018, p. 936). Moreover, the judge stressed that NT1 had not accepted his guilt, had misled the public and the court, and shown no remorse (*NT1 and NT2 v Google LLC,* para 170). Also, Justice Warby held that NT1's case suffered from a lack of causation in establishing whether the interference with his privacy rights would have occurred irrespective of Google's actions (*NT1 and NT2 v Google LLC,* para 151). In other words, NT1 did not make a compelling case regarding the harm suffered as a result of the continued processing of the links. This is significant considering *Google Spain's* finding that *if* existing evidence shows that the availability of a search result is causing prejudice to the claimant's rights, this would be a strong factor in favour of delisting. However, due to the facts of this decision, NT1's delisting claim was dismissed. Subsequently, the claimant appealed to the Court of Appeal but eventually agreed to settle the case before it was heard (Corfield 2018).

In contrast, the High Court upheld NT2's claim for misuse of private information in addition to his delisting claim. According to the judge, NT2's situation was different because his criminal conviction did not involve dishonesty and his punishment had been based on a plea of guilt (*NT1 and NT2 v Google LLC,* para 203). Thus, Justice Warby argued that NT2's role as a public figure had been much less

---

[69] Which would speak in favour of delisting.

significant as his previous criminal conduct was of little relevance to his current activities and there was no plausible risk of his re-offending (FitzPatrick et al. 2018, p. 937). In contrast to NT1, he also established a stronger case on the harm suffered both to his business and family as a result of the continued availability of the links (*NT1 and NT2 v Google LLC,* paras 221–222). Hence, applying the guidance provided by Google Spain and the Working Party, the court found that information about the crime and its punishment had become out of date, irrelevant, and of no sufficient legitimate interest to users of Google to justify its continued availability (*NT1 and NT2 v Google LLC,* para 223). However, despite upholding NT2's claim for misuse of private information, the court held that he was not entitled to damages or compensation. The judge took this view because he found that Google had taken reasonable care (*NT1 and NT2 v Google LLC,* paras 227, 228, 230).

In addition to the helpful commentary as to the application of the Working Party criteria, this judgment contains significant conclusions as to how UK courts should balance out the different interests involved in a right to be forgotten case *post-Google Spain*. As noted above, such a balancing exercise is also required under article 17 of the GDPR. One of the main findings resulting from the High Court judgment is therefore that delisting claims will primarily depend on the underlying facts as the balancing exercise will turn on the particular circumstances of the individual claimant (FitzPatrick et al. 2018, p. 937). Further, in cases concerning criminal offences, the claimant's performance as a witness and attitude to their previous conduct is likely to be taken into account.[70] In essence, the ruling offers a tentative first step towards clarifying the criteria for a delisting order in cases involving criminal convictions and offers a significant endorsement of the right to be forgotten in such cases (Costello 2018, p. 281). Thus, if traditionally, the UK seemed to prioritise freedom of expression over privacy rights, post-*Google Spain*, the current signs emerging is that UK courts have demonstrated a willingness to accept the notion of digital 'forgetting'. This argument can be underpinned by the High Court's finding that there is no presumption in favour of either party when engaging in the balancing exercise between the data subjects' 'right to be forgotten' and the rights of internet search engines in delisting claims.

4 Concluding Remarks: Strengthening the Right to Be Forgotten Going Forward?

This chapter aimed to illustrate the trajectory of the present status of UK and (still) relevant EU law relating to the right to be forgotten, and particularly, attempted to unveil the significance of the changes caused by the GDPR. It has shown that a wide range of remedies is available to allow for the respect of the proportionality principle and striking a fair balance between the various competing rights. Such remedies

[70]For further detail see FitzPatrick et al. (2018), p. 937.

range from injunctions to retain the defendant from further publication of the infringing data to pecuniary damages, anonymisation of the data, de-listing, removal, blocking or destruction of the data.

Overall, the EU's measures to strengthen the right to be forgotten are to be welcomed. However, additional efforts from legislators and courts are required to define the outer boundaries of this right to ensure it strikes an appropriate balance between freedom of expression (including the freedom to receive information) on the one hand and digital privacy interests on the other. Also, clarification is necessary regarding the right to freedom to conduct a business, which intermediaries should enjoy.

The right to be forgotten raises important concerns as to how private entities, such as Google, are competent to strike a fair balance in these cases where fundamental rights are in conflict. It is suggested that data controllers should be provided with further guidelines based upon developing ECtHR jurisprudence to ensure for example, that the proportionality principle is respected. For example, it is not self-evident how the outcome should depend upon whether the data subject is a politician, a celebrity, or a private individual? What distinction should be made between the publication of true or false data? Can we trust such nuanced and difficult balancing to a private entity? The obvious first step is to ensure transparency in their decision-making process, combined with the ability to have courts oversee and review particular cases. It must also be kept firmly in mind that private commercial entities are not independent. It cannot be assumed that their interests are aligned with those of data subjects or the public interest.

The potentially chilling effects on journalism should not be overlooked either. Despite the exemption provided under the updated data protection legislation, there might be uncertainties surrounding qualification as 'a journalist', which the legislation does not attempt to define. Given the growing prominence of 'citizen journalism', this represents a potential barrier to the efficient operation of the right to be forgotten. Furthermore, it would seem impossible for a data controller to predict which information might, in future, prove to be of historical interest despite this being significant in the decision whether, or not, to remove it.

This leads to a further concern linked to the sheer number of requests which data controllers must process on a daily basis. This suggests that the freedom to conduct a business ought to feature more predominantly in the debate. Search engine companies are required to determine an enormous number of requests, but this task is far from straightforward since it requires assimilation of potentially complex body facts and exercise of considered judgment against a background of ever-evolving law. If private entities are to go discharge these duties with more than a cursory assessment, then it is essential that they are provided with additional guidance. This applies equally in relation to the data controllers' role of establishing whether data should remain because it falls within one of the purposes of article 17(3) GDPR. In particular, the extent to which algorithmic, or other automated anti-piracy systems, is fit for this purpose should be clearly spelt out.

Overall, there exists a broad academic consensus in the UK academic landscape that Google should release more information surrounding its day-to-day handling of right to be forgotten requests.[71] In practice, it remains unclear how this private actor actually decides where the balance of fundamental rights lies in specific cases when determining whether to delist a search engine result or not. Indeed, Google's own Advisory Council on the right to be forgotten advocates more transparency (The Advisory Council to Google on the Right to be Forgotten 2015, p. 20).

However, it should be acknowledged that the Google Transparency Report has been extended to include data concerning requests seeking to exercise the right to be forgotten in the EU. This first step towards greater transparency is welcomed. For example, based on this information, we have learnt that between 29th May 2014 and 18th May 2017, Google has received a total of 724,269 requests and has evaluated 2,041,921 URLs for possible removal (Google 2017). The report further revealed that Google has received a total of 107,840 requests from the UK within that timeframe (Google 2017). This represents approximately 14% of the total of requests submitted from all the EU member states.

In this context, it should be noted that the old data protection laws were criticised for being little known amongst the general public.[72] O'Callaghan and de Mars (2016, p. 45) argue that the apparent under-use of data protection provisions was attributable to the fact that the domestic data protection laws derived from international obligations rather than domestic initiatives. Ghezzi et al. (2014, p. 72) also stress that further clarification as to the scope of the exemptions would have rendered the Directive, and hence the DPA 1998, more effective. As they say:

> There is an urgent need to clarify the rules applying to data processing by individuals for private purposes and, moreover, its compatibility with the data protection rules derogation for the processing of data carried out solely for journalistic purposes and protected by the right to freedom of expression (Art. 9 DPD). What is to be understood by 'purely personal purposes'? Does the posting of information on a social networking site equate to the disclosure of information for private purposes, that is, to our private (although admittedly large) group of selected contacts? Or does it equate to disclosure of information to the public? If so, then there is another 'twist' one should take into account. Private individuals who disclose information, opinions or ideas to the public – for example through SNS, blogs, *YouTube* or *Twitter* – would then be protected by the freedom of expression, receiving the same treatment as media professionals processing data 'solely for journalistic purposes or the purpose of artistic or literary expression' (Ghezzi et al. 2014, p. 72).

It is yet to be determined whether the GDPR and the UK DPA 2018 are likely to change this status quo. However, considering the high media profile of regulatory or court decisions concerning the GDPR,[73] it seems reasonable to assume that the general public is more likely to be aware of their rights originating from this (international) legislation than they were before.

[71]See the letter sent by 80 academics and shared with The Guardian in Kiss (2015). The letter is available at https://medium.com/@ellgood/open-letter-to-google-from-80-internet-scholars-release-rtbf-compliance-data-cbfc6d59f1bd.

[72]See for example Leveson (2012), p. 999; Hurst (2015), p. 187.

[73]As an example for the recent international media coverage of CNIL's 'record fine' against Google, see Hern (2019).

In any case, it is welcomed that the right to be forgotten is enshrined in a regulation rather than a directive as this engenders greater uniformity. Under the old legal framework, harmonisation was hampered as some data type is subject to removal in one member state, but not in another. These differences in the interpretation of the right to be forgotten were inevitable and raised concerns.

Further, one should consider the tensions created between the EU and the US on this issue. The EU right to be forgotten is likely to cut across the US First Constitutional Amendment, which interferes with combating cross-jurisdictional Internet-based crime.[74] The *Google Spain* decision underlines a pressing need to specify the territorial scope of the right to be forgotten. Part of EU law has extra-territorial reach, yet given the lack of unanimous consensus on the right to be forgotten, this is likely to create difficulties of enforcement in the US. The goals sought by this right might be thwarted in a global and digital world as a result of differences in legal approaches to the interaction of freedom of expression with other fundamental rights.

Regard must be had to the complementary objectives of improved data security and reduction of data vulnerability. These seem to go hand in hand with the goals sought by the right to be forgotten. After all, ensuring individuals are more in control in data shared online and have more say on how it is processed by data controllers underlies the right to be forgotten in recognition of the sheer volume of data available online and its increasing vulnerability. Not only can data be hacked by criminals, shared with the government and subject to commercial exploitation, but it can be misused in other ways. It is worth considering whether data vulnerability should be tackled by better enforcement mechanisms, including larger penalties, as Bernal (2018) has suggested.

Finally, a further consideration in the hierarchy of legal norms is required, particularly in respect of the GDPR, the E-Commerce Directive (and the safe harbour provisions),[75] and the Copyright Directive.[76] Whilst some of this is considered to some extent in relation to the E-Commerce Directive by virtue of recital 21 and article 2(4), it seems likely that Stalla-Bourdillon (2017) is correct in saying that this treatment is over-simplistic.

In conclusion, we do not believe the 'right to be forgotten' should be further strengthened *before* certain existing issues are resolved. Namely, the international interaction between the GDPR and the US legislation, additional guidance for search engines to comply with human rights consideration and preserve freedom of expression/freedom of receiving information adequately, and the interplay with other pieces of legislation such as the E-Commerce Directive and the proposed new copyright Directive (McCarthy 2016, p. 360).

From a UK perspective, it will be interesting to see how the jurisprudence relating to the right of to be forgotten is going to develop in light of the latest reforms in both domestic and international law. Particularly, balancing out the different fundamental

[74]See for example Rosen (2012).

[75]See Directive 2000/31/EC (2000), pp. 1–16.

[76]See Directive 2001/29/EC (2001), pp. 10–19; see also European Commission (2016).

rights and interests at stake is likely to remain a difficult task and, according to the latest High Court judgment, will be subject to a case-by-case analysis.

References

Advocate General (2019) C-507/17 *Google v CNIL* Opinion of AG [2019] ECLI:EU:C:2019:15, Opinion of AG Szupnar

Aplin T (2007) The future of breach of confidence and the protection of privacy. Oxf Univ Commonwealth Law J 7(2):137–173

Article 29 Data Protection Working Party (2014) Guidelines on the implementation of the CJEU judgment on "Google Spain and Inc v. Agencia Espanola de Proteccion de Datos (AEPD) and Mario Costeja Gonzalez" C-131/12. https://ec.europa.eu/justice/article-29/documentation/opinion-recommendation/files/2014/wp225_en.pdf. Accessed 20 Feb 2019

Bernal P (2018) A right to delete? Eur J Law Technol 2(2). http://ejlt.org/article/view/75/144. Accessed 20 Feb 2019

Boulanger M (2010) United Kingdom – implementation of Directive 95/46/EC (European Commission DG JFS, Unit C3: Data Protection, 2010). http://amberhawk.typepad.com/files/dp_infraction_reasons.pdf. Accessed 20 Feb 2019

Carey P (2018) Introduction. In: Carey P (ed) Data protection – a practical guide to UK and EU law. Oxford University Press, New York, p xxxiv

CJEU (Press release) (2019) Advocate General Szpunar proposes that the Court should limit the scope of the de- referencing that search engine operators are required to carry out to the EU. https://curia.europa.eu/jcms/upload/docs/application/pdf/2019-01/cp190002en.pdf. Accessed 20 Feb 2019

Corfield G (2018) Google settles Right to Be Forgotten case on eve of appeal hearing. Available via The Register. https://www.theregister.co.uk/2018/12/20/google_settles_nt1_right_to_be_forgotten_lawsuit/. Accessed 20 Feb 2019

Costello R (2018) The Right to be forgotten in cases involving criminal convictions NT1 and NT2 v Google and The Information Commissioner [2018] EWHC 799 (QB). Eur Hum Rights Law Rev 3:268–282

Council of Europe (1981) Convention for the protection of individuals with regard to automatic processing of personal data. http://www.coe.int/en/web/conventions/full-list/-/conventions/treaty/108. Accessed 20 Feb 2019

de Baets A (2016) A historian's view on the right to be forgotten. Int Rev Law Comp Technol 30 (1-2):57–67

European Commission (2016) Proposal for a Directive of the European Parliament and of the Council on copyright in the Digital Single Market COM/2016/0593 final – 2016/0280 (COD)

European Data Protection Board (2018) Guidelines 3/2018 on the territorial scope of the GDPR (Article 3) - Version for public consultation. https://edpb.europa.eu/sites/edpb/files/consultation/edpb_guidelines_3_2018_territorial_scope_en.pdf. Accessed 20 Feb 2019

European Union Select Committee (2014) Inquiry on the right to be forgotten evidence Session No. 3. http://data.parliament.uk/writtenevidence/committeeevidence.svc/evidencedocument/eusub-f-home-affairs-health-and-education-committee/the-right-to-be-forgotten/oral/11381.html. Accessed 20 Feb 2019

Finck M (2018a) Google v CNIL: defining the territorial scope of European Data Protection Law. Available via Oxford Business Law Blog. https://www.law.ox.ac.uk/business-law-blog/blog/2018/11/google-v-cnil-defining-territorial-scope-european-data-protection-law. Accessed 20 Feb 2019

Finck M (2018b) Blockchains and data protection in the European Union. Eur Data Protect Law Rev 18(1):17–35

FitzPatrick P et al (2018) High Court of England and Wales considers 'right to be forgotten' for the first time. J Intellect Prop Law Pract 13(12):935–937

Garstka K, Erdos D (2017) Hiding in plain sight: right to be forgotten and search engines in the context of international data protection frameworks. In: Belli L, Zingales N (eds) Platform regulators – how platforms are regulated and how they regulate US. FGV Direito Rio Edition, Rio de Janeiro, pp 127–146

Ghezzi A, Pereira AG, Vesnić-Alujević L (2014) The ethics of memory in a digital age: interrogating the right to be forgotten. Palgrave Macmillan, Basingstoke, p 17, 35, 67, 72, 95

Google (2017) Transparency report. https://www.google.com/transparencyreport/removals/europeprivacy/?hl=en. Accessed 20 Feb 2019

Hern A (2019) Google fined record £44m by French data protection watchdog. https://www.theguardian.com/technology/2019/jan/21/google-fined-record-44m-by-french-data-protection-watchdog. Available via The Guardian. Accessed 20 Feb 2019

House of Lords European Union Committee (2014) EU data Protection law: a 'right to be forgotten'? (2nd Report of Session 2014-15 of the European Union Committee). https://publications.parliament.uk/pa/ld201415/ldselect/ldeucom/40/40.pdf. Accessed 20 Feb 2019

Hurst A (2015) Data privacy and internet liability: striking a balance between privacy, reputation, innovation and freedom of expression. Entertain Law Rev 26(6):189

Information Commissioner's Office (2014) Data protection and journalism: a guide for the media: pp 27–39. https://ico.org.uk/media/for-organisations/documents/1552/data-protection-and-journalism-media-guidance.pdf. Accessed 20 Feb 2019

Information Commissioner's Office (2016) Information Commissioner's Annual Report and Financial Statements 2015/16. https://ico.org.uk/media/about-the-ico/documents/1624517/annual-report-2015-16.pdf. Accessed 20 Feb 2019

Information Commissioner's Office (2018) Exemptions. https://ico.org.uk/for-organisations/guide-to-data-protection/guide-to-the-general-data-protection-regulation-gdpr/exemptions/. Accessed 20 Feb 2019

Kiss J (2015) Google must be more open on 'right to be forgotten, academics warn in letter. Available via The Guardian. https://www.theguardian.com/technology/2015/may/14/google-right-to-be-forgotten-academics-letter. Accessed 20 Feb 2019

Leveson B (2012) An inquiry into the culture, practices and ethics of the press. Stationary Office, London, p 999

Lloyd-Jones H, Carey P (2018) The rights of individuals. In: Carey P (ed) Data protection – a practical guide to UK and EU law. Oxford University Press, New York, p 146

Mc Cullagh K (2018) The UK Data Protection Act 2018. In: E-conference, National Adaptations of the GDPR. Available via Blog Droit Europeen. https://blogdroiteuropeen.files.wordpress.com/2018/06/karen.pdf. Accessed 20 Feb 2019

McCarthy HJ (2016) All the World's a stage: the European right to be forgotten revisited from a US perspective. J Intellect Prop Law Practice 11(5):360

McGoldrick D (2013) Developments in the right to be forgotten. Hum Rights Law Rev 13 (4):761–776

Moreham N (2005) Privacy in the common law: a doctrinal and theoretical analysis. Law Q Rev 121(Oct):628–656

O'Callaghan P, de Mars S (2016) Narratives about privacy and forgetting in English law. Int Rev Law Comp Technol 30(1–2):42–56

Peers S (2014) The CJEU's Google Spain Judgment: failing to balance privacy and freedom of expression. Available via EU Law Analysis. http://eulawanalysis.blogspot.co.uk/2014/05/the-cjeus-googlespain-judgment-failing.html. Accessed 20 Feb 2019

Powles J, Larsen R (2015) Academic commentary: Google Spain - Compiled by Julia Powles and Rebekah Larsen. http://www.cambridge-code.org/googlespain.html. Accessed 20 Feb 2019

Rosen J (2012) The Right to Be Forgotten. Stanford Law Review Online. https://www.stanfordlawreview.org/online/privacy-paradox-the-right-to-be-forgotten/. Accessed 20 Feb 2019

Select Committee on Communications (2015) Press regulation: where are we now? (HL 2014–15, 135). https://publications.parliament.uk/pa/ld201415/ldselect/ldcomuni/135/135.pdf. Accessed 20 Feb 2019

Stalla-Bourdillon S (2017) The GDPR, the proposed copyright directive and intermediary liability: one more time! Available via INFORRM. https://inforrm.wordpress.com/2017/03/16/the-gdpr-the-proposed-copyright-directive-and-intermediary-liability-one-more-time-sophie-stalla-bourdillon/. Accessed 20 Feb 2019

The Advisory Council to Google on the Right to be Forgotten (2015) Report of the Advisory Committee to Google on the Right to be Forgotten. https://drive.google.com/file/d/0B1UgZshetMd4cEI3SjlvV0hNbDA/view. Accessed 20 Feb 2019

The European Parliament and The Council of the European Union (1995) Directive 95/46/EC (1995) of the European Parliament and of the Council of 24 October 1995 on the protection of individuals with regard to the processing of personal data and on the free movement of such data OJ L 281/31

The European Parliament and The Council of the European Union (2000) Directive 2000/31/EC of the European Parliament and of the Council of 8 June 2000 on certain legal aspects of information society services, in particular electronic commerce, in the Internal Market ('Directive on electronic commerce') OJ L 178, pp 1–16

The European Parliament and The Council of the European Union (2001) Directive 2001/29/EC of the European Parliament and of the Council of 22 May 2001 on the harmonisation of certain aspects of copyright and related rights in the information society OJ L 167, pp 10–19

UK Government (2013) Defamation Act 2013 Explanatory Notes. http://www.legislation.gov.uk/ukpga/2013/26/pdfs/ukpgaen_20130026_en.pdf. Accessed 20 Feb 2019

UK Government (2018a) Data Protection Act 2018 Explanatory Notes. http://www.legislation.gov.uk/ukpga/2018/12/pdfs/ukpgaen_20180012_en.pdf. p 11. Accessed 20 Feb 2019

UK Government (2018b) Policy paper digital charter. https://www.gov.uk/government/publications/digital-charter. Accessed 20 Feb 2019

Walden I (2011) Privacy and data protection. In: Reed C (ed) Computer law. Oxford University Press, Oxford, pp 584–585

Welfare D, Carey P (2018) Territorial scope and terminology. In: Carey P (ed) Data protection – a practical guide to UK and EU Law. Oxford University Press, New York, p 6

Wilson I (2018) NT1 and NT2 v Google Inc: How to seek the delisting of search engine results following the first English decision on the "right to be forgotten". Available via INFORRM. https://inforrm.org/2018/04/20/nt1-and-nt2-v-google-inc-how-to-seek-the-delisting-of-search-engine-results-following-the-first-english-decision-on-the-right-to-be-forgotten/. Accessed 20 Feb 2019

A Turkish Law Perspective on the *"Right to Be Forgotten"*

Kadir Berk Kapancı and Meliha Sermin Paksoy

Abstract Since the rule is changed from easily forget to easily remember thanks to the digital Internet platforms, the question if there should be a right to be forgotten consecrated to individuals has become a frequently asked question with—of course—no "one correct answer" in different legal systems all around the world. Turkish legal practice has also encountered different cases where the question if a right to be forgotten should be recognized/accorded or not. Thus, the concept itself and discussions thereon prove to be heated nowadays among legal scholars. Accordingly, this article mainly aims to legally analyze the newly emerging concept "*the right to be forgotten*" and its potential practical impacts to the existing or future Internet technologies, in light of the Turkish legislations and existing case law.

1 Introduction

Turkish Law does not explicitly and separately regulate "*the right to be forgotten*" under a specialized body of legislation however it still exists in the legal practice.[1] This can be deduced from the general body of legislation and with a broad interpretation angle, from data protection law legislations; furthermore, the existence of this right is also confirmed by different high court decisions both handed in the afterwards of European Court of Justice (ECJ) ruling with regard to the right to be forgotten. Moreover, in Turkey the existence and the practice of the right to be

[1]Develioğlu (2017), p. 93; Korkmazer (2016), p. 127; Akkurt (2018), p. 85; Bozkurt (2014), p. 97; Karan (2018), p. 4220.

K. B. Kapancı (✉)
MEF University (Faculty of Law, Department of Civil Law), İstanbul, Turkey
e-mail: kapancib@mef.edu.tr

M. S. Paksoy (✉)
Altinbas University (Faculty of Law, Department of Civil Law), İstanbul, Turkey
e-mail: sermin.paksoy@altinbas.edu.tr

F. Werro (ed.), *The Right To Be Forgotten*, Ius Comparatum – Global Studies in Comparative Law 40, https://doi.org/10.1007/978-3-030-33512-0_11

forgotten is not limited to the Internet, but it is applicable in any type of publicly available source such as printed publications.[2]

2 Legal Framework of the Right to Be Forgotten in Turkey

Under Turkish law, protection of the right to be forgotten is threefold. Before any specific regulation was enacted, scholars argued that the right to be forgotten could be derived from articles of Turkish Civil Code protecting general right of personality.[3] The right to be forgotten is accepted as a part of general right of personality since it is a manifestation of the right to informational self-determination and also of the right to privacy.[4] Furthermore, widespread usage of Internet as a violation tool has triggered enactment of Law No. 5651 on Regulating Broadcasting in the Internet and Fighting against Crimes Committed through Internet Broadcasting. This law sets accelerated existing protection mechanisms for those whose personality rights have been infringed through publications in Internet and envisages criminal punishment for those who do not comply with court orders.[5] So far, the right to be forgotten has been protected under the umbrella of general right of personality.

Eventually, after a very long drafting process Turkish Data Protection Law No.6698 was enacted in the year 2016. This Law, based on the European Union's Data Protection Directive 95/46/EC,[6] includes a more specific regulation on this issue. However, in its text "*the right to be forgotten*" is not explicitly mentioned.

Article 7 of this Law is as follows: "*Personal data which are processed in accordance with this law or relevant other laws shall be deleted, destroyed or anonymized either ex officio or upon request by the data subject in case the reasons necessitating their processing cease to exist*".[7] Parallel to this, Article 11 states that "*Data subject has a right to request deletion or destruction of personal data within the framework of the conditions set forth under Article 7*".[8]

---

[2]See Sect. 6.2 below.

[3]Sözüer (2017), p. 167; Akkurt (2016), p. 2613; Sağlam (2017), p. 694.

[4]Aksoy (2008), p. 241; Önok (2017), p. 159; Şerefoğlu Henkoğlu (2017), p. 193; Akkurt (2016), pp. 2614–2615; Akkurt (2018), p. 90; Çelik (2017), pp. 396–397.

[5]Akkurt (2018), p. 85; Elmalıca (2016), p. 1624; Bozkurt (2014), p. 97; Karan (2018), p. 4221.

[6]Directive 95/46/EC of the European Parliament and of the Council of 24 October 1995 on the protection of individuals with regard to the processing of personal data and on the free movement of such data.

[7]Some turkish scholars argue that this provision regulates the right to be forgotten, although the name of it is not directly mentioned (See Çırak 2018, p. 164; Duman 2017, p. 235; Bozkurt Yüksel 2016, pp. 36–37). According to the opposing view, the said provision rather foresees a right to erasure which is more different than the former (See Develioğlu 2017, p. 93; Sözüer 2017, pp. 38 et seq.).

[8]Sözüer (2017), pp. 181–182; Karan (2018), p. 4221.

Under the umbrella of this law, it is the Board of Protection of Personal Data who decides which data should be deleted. This board consists of nine members, five of which are appointed by the Turkish parliament, The President and the cabinet appoint remaining members by naming two members each.[9]

Regulation (relevant to the Law No. 6698) on Protection of Personal Data has also a very similar article on the right to be forgotten. According to Article 9 of this Regulation, *"personal health data which are processed in accordance with Law No. 6698 or relevant other laws should be deleted or anonymized upon request by the data subject in case the reasons necessitating their processing cease to exist".*

Though it has no direct application, Article 20 of Turkish Constitution, regulating the right to private life and privacy, contains the very foundations of the right to be forgotten as it states that *"everyone has the right to request the protection of his/her personal data. This right includes (. . .) requesting the deletion of his/her personal data".*[10]

Thus, in Turkish law, the data subject can enjoy a threefold protection as stated above. There are specific remedies granted by these protection options.

3 Available Remedies for the Enforcement of the Right to Be Forgotten Under Turkish Law

In Turkish law, there are several compensatory and coercive remedies to enforce the right to be forgotten.

According to Article 11 of Turkish Data Protection Law, the data subject may request compensation for the damages in case he/she incurs damages due to unlawful processing of personal data.[11] At the same time, as it is indicated in Article 14 paragraph 3 of this law, the data subject whose personality rights are infringed through the rejection of the erasure request may ask for immaterial and material damages according to Turkish Civil Code Article 25 and Turkish Code of Obligations Article 58 (only immaterial damages). Under the Article 25, the data subject may also ask for disgorgement of profits (if there is any).[12]

Turkish Civil Code Article 25 and Turkish Data Protection Law enable the plaintiff to receive material damages. Nevertheless, it is very difficult for the plaintiff to prove the amount of his/her material damage and the causality between the damages claimed and the violation of the defendant.[13] That is why this option is

[9]See also Develioğlu (2017), p. 152 et seq. for more detailed information with regard thereof.

[10]Şerefoğlu Henkoğlu (2017), p. 196; Küpeli (2016), p. 232; Akkurt (2016), p. 2613; Çelik (2017), p. 395; Sağlam (2017), p. 694; Nair and Balta (2017), p. 122; Sözüer (2017), p. 164 et seq.; Karan (2018), p. 4222; Ahi (2014), Bozkurt Yüksel (2016), p. 36; Sumer (2016), p. 777.

[11]Karan (2018), pp. 4221–4222.

[12]Develioğlu (2017), pp. 135–136; Sağlam (2017), p. 698.

[13]Develioğlu (2017), p. 134.

almost never used. Additionally, fault, be it in form of willful conduct or negligence (at any level), must exist. On the other hand, it is also possible for the plaintiff to ask compensation for his/her immaterial damages according to Article 25 of Turkish Civil Code and Article 58 of Turkish Code of Obligations.[14] Compared to the material damages, it is easier to obtain immaterial damages. However in this case, the plaintiff has also to prove that resistance/refusal to his/her request of erasure constitutes a violation of his/her right to be forgotten which is a manifestation of his/her general right of personality. Moreover, in a similar way, existence of the fault (willful conduct or negligence) of the defendant is required which means the defendant should have resisted to the lawful erasure request of the plaintiff.[15] For example, in a decision of General Assembly of the Civil Chambers of the Court of Cassation (N. 2014/4-56 - 2015/1679 and dated 17/05/2015) plaintiff could recover her immaterial damages.[16]

If the data controller rejects request of erasure, the data subject may consult to several coercive remedies. At this point, the data subject may file a complaint to the Board of Data Protection to have an inspection started. As a result of this inspection if the Board decides that a violation exists, it will order the data controller to eliminate the identified illegalities. Only in extraordinary cases when serious or irreparable losses may occur and illegality clearly exists, the Board may decide processing of data to be ceased. If the data processor rejects the data subject's request and then does not comply with the order of the Board, he/she may receive up to two years of incarceration punishment according to Article 138 of Turkish Criminal Code.[17]

In a similar way, if personal data of the data subject are processed or published in internet, according to Article 9 of Law No. 5651 on Regulating Broadcasting in the Internet and Fighting against Crimes Committed through Internet Broadcasting, the data subject may ask the Criminal Court of Peace to order limitation of the access to the relevant data. Content and service providers who do not comply with the order of the criminal court of peace will face criminal punishment.[18]

As an alternative, the data subject may ask the civil courts to prohibit a threatened infringement or to order cessation of the existing infringement under the Article 25 of Turkish Civil Code. In particular, he/she may also request that the correction or the judgment be notified to third parties or published.[19]

[14]Develioğlu (2017), p. 135; Sağlam (2017), p. 698.

[15]Küpeli (2016), p. 233.

[16]See Sect. 6.2 below.

[17]Küpeli (2016), p. 233; Şerefoğlu Henkoğlu (2017), p. 196; Nair and Balta (2017), p. 122; Develioğlu (2017), p. 125; Sağlam (2017), p. 695; Ahi (2014), Bozkurt Yüksel (2016), p. 36.

[18]Küpeli (2016), p. 234; Nair and Balta (2017), p. 122; Karan (2018), p. 4221.

[19]Develioğlu (2017), pp. 130–134.

4 Balance of Different Interests: Limits of the Right to Be Forgotten in Turkey

The right to be forgotten is firstly limited by specific provisions enabling continuous processing of individual data. Since there is not any specific provision, then in each concrete case a balance has to be struck between the public interest, freedom of expression, freedom of the media and the data subject's right to be forgotten.[20] If in the concrete case public interest, freedom of the media and freedom of speech outweighs the right to be forgotten of the data subject, courts will abstain from ordering delinking of the relevant information. In a similar way, for example if the data subject is a public figure and if the public has an undeniable interest to know the relevant data, courts will not grant the right to be forgotten.[21]

To balance these conflicting interests, courts make sure the relevant processing of data is accurate, adequate, relevant and not excessive to the purposes for which they are collected.[22] According to Turkish Constitutional Court, in their case-by-case assessment, courts must also consider content, accuracy, up-to-dateness, date and duration of this publication, historical, scientific or statistical value of the relevant news, contribution of the relevant news to the public interest (value of the news for the public and its relevancy for the future), the data subject's social position, fame and job, subjectivity and objectivity of the publication and interest of the public for the relevant publication before they accept or reject the data subject's request of erasure or delinking.[23]

Very recently Turkish Data Protection Board has published short summary of a relevant case (the full and detailed version of the decision is not provided yet) in which Board has contemplated on the limits of the right to be forgotten. In this case applicant asked the erasure of an opinion written by a columnist. The Board has rejected the request of the applicant since he is still in a position of public interest and therefore opinion expressed by the columnist should the respected as a reflection of the freedom of expression.[24]

In another recent case, which has been publicly known via newspapers, a model asked the court to delist the news on her previous marriage and her other 'by the media made-up' relations on the grounds that in her profession she would like to be known only with her work. Additionally, she has claimed made-up news on her

[20] Akgül (2016), p. 22; Yılmaz (2018), p. 119; Ocak (2018), p. 520; Akkurt (2016), p. 2617 et seq.; Akkurt (2018), p. 88; Çırak (2018), p. 170; Sözüer (2017), p. 71 et seq.; Karan (2018), p. 4241; Duman (2017), p. 239.

[21] Ocak (2018), p. 520.

[22] Yavuz (2016), pp. 135–136; Ahi (2014), Akkurt (2016), p. 2617 et seq.

[23] See Sect. 6.1 below.

[24] Turkish Data Protection Authority (2018) https://www.kvkk.gov.tr/Icerik/4214/Kurul-Kararlari. Accessed 1 Nov 2018.

private life violated her right of personality.[25] We think that the verdict of this pending case will most probably further clarify limits of the right to be forgotten in Turkey.

5 Assessment of the Implementation of the Right to Be Forgotten Under Turkish Law

In Turkish law, the right to be forgotten has just recently become a hot topic among scholars and in the media. There are a lot of new academic publishing on the subject.[26] Again there are also two relatively new, landmark high court decisions. The first case dates back to 2015, it was a decision of the Turkish Court of Cassation. Moreover, very recently in 2016 Turkish Court of Constitution has also given another important decision on the right to be forgotten. Definitely, this decision, which has also enjoyed wide-spread media coverage, will set the fundamental standards and the relative criteria for the sound implementation of this right and ensure effective functioning of it.[27]

Additionally, most of the scholars in Turkey[28] expressly welcome ECJ ruling on Google Spain.[29] Usually, rather than criticizing Google Spain case, some of them express their concern because of the ambiguity of the limits of the right to be forgotten. Notwithstanding this fact, almost no opinion is expressed against Google Spain case so far, yet nearly all of the scholars agree that the limits and conditions of this right have to be more specific.[30]

However, so far, despite the intense media and academic interest, there is no massive influx of erasure requests. That is why the right to be forgotten will appear to be implemented very slowly and in the long run. The biggest obstacle is created by Google which does not respond Turkish applicants' requests of right to be forgotten properly. In one instance, a Turkish woman has repeatedly requested Google to remove the newspaper links on a prostitution gang case, in which she was also accused for being a gang member. Even though relevant news dates back 2007 and she got acquitted of relevant offences at the end, Google stayed indifferent to her persistent requests by denying her even a grounded rejection.

---

[25]Hurriyet Turkish Newspaper (2018) http://www.hurriyet.com.tr/gundem/eski-model-iliskileriyle-hatirlanmak-istemiyor-unutulma-hakkini-istedi-40941797. Accessed 1 Nov 2018.

[26]See References.

[27]For a detailed information about these decisions please see Sect. 6 below.

[28]Akgül (2016), p. 35; Şerefoğlu Henkoğlu (2017), p. 199; Yılmaz (2018), p. 119; Korkmazer (2016), p. 126; Sumer (2016), p. 790.

[29]Google Spain SL, Google Inc. v. Agencia Española de Protección de Datos (AEPD), M. C. Gonzalez, http://curia.europa.eu/juris/document/document_print.jsf?doclang=EN&docid=152065.

[30]See for instance Ocak (2018), p. 521; Ahi (2014).

Another obstacle might be the reluctance of the Turkish courts to apply criminal punishment to sanction denial of this right. Additionally the desire of the public authorities to process as much personal data as possible, arbitrarily and without meeting transparency standards might constitute another hindrance.

At the moment, the concrete practical impacts of the Turkish Constitutional Court's decision[31] are unknown because decisions of courts of first instance are not publicly available until they get finalized through appeal process. So, it will take at least 2 of 3 years to gather more reliable information on the implementation of the right to be forgotten.

6 Judicial Applications with Regard to the Right to Be Forgotten in Turkey

In Turkey, as it is mentioned above, there are two recent landmark court decisions which recognize and practice the right to be forgotten following the ECJ ruling. The last decision dated 2016 is the one of the Turkish Constitutional Court (Sect. 6.1) while the first decision dated 2015 is the one of the Turkish Court of Cassation (Sect. 6.2).

6.1 Decision of the Turkish Constitutional Court (Numbered 2013/5653, Dated 03/03/2016, Rendered as a Result of an Individual Application)

In this decision right to be forgotten in Internet is explicitly recognized and practiced.

The case is about a request for erasure (removal) of past news from the web archive of a nation-wide published newspaper. The applicant served with fine due to the illegal drug use, and a related newspaper reported about his lawsuit as news three times in the years of 1998 and 1999. In 2013, the applicant requested the erasure of the news from the web archive of the newspaper with the grounds that continuous availability of such news infringe his rights to privacy and negatively affect his social, family and business relations. However, the newspaper did not comply with this request. So, the applicant filed a criminal lawsuit against the newspaper. The court accepted the request, but the newspaper objected to this decision on the higher court. The latter decided to the rescission of the former's decision. This time, using his right of individual application, the applicant applied to the Constitutional Court on the grounds that his fundamental right of personal honor (dignity) and public reputation is infringed. Constitutional Court accepted this applicant but the second

[31]See Sect. 6.1 below.

section of the Constitutional Court sent the matter to the attention of the General Assembly due to its importance.

Referring to the European Court of Justice relative ruling, the General Assembly of the Constitutional Court, in its decision dated 2016, explicitly mentioned the right to be forgotten and decided that the applicant's right of personal honor (dignity) and public reputation which is guaranteed by the Article 17 of the Constitution is infringed in the present case *(paragraph 46).*

According to the Court, although there is no explicit provision (at that point Law N. 6698 was not also yet enacted and entered into force) with regard to the right to be forgotten in Turkish Law, this right must be accepted as direct outcome of the Articles 5 (*Fundamental aims and duties of the state*), 17 (*Personal inviolability, corporeal and spiritual existence of the individual*) and 20 (*Privacy and private life*) of the Turkish Constitution *(paragraph 47).* This right creates a positive duty on the state, in order to give an opportunity to individuals to improve their spiritual existence *(paragraph 49).* Thus, considering this duty, the state should take the required measures in order to protect the personality rights of its citizens.[32] Apparently, to establish an environment where the right to be forgotten can be easily practiced fall under this set of measures—of course only if the requirements thereof are duly fulfilled.

However, in the practice of this right, in order to remove the news that are related to personal data, one should thoroughly take into account the relevant factors such as duration of the publication, content, accuracy, up-to-dateness, characteristic of the news as an historical data, public interest *(especially whether it is related to a publicly known person or not, the interest of the general public etc.)* and evaluate the subject matter accordingly.

In its evaluation of the present case, the Court concluded that although the content of news is authentic, it is in a distant, remote past (14 years passed); furthermore, it is of no scientific, historical or statistical interest and therefore its direct accessibility is not of importance or required; moreover, the applicant is not a public figure. Thus, there is no public interest in its direct accessibility and the newspaper should comply with the data subject's erasure request *(paragraphs 73–74).* In short, the Court held that the said news must be evaluated in light of the right to be forgotten of the applicant, and the personality right of the applicant must be preferred over the freedom of expression and media (press). Whereas the existed equilibrium between the freedom of media and protection of personality rights are unbalanced due to the digital platforms, to give a right to be forgotten to individuals will hopefully recreate it. However in this practice, one must be gentle and the core of the freedom of media should never be touched or affected negatively.[33]

As seen, it can be said that Turkish Constitutional Court has embraced Google Spain ruling of ECJ by giving a decision in line with it and by citing it in its decision.

---

[32]See also Çelik (2017), p. 402.

[33]See also Çelik (2017), p. 401; Çırak (2018), pp. 183–184; Ocak (2018), p. 518; Sözüer (2017), p. 199.

However, it should also be added that Turkish Constitutional Court has diverged from Google Spain case in one aspect by ordering a nation-wide published newspaper to take down the news from its web-archive.[34] Whether it is speculated, this may have happened because of the fact that the applicant did not ask any search engine to comply with his right to be forgotten.

6.2 Decision of the General Assembly of the Civil Chambers of the Turkish Court of Cassation (Numbered 2014/4-56 - 2015/1679 and dated 17/05/2015)

In this decision the right to be forgotten in Internet is indirectly recognized.[35] Nevertheless it is appreciated by the legal doctrine, since it has firstly discussed the right and also has given its first jurisprudential definition.[36]

The case is about the request for damages due to the infringement of personality rights of the claimant who is the victim in an earlier adjudicated (in the year of 2009) sexual assault case. The text of this decision is published in 2010 in a book with the claimant's full name. Following that, the claimant filed a lawsuit against the writer on the ground that her personality rights are infringed, and requested the books to be pulled off the shelves and damages for pain and suffering. There against, the respondent asserted that, firstly the book has a scientific quality and therefore it is not addressed to large mass of the population; secondly that the referred penal court decision is already published and therefore accessible from the official information system of the national judicial network; thirdly the case is also reflected in the press and known judicial locality of the deciding penal court and finally her name is fully referenced only in one section of the book as a result of typesetting error (which is in the afterwards duly corrected and the corrected version of the book is also provided to the claimant) and not in the other sections and for all these reasons demanded the rejection of the claimants claims. The court of first instance, decided that there is an infringement of the personality rights of the claimant considering the structure and values of the Turkish society and the book is sold countrywide and accepted the claims. Following the appeal of the decision by the respondent, the Court of Cassation examined the case and reversed the decision by plurality of the votes on the grounds that the scientific quality of the book, and the publicly-known character of the referenced decision. In his dissenting vote, one of the judges implied that there is a definite infringement of the claimant's personality rights, taken into account the current developments with regard the human rights practices and especially the ones

[34]Ocak (2018), p. 520.

[35]The term "*indirect recognition*" indicates how the court reacts towards the right to be forgotten in Internet. In this case, court recognizes plaintiff's right to be forgotten in printed (published) materials. By doing so, it indirectly recognizes as well the right to be forgotten in Internet.

[36]Çelik (2017), p. 400; Elmalıca (2016), p. 1629; Ocak (2018), p. 517; Sumer (2016), p. 775.

related to the right to be forgotten. Following this decision of reversal of the Court of Cassation, on its side, the court of first instance insisted on its first decision and the conflict is re-appealed by the respondent.

The General Assembly of Civil Chambers of the Court of Cassation which examined the case, emphasized the importance of the personal data protection and decided by plurality of votes that the reference to the full name of the claimant (via the said court decision) in the book constitutes an infringement of her personality rights, especially her right to be forgotten and to the confidentiality of her private life and accordingly approved the decision of the court of first instance. According to the Court, although the right to be forgotten is mainly related to digital personal data, it is a necessity to accept such right in other domains where the public has a direct and easy access in light of its particularities and its relation to human rights. The decision referred to the ECJ ruling as well.

As seen this decision is going a bit further in its approach, if ECJ ruling is taken into account as well, in so what it accepts an application of the right to be forgotten outside of the Internet domain. Such a broad approach of the concept may be criticized, since it may risk leaving a very restricted place to the freedom of expression if it is brought to far.[37] However some Turkish scholars state that there is no harm to follow this decision, at least for the other platforms (outside the Internet) where personal data are easily accessible by the public (as emphasized in the decision), since there is a high level of resemblance.[38] Moreover, some scholars seem also concerned about the *ratio decidendi* and the potential effect of this decision, on the account of the fact that it does not properly illuminate the scope of application of the right (to be forgotten) with relation to other rights such as right to privacy and right to protection of the personal data.[39] Nevertheless, as it has been already told, it is the first decision admitting the existence of the right to be forgotten under Turkish Law and thus it has a special value.

7 Implementation by Google: What About Turkey?

As it is known Google formulated user application forms in order to correctly implement the right to be forgotten. Even though there is also a Turkish application form, Turkey is not listed in the search bar regarding applicant's country. From there it may be easily deduced the specific procedure provided by Google, is not yet available in the Republic of Turkey. As stated above mentioned example, Google denies this right to Turkish citizens without any explanation.[40] The most probable

[37] Yavuz (2016), p. 99; Ocak (2018), pp. 517–518; Çırak (2018), p. 180; Elmalıca (2016), p. 1629.

[38] For an embracing view of this last approach in the Turkish legal doctrine see Akgül (2016), p. 34; Küpeli (2016), p. 236; Bozkurt Yüksel (2016), p. 37; Sözüer (2017), p. 188; Sumer (2016), p. 790.

[39] Yavuz (2016), p. 99; Duman (2017), p. 235; Sumer (2016), p. 791.

[40] See Sect. 5 above.

reason in behind this behavior is that Turkey is not a party of the European Union and within the legal realm of European Court of Justice, although it is a country still in the processes of entering thereto. However, it should be noted that it is still possible to individually apply to Google, in a more general fashion; after all Google, operating in Turkey, should comply with the mandatory Turkish regulations regarding the issue. Whether Google does not accept the request and does not delete or delink the related data—while there are valid grounds for doing the contrary-, one should apply then to the Turkish Courts and seek for the required remedy via court proceedings. This is what has happened in above mentioned prostitution gang case—mentioned above; indifference and abstention of Google have forced the relevant applicant to initiate Turkish court proceedings to enforce her right to be forgotten.

8 Some General Suggestions with Regard the Practice of the Right to Be Forgotten and Conclusions

To begin with, for a sound practice of the right to be forgotten, we think that search engines should be more transparent about the ways they use to implement the right to be forgotten. After all, according to Articles 5 and 12 of General Data Protection Regulation of European Union, data processing must be transparent. In our opinion, as data processors, search engines have to explain on what grounds it rejects erasure request and keeps relevant data available in its search engine.[41]

As a matter of fact, granting an unlimited discretion to search engines may lead to problematic and dubious practices.[42] Therefore, in an ideal practice, the role of the search engines in the processes of the right to be forgotten should be rendered into a more "passive" one. Countries should definitely provide guidelines for relevant data processors. Otherwise, in the practice of the said right, if the balancing mechanism of the rights to freedom of expression and media and to protection of personal data (in a more general way personality rights) will be carried out by a private law entity alone, this may cause ambiguity and arbitrariness, more specifically excessive or insufficient protection of right to be forgotten. Especially, if the search engines abuse this power and act as censors, important information for the collective memory of humanity may get lost under their sole discretion.[43] That way, entire human history may be changed and rewritten in a similar manner of dystopia scenarios (i.e. such as in the well-known novel "1984" of George Orwell) and the future may be easily controlled in an unwanted manner for the entire humanity.[44] Besides, if small scale

[41]Yılmaz (2018), p. 120.

[42]Elmalıca (2016), p. 1621; Önok (2017), pp. 173–174; Sözüer (2017), p. 138 et seq.

[43]For a similar criticism see Önok (2017), p. 175; Laçin (2014), p. 406 et seq.; Şerefoğlu Henkoğlu (2017), p. 206.

[44]See and compare Yılmaz (2018), p. 119; Sözüer (2017), p. 139; Duman (2017), p. 238.

search engines are also taken into consideration, it is dubious to leave this responsibility only to private entities.[45] Namely, under political or economical pressure, they might be more willing to erase data. Moreover how they will manage with relevant costs is another problem.

Separately from that, to be realistic, if the national courts or national data protection agencies are empowered as entities of first instance, the amount of workload and charges that may arise will be immense, maybe even beyond imaginable extent that they could not effectively carry out the practice.[46]

Furthermore, thinking geographical—and technological-wise, it is clear that the online blocking or locking hazardous information shall be insufficient or inadequate most of the times, since there are always other ways to reach them.[47]

In any case, the overall transparency of the practice is of utmost importance and three main criteria should be at all times considered, these are the following: the transparency of the individual applications (i); the clarification and certification of the general policy and the publicly sharing of the statistical information with regard the practice (ii); the thorough explanation of the grounds of acceptance/refusal of the application to each individual applicant (iii).[48]

Furthermore, no matter how transparent the said practices relevant to the right to be forgotten are to be carried out, either by the search engines or by the national data protection agencies or courts, it should always be confidential. This is a direct/ rational consequence of the needs of the injured party exercising his/her right to be forgotten. His/her data should be kept confidential the whole time/and in the afterwards of the related processes. Otherwise, this may cause a second injury while remedying the first one.[49] Paradoxically, Gonzalez is now a famous public figure and his inability to pay his debts once upon a time is a widespread information because of a case he has won, while what he asked for in the first place was solely to be forgotten.

In our opinion, as a further step, a uniform international practice just like in European Union (and arguably but maybe a supra-national unifying authority like European Court of Justice or European Court of Human Rights) is highly needed especially for extra-territorial reach and enforcement of the right to be forgotten.[50] Otherwise, from a geographical point of view one cannot make reference to a uniform application thereof and it causes incoherent practice. Taken also in account different approaches to the right to be forgotten of different countries, it will not be

[45] Yavuz (2016), pp. 177–178; Önok (2017), p. 174; Sözüer (2017), p. 139.

[46] See also Yavuz (2016), p. 179; Elmalıca (2016), p. 1621; Gülener (2012), p. 235; Sözüer (2017), p. 139.

[47] See also Elmalıca (2016), p. 1630; Sözüer (2017), p. 141.

[48] See and compare Yavuz (2016), pp. 149–150; The Advisory Council to Google on the Right to be Forgotten (2015), p. 21.

[49] Yavuz (2016), p. 180.

[50] See also Elmalıca (2016), p. 1621; Bozkurt Yüksel (2016), p. 38.

easy to obtain a consistent application to solve similar issues thinking on a country-wise basis.[51]

Until that (further) step is taken, which is obviously needing time, encumbering the search engines with the responsibility on first hand, will be the best way to create a solution to the existing problem. However, this practice should be at all times monitored to and controlled by the national authorities (data protection agencies/ boards or courts). Additionally, search engines should be provided with binding guidelines. This will function as a filtering mechanism.[52] Thus, search engines will carry out the main inspection however always along with the assistance and the supervision (a sort of cooperation) of the data protection agency and/or court.[53]

In this regard, the questions of who is competent in the right's implementation (whether the search engines of different sorts or data protection agencies or national courts or supra-national and binding authorities), and how this implementation should be carried out (under which principles the data shall be evaluated, whether the data is subject to be erased or rendered unavailable/inaccessible, what are the collective components of the balance test between the freedom of expression/media and the protection of personal data should [i.e. the public character of the data, its source, its existence in terms of time] etc.) need to be clarified.

Conclusively, we maintain that one should always bear in mind that on one hand the exaggerated/extreme practice of the right to be forgotten shall negatively affect the freedom of expression, on the other hand its inefficient practice shall lead to infringe the personality rights. Therefore, this point should be duly/thoroughly clarified at all times.

References

Ahi Ş (2014) Unutulma Hakkı (The Right to be Forgotten). http://www.bilisimhukuk.com/2014/02/unutulma-hakki-the-right-to-be-forgotten/. Accessed 1 Nov 2018

Akgül A (2016) Kişisel Verilerin Korunmasında Yeni Bir Hak: "Unutulma Hakkı" ve AB Adalet Divanı'nın "Google Kararı". Türkiye Barolar Birliği Dergisi 116:11–38

Akkurt SS (2016) 17.06.2015 Tarihli, E. 2014/4-56, K. 2015/1679 Sayılı Yargıtay Hukuk Genel Kurulu Kararı ve Mukayeseli Hukuk Çerçevesinde "Unutulma Hakkı". Ankara Üniversitesi Hukuk Fakültesi Dergisi 65(4):2605–2635

Akkurt SS (2018) "Unutulma Hakkı"nın Kişilik Hakkı Kapsamındaki Kişisel Değerlerle İlişkisi. Konya Ticaret Odası Karatay Üniversitesi Hukuk Fakültesi Dergisi 3(1):81–104

Aksoy HC (2008) The right to personality and its different manifestations as the core of personal data. Ankara Law Rev 5(2):235–249

Bozkurt A (2014) Sanal Ortamda "Unutulmak", Bir Hak mı? Bilişim - Türkiye Bilişim Derneği Aylık Bilişim Kültürü Dergisi 42(161):94–96

---

[51] See also Gülener (2012), p. 226 et seq.

[52] Yavuz (2016), p. 179.

[53] Yavuz (2016), p. 181; The Advisory Council to Google on the Right to be Forgotten (2015), p. 35.

Bozkurt Yüksel AE (2016) İnternet ve Unutulma Hakkı. In: 4. Uluslararası Bilişim Hukuku Kurultayı 2016 Bildiriler Kitabı, İzmir, pp 23–43

Çelik Y (2017) Özel Hayatın Gizliliğinin Yansıması Olarak Kişisel Verilerin Korunması ve Bu Bağlamda Unutulma Hakkı. Türkiyet Adalet Akademisi Dergisi 8(32):387–406

Çırak E (2018) Dijital Çağda Sonsuza Kadar Hatırlanmaya Karşı: Unutulma Hakkı. Ceza Hukuku Dergisi 36:161–189

Develioğlu M (2017) 6698 sayılı Kişisel Verilerin Korunması Kanunu ile Karşılaştırmalı Olarak Avrupa Birliği Genel Veri Koruma Tüzüğü Uyarınca Kişisel Verilerin Korunması Hukuku. Oniki Levha, İstanbul

Duman Ö (2017) Özel Yaşama Saygı Hakkı Çerçevesinde Unutulma Hakkına İlişkin Karar Analizi. İstanbul Barosu Dergisi 91(4):227–240

Elmalıca H (2016) Bilişim Çağının Ortaya Çıkardığı Temel Bir İnsan Hakkı Olarak Unutulma Hakkı. Ankara Üniversitesi Hukuk Fakültesi Dergisi 65(4):1603–1636

Google Spain SL, Google Inc. v. Agencia Española de Protección de Datos (AEPD), M. C. G., http://curia.europa.eu/juris/document/document_print.jsf?doclang=EN&docid=152065. Accessed 1 Nov 2018

Gülener S (2012) Dijital Hafızadan Silinmeyi İstemek: Temel Bir İnsan Hakkı Olarak "Unutulma Hakkı". Türkiye Barolar Birliği Dergisi 102:219–240

Hurriyet Turkish Newspaper (2018). http://www.hurriyet.com.tr/gundem/eski-model-iliskileriyle-hatirlanmak-istemiyor-unutulma-hakkini-istedi-40941797. Accessed 1 Nov 2018

Karan U (2018) İnternette Haber Arşivlerinde Unutulma Hakkı: İfade Özgürlüğü İhlali mi, Özel Yaşama Saygı Hakkının Doğal Sonucu mu? Legal Hukuk Dergisi 16(189):4201–4254

Korkmazer G (2016) Unutulma Hakkına Dair Avrupa Birliği Adalet Divanı Google Kararının Değerlendirilmesi. Türk Hukuk Araştırmaları Dergisi 1(2):125–129

Küpeli C (2016) Şiddet Mağduru Kadınların Unutulma Hakkı. Marmara Üniversitesi Hukuk Fakültesi Hukuk Araştırmaları Dergisi 22(1):229–238

Laçin İ (2014) Unutulma Hakkı. Galatasaray Üniversitesi Hukuk Fakültesi Dergisi 2014 (1):391–419

Nair G, Balta E (2017) Bilgi İletişim Teknolojileri Kullanımında Sınırları Aşan Bir Sosyal Sorun Alanı Olarak Unutulma Hakkı. Cumhuriyet Üniversitesi Sosyal Bilimler Dergisi 41(2):113–126

Ocak A (2018) Hakları Dengelemek: Unutulma Hakkı İfade Özgürlüğüne Karşı, Balancing rights: the right to be forgotten v. freedom of expression. Türkiye Adalet Akademisi Dergisi 9 (33):507–535

Önok M (2017) Kişisel Verilerin Korunması Bağlamında "Unutulma Hakkı" ve Türkiye Açısından Değerlendirmeler. İstanbul Kültür Üniversitesi Hukuk Fakültesi Dergisi 16(1):155–188

Sağlam İ (2017) Unutulma Hakkı. In: Akipek Öcal Ş, Uzun Kazmacı Ö, Ayar A, Sayın ZG, Özçelik NŞ (eds) 1926'dan Günümüze Türk – İsviçre Medeni Hukuku, vol 1. Seçkin, Ankara, pp 683–700

Şerefoğlu Henkoğlu H (2017) Unutulma Hakkı: Dijitalleşme Sürecinde Bilgiye Erişim Özgürlüğünü Tehdit Eder mi? In: Özdemirci F, Akdoğan Z (eds) Bilgi Sistemleri ve Bilişim Yönetimi: Beklentiler ve Yeni Yaklaşımlar. T.C. Ankara Üniversitesi Bilgi Yönetim Sistemleri Belgelendirme Merkezi (Bil-Bem), Ankara, pp 191–212

Sözüer E (2017) Unutulma Hakkı – İnsan Hakları Perspektifinden Bir İnceleme. Oniki Levha, İstanbul

Sumer O (2016) "Unutulma Hakkı"nın Türk İçtihatına Girişi. Legal Hukuk Dergisi 14 (158):775–792

The Advisory Council to Google on the Right to be Forgotten – Final Report (2015). https://static.googleusercontent.com/media/archive.google.com/tr//advisorycouncil/advisement/advisory-report.pdf. Accessed 1 Nov 2018

Turkish Data Protection Authority (2018). https://www.kvkk.gov.tr/Icerik/4214/Kurul-Kararlari. Accessed 1 Nov 2018

Yavuz C (2016) İnternet'teki Arama Sonuçlarından Kişisel Verilerin Kaldırılması: Unutulma Hakkı. Seçkin, Ankara

Yılmaz SS (2018) Kişisel Verilerin Korunması Regülasyonu ve Unutulma Hakkı. Terazi Hukuk Dergisi 13(142):116–121

Part II
Americas

Argentina: The Right to Be Forgotten

Judge Marcelo López Alfonsín

Abstract There is neither specific Argentine legislation guaranteeing the right to be forgotten nor specific remedies for infringement of such a right. Argentina finds similar values and protections in the constitutional rights of privacy and dignity, but has not followed the ECJ ruling in the Google Spain case in accordance with the European approach to the right to be forgotten.

1 How Is the Right to Be Forgotten Protected Under Your Law? Does Your Law Specifically Grant a Right to Be Forgotten or Does This Right Derive from a More General Framework?

In recent years, the right to be forgotten has been placed in the spotlight around the world especially in Europe. In this context, Argentina is dealing with the complex issue of internet intermediaries' liability and the debate about the protection of privacy in the digital age has also gained prominence.

Although there is no specific legislation in our country that guarantees the right to be forgotten, it is possible to find a more general framework from which it is possible to derive the grant of this right.

The argentine legal system recognizes the right to honor, to privacy and to one's own image as spheres of protection of human dignity and there is as well an expressed safeguard of the protection of personal data.[1] Moreover, the national framework incorporates international human rights treaties[2] in which it is also

[1]Argentine Constitution, articles 18, 19, 33 & 43; Personal Data Protection Law 25.326.

[2]Argentina's Constitution provides, when it incorporates the human rights treaties mentioned in article 72, subsection 22, that those provisions acquire constitutional status, meaning that in case of conflict judges must declare the unconstitutionality of the national laws.

J. M. L. Alfonsín (✉)
Universidad de Buenos Aires, Buenos Aires, Argentina
e-mail: mlalfonsin@jusbaires.gov.ar

F. Werro (ed.), *The Right To Be Forgotten*, Ius Comparatum – Global Studies in Comparative Law 40, https://doi.org/10.1007/978-3-030-33512-0_12

possible to find legal provisions on the matter.[3] In this connection, the guidelines developed by international human rights law should be taken into account as it is recognized that the protection of the right to privacy must be encouraged by states in the digital age.[4]

In addition, it is important to highlight that the Argentine Supreme Court of Justice in the *Rodriguez v. Google*[5] case pronounced a relevant sentence about the civil liability of Internet search engines. In this case, professional model Belén Rodríguez claimed that the search engines were violating her right to her own image and that they were engaging in defamation, as her name was linked to web pages with sexual content. She requested to order Google Argentina and Yahoo to remove and permanently block all search results associating her name to websites having sexual, pornographic content; to remove all thumbnails using her image from the search results; and to be compensated for the monetary and moral harms suffered. The ruling of *Belen Rodriguez* has become a leading case in the subject, as it was the first time the Supreme Court have established some guidelines about the role played by internet search engines.

Although the court rejected the claimant's demand,[6] it was stated that there is no strict liability for third-party violations because search engines are not responsible of the published contents unless and until they are notified of the harm caused. According to this, justices detailed that Internet intermediaries become liable only upon obtaining "effective knowledge" of the illegal content involving the notification by a court or other competent authority that could place intermediaries on notice. On the contrary, the court conceded that in a few cases of "gross and manifest" harm, involving content whose illegality is beyond doubt, a proper notification by the affected party might be sufficient to require intermediaries to act. The Court stated that such categories may include child pornography, speech

[3]American Declaration of the Rights and Duties of Man, article 5 ("Every person has the right to the protection of the law against abusive attacks upon his honor, his reputation, and his private and family life."); Universal Declaration of Human Rights, article 12 ("No one shall be subjected to arbitrary interference with his privacy, family, home or correspondence, nor to attacks upon his honour and reputation. Everyone has the right to the protection of the law against such interference or attacks."); International Covenant on Civil and Political Rights, article 17 ("No one shall be subjected to arbitrary or unlawful interference with his privacy, family, home or correspondence, nor to unlawful attacks on his honour and reputation."); American Convention on Human Rights, article 11, para. 2 ("No one may be the object of arbitrary or abusive interference with his private life, his family, his home, or his correspondence, or of unlawful attacks on his honor or reputation.").

[4]UN Council on Human Rights, "*Promotion, Protection and Enjoyment of Human Rights on Internet,*" June 29th 2012; Office of the Special Rapporteur for Freedom of Expression, Inter-American Commission on Human Rights "*Freedom of Expression and the Internet.*" June 1st 2011.

[5]Argentina Supreme Court of Justice, "*Rodríguez, María Belén c/Google Inc. s/daños y perjuicios", sentence of the 28th October 2014.*

[6]The Supreme Court followed the ruling of the case Rodríguez in subsequent cases (Argentina Supreme Court of Justice, such as "*Da Cunha, Virginia c/ Yahoo de Argentina S.R.L. Y otro s/ daños y perjuicios*", and "*Lorenzo, Bárbara cl Google Inc. si daños y perjuicios*", sentences of the 30th December 2014).

that directly endangers the life or physical integrity of others, clear incitement to violence or discrimination, or clearly unlawful publications that grossly violate individual privacy or cause deliberate harm to one's reputation.[7] Moreover, the Court has highlighted the importance of freedom of expression as well as the need to protect the individual right to honor and to privacy.

In the light of the foregoing, it should be pointed out that, although there is no specific regulation that guarantees the right to be forgotten, I believe that the argentine framework concerning the right to privacy and the protection of personal data must be understood as a safeguard against Internet harmful contents.

2 What Are the Limits to the Right to Be Forgotten Under Your Law?

Even though there are no legal regulations that guarantee the right to be forgotten in Argentina in the *Rodríguez* case the Supreme Court recognized that there is a limit to the protection of the right to privacy, which is derived directly from the safeguard of the freedom of expression and the prohibition of prior censorship, established in article 13 of the American Convention on Human Rights.[8] In this connection, the Law 26.032 established that the search, reception and dissemination of information and ideas through the internet service is considered to be included within the constitutional guarantee that protects the freedom of expression.

Based on this standard, the Supreme Court ruled in favor of subjective liability because it recognized that holding search engines responsible for contents not created by them would imply a restriction of the freedom of expression on the internet. Thus, the actual limit according to the Court is the identification of the harm and the following notification to the search engines.

According to the *Rodriguez* case ruling, there is another limit to the right to be forgotten, as the Supreme Court stated that it is not allowed to enable a private actor to establish if there is violation of the right to privacy, as it could affect the freedom of expression and the right to access to public information, which means that the affected person, unless the harm is "gross and manifest", must claim to the competent authority.

---

[7]Argentina Supreme Court of Justice, *"Rodríguez, María Belén c/Google Inc. s/daños y perjuicios", sentence of the 28th October 2014, parag.17.*

[8]The Supreme Court has recalled in the *Rodriguez* case that any restriction or limitation on free expression should be interpreted restrictively and that all censorship has a strong presumption of unconstitutionality.

3 What Are, in Your Law, the Legal Remedies Available to Enforce the Right to Be Forgotten?

At present, in Argentina there are no available remedies to enforce the right to be forgotten.

Firstly, in the aforementioned *Rodriguez* case justices stated that the right to judicial protection in accordance with international human rights law involves the right to a simple and prompt recourse to a competent court or tribunal for protection against acts that violate fundamental rights. Consequently, they have emphasized the need for legal remedies to demand the protection of the personal data and to ensure the right to privacy of Internet users.

Secondly, it should be mentioned that in the *Gil Dominguez* case, a citizen promoted a class action against the Government of the City of Buenos Aires given the lack of available legal remedies to protect the privacy rights of the consumers in Buenos Aires. The ruling of this case recognized the right to be forgotten and, ordered to the local authority to incorporate an administrative procedure in order to protect the right of privacy of internet users.[9] Even though the protocol has not been implemented,[10] the proposal seeks to allow people to request easily the deletion of personal data that might be considered harmful of their privacy on the internet.

Finally, it should be mentioned that the argentine legal framework recognized a remedy called *habeas data*,[11] allowing any person to obtain information on the data about himself registered in public records or private databases and, in case of false data or discrimination, this action may be filed to request the suppression, rectification, confidentiality or updating of said data. Nowadays, as the right to be forgotten is not expressly protected under law, this remedy should be considered the only one available for people whose privacy is being damaged on the Internet.[12]

---

[9]City of Buenos Aires, Juzgado Contencioso Administrativo y Tributario de la Ciudad de Buenos Aires N°18, Secretaria N° 36 [Court of first instance] "*GIL DOMÍNGUEZ ANDRÉS FAVIO CONTRA DIRECCIÓN GENERAL DE DEFENSA Y PROTECCIÓN DEL CONSUMIDOR DEL GCBA S/AMPARO",Expte.A352-2014/0*, sentence of October 10th 2014. Press release: http://www.lanacion.com.ar/1736360-el-derecho-al-olvido-en-internet-se-debera-aplicar-tambien-en-la-capital-federal.

[10]The protocol could not be implemented because the ruling was reversed on appeal.

[11]Argentine Constitution, article 43 and Personal Data Protection Law 25.326.

[12]Basterra M, "Definiendo el alcance y los límites de la responsabilidad de los buscadores de Internet." http://marcelabasterra.com.ar/wp-content/uploads/2016/04/DEFINIENDO_EL_ALCANCE_Y_LOS_LIMITES_DE_LA_RESPONSABILIDAD_DE_LOS_BUSCADORES_EN_INTERNET.pdf https://es.scribd.com/document/247961027/Articulo-LA-LEY-Responsabilidad-de-los-buscadores-de-Internet-Marcela-Basterra-pdf/.

4 As a Follow-up to the Previous Question, Does Your Law Allow the Plaintiff to Receive Material or Immaterial Damages? If Yes, Is Such Remedy Realistic in Practice?

The lack of specific regulation that enshrines the right to be forgotten and the absence of legal framework about the liability of internet intermediaries, requires applying traditional civil liability rules in order to enable internet users to receive material or immaterial damages. In this connection, the National Civil and Commercial Code established that a human person injured in his or her personal or family privacy, honor or reputation, image or identity can claim the repair of the damages.[13] However, this framework does not expressly refer to the harm derived from the use of the internet.

As explained above, in the *Rodríguez* case the Supreme Court ruled in favor of subjective liability of internet search engines after they are notified by the affected person or a competent authority, depending on the type of harm. Therefore, the only way to receive material or immaterial damages is to prove the negligence of the search engines after they are asked to remove a content that could be harmful to an internet user.

Notwithstanding, the lack of legislation on the specific obligations of internet intermediaries is an obstacle in practice to allow this type of lawsuits.

5 In General, How Do You Assess the Implementation of the Right to Be Forgotten in Your Law? Is It Effective? Is It Used in Practice? Are There Particular Obstacles in the Implementation of This Right?

In Argentina, the protection of the right to be forgotten is not effective because of the lack of legislation on the obligations of internet intermediaries. In this context, individual claims and litigation have provided some judicial standards. However, these are not binding on other courts and do not establish clear rules of procedure for individuals.

As it was previously stated, unless there is a huge legal basis to protect the right to privacy and personal data, the national framework does not provide prompt and effective remedies in accordance with the speed at which data on the internet circulate.

Moreover, from the point of view of the Argentine civil liability rules it is not possible to held responsible internet intermediaries as there are no laws that regulate their obligations. In this context, there are particular obstacles for the fulfillment of the right to be forgotten. Therefore, those affected in their right to privacy do not

[13] Argentine National Civil and Commercial Code, article 52.

have legal tools and, consequently, there are no available remedies at present to the implementation of this right properly.

Finally, it should be pointed out that the speed of dissemination of the data produced by the internet and its immediately resulting damages require procedures with a certain degree of fastness and prompt response, which does not exist in Argentina.

6 How Did Courts and Commentators in Your Country Welcome the ECJ Ruling on Google v González?

The ECJ ruling on *Google v. Gonzalez* was not followed by argentine courts and, in general terms, commentators do not agree with the statement of the European Court. In this regard, according to Basterra the absence of legal obligations of the search engines for the contents on the internet in Argentina, is the principal obstacle to hold them responsible.[14]

7 For Those Who Are from a Country That Is Not Part of the European Union, Did Your Courts Follow the ECJ Ruling on the Right to Be Forgotten? Is It Likely Do That They Will Follow It?

In the absence of explicit recognition in Argentina of the right to be forgotten, the *Google v. González* case and comparative law were taken into account by the Supreme Court as a subsidiary source of law in the *Rodriguez* case.[15] Indeed, Lorenzetti and Maqueda, two of the five judges of the Court which are at present in exercise, left open the possibility of a future recognition of the right to be forgotten based on the ECJ ruling as an *obiter dictum*. However, the final judgment did not follow the European doctrine. The principal difference is that the Court established, in accordance with the protection of the right to freedom of expression in the Inter-American system, that it is not appropriate to hold search engines as objectively liable for data processing. Liability would be attributed to them only if certain information causes damages and there is culpable conduct on the part of internet intermediaries.

Thus, the application of *Google v. González* case in Argentina should take into account the differences between the European framework for the protection of

[14] Basterra M, "*Definiendo el alcance y los límites de la responsabilidad de los buscadores de Internet.*"

[15] In the *Rodriguez* case the Supreme Court quoted the legislation about internet intermediaries' liability of Portugal, Brasil, United States, Canada, Italy, United Kingdom and Spain.

personal data and the strong emphasis on the right to freedom of expression in the Inter-American system. As it was already explained, applying the ECJ ruling in Argentina could affect the prohibition of prior censorship contemplated in article 13 of the American Convention on Human Rights.

Unlike the European ruling, it would not be allowed in Argentina to enable Internet intermediaries to decide whether or not to block contents on the internet. The Court has already stated that, unless the damage is "gross and manifest", a court or a competent authority must get involved. In this vein, in the *Gil Dominguez* case it was ruled in favor of the involvement of an administrative authority to incorporate an administrative procedure to enforce the right of privacy of internet consumers.

Finally, as it will be explained later, drafts law on the liability of internet intermediaries also proposed the intervention of a state body.

8 Did Your Law Already Grant a Similar Right to Be Forgotten Than the One Stated in the ECJ Ruling?

The Argentine legal system did not expressly grant a right to be forgotten similar to the one stated in the ECJ ruling.

9 To Implement the ECJ Ruling, Google Has Created a Form in Which Anyone Interested Can Submit a Request to Have Information About Him-or Herself Be Delisted. Based on This Request, Google Will Weigh Between the Private Interest of the Petitioner and the Public Interest to Be Informed. Google Does Not Disclose the Ways in Which It Deals with Requests. In Particular, Google Does Fully Not Disclose, the Category of Requests That Are Excluded or Accepted, the Proportion of Requests and Successful De-listings and, Among Others, the Reason for the Denial of Delisting. Do You Think That Google Should Be More Transparent About the Ways It Uses to Implement the Right to Be Forgotten?

I strongly believe that the procedure established by Google from the ECJ ruling must comply with the standards of transparency and accountability concerning the publication of contents and restriction policies and practices.

In this regard, it is possible to recall that the Article 29 Data Protection Working Group (created under Directive 95/46/EC) encouraged the search engines to provide the de-listing criteria they use, and to make more detailed statistics available.[16]

According to this, the advertising of the content restriction criteria that the search engines should comply with derives from the "*Manila Principles on Intermediary Liability*", adopted by civil society organizations in 2015. These principles stated that intermediares should *"publish their content restriction policies online, in clear language and accessible formats, and keep them updated as they evolve, and notify users of changes when applicable*" and "*publish transparency reports that provide specific information about all content restrictions taken by the intermediary, including actions taken on government requests, court orders, private complainant requests, and enforcement of content restriction policies"*.[17]

10 Is the Procedure Prepared by Google Used in Your Country?

In Argentina the procedure prepared by Google is not used because the ECJ ruling is only applicable to countries from the European Union. Moreover, as it was already explained, there is no specific procedure to request the removal or modification of personal data on the internet.

As it was explained, in the *Gil Dominguez* case a judge ordered the implementation of an administrative procedure for the protection of the right to privacy of internet consumers, which could be considered similar to the procedure established by Google. However, taking into account the argentine framework, the court ordered the involvement of a local authority.[18]

[16]Article 29 Data Protection Working Party, "*Guidelines on the implementarion of the cour of justice of the european union judgment on "Google Spain and Inc v. Agencia Espanola de Proteccion de Datos (AEPD) and Mario Costeja González" C/131/12*", Adopted on November 26th 2014. Available at: http://ec.europa.eu/justice/data-protection/article-29/documentation/opinion-recommendation/files/2014/wp225_en.pdf.

[17]Manila Principles on Intermediary Liability, *Best Practices Guidelines for Limiting Intermediary Liability for Content to Promote Freedom of Expression and Innovation*, March 24th 2015, Principle VI. Available at: https://www.manilaprinciples.org/.

[18]The protocol could not be implemented because the ruling was reversed on appeal.

11 Is There Any Upcoming Legal Reform in Your Country Whose Purpose Is to Reinforce or Modify the Right to Be Forgotten?

After the aforementioned *Rodríguez* case, the right to be forgotten has been placed in the spotlight in Argentina. Although they were not finally approved, two draft laws were submitted in the National Congress seeking the regulation of internet intermediaries' liability.

One of them seeks to eliminate or limit the access to personal data that are contained on the internet and that are likely to impair the right to privacy or honor, through a procedure similar to the one that Google established after the ECJ ruling. It proposes allowing individuals to notify directly the search engines to remove or eliminate the contents on the internet.[19]

The other one aims to prevent the dissemination of messages with discriminatory content on the internet and proposes to establish a procedure at the head of an administrative authority to sanction internet intermediaries, and it grants the right to request material or immaterial damages for individuals.[20]

12 In Your Opinion, What Should Be the Next Step in the Protection of the Right to Be Forgotten? Do You Think That One Must Go Further and Strengthen the Right to Be Forgotten? Do You Think That the European Union Should Modify or Adapt Its Legislation on the Right to Be Forgotten?

At present, the great development of digital media has triggered a new way of information exchange that represents a real threat to users' control over their privacy. It is undeniable that rapid technological evolution and globalization brought with it new challenges for the protection of personal data since today the right to privacy can be seriously affected by the improper use of information on the internet. Therefore, it is necessary to rethink the actual legal framework to strengthen the protection of the right to privacy of individuals.

[19]Draft Law Number 7989-D-2014. Available at: http://www1.hcdn.gov.ar/proyxml/expediente.asp?fundamentos=si&numexp=7989-D-2014.

[20]Draft Law Number 7379-D-2014. Available at: http://www1.hcdn.gov.ar/proyxml/expediente.asp?fundamentos=si&numexp=7379-D-2014.

In this context, the judges or the legislators should begin to remedy these matters.[21] The legal framework of data protection is undoubtedly a starting point for developing the discussion in Argentina, but it does not seem to be sufficient.

Consequently, the next step not only in Argentina, but also in the European Union, should involve the regulation by law of the internet intermediaries liabilities and the enforcement of judicial remedies, in order to strengthen the right to be forgotten of individual without undermining the freedom of expression.

References

American Convention on Human Rights, article 11, para. 2

American Declaration of the Rights and Duties of Man, article 5

Argentine Constitution, articles 18, 19, 33 & 43

Argentine National Civil and Commercial Code, article 52

Article 29 Data Protection Working Party, "*Guidelines on the implementarion of the cour of justice of the european union judgment on "Google Spain and Inc v. Agencia Espanola de Proteccion de Datos (AEPD) and Mario Costeja González" C/131/12*", Adopted on November 26th 2014

Basterra M, "Definiendo el alcance y los límites de la responsabilidad de los buscadores de Internet." http://marcelabasterra.com.ar/wp-content/uploads/2016/04/DEFINIENDO_EL_ALCANCE_Y_LOS_LIMITES_DE_LA_RESPONSABILIDAD_DE_LOS_BUSCADORES_EN_INTERNET.pdf https://es.scribd.com/document/247961027/Articulo-LA-LEY-Responsabilidad-de-los-buscadores-de-Internet-Marcela-Basterra-pdf

City of Buenos Aires, Juzgado Contencioso Administrativo y Tributario de la Ciudad de Buenos Aires N°18, Secretaria N° 36 [Court of first instance] "*GIL DOMÍNGUEZ ANDRÉS FAVIO CONTRA DIRECCIÓN GENERAL DE DEFENSA Y PROTECCIÓN DEL CONSUMIDOR DEL GCBA S/AMPARO",Expte.A352-2014/0*, sentence of October 10th 2014

Draft Law Number 7379-D-2014. Available at: http://www1.hcdn.gov.ar/proyxml/expediente.asp?fundamentos=si&numexp=7379-D-2014

Draft Law Number 7989-D-2014. Available at: http://www1.hcdn.gov.ar/proyxml/expediente.asp?fundamentos=si&numexp=7989-D-2014

International Covenant on Civil and Political Rights, article 17

Manila Principles on Intermediary Liability, *Best Practices Guidelines for Limiting Intermediary Liability for Content to Promote Freedom of Expression and Innovation*, March 24th 2015, Principle VI

Office of the Special Rapporteur for Freedom of Expression, Inter-American Commission on Human Rights "*Freedom of Expression and the Internet.*" June 1st 2011

Personal Data Protection Law 25.326

Rodriguez v. Google, Argentina Supreme Court of Justice, "*Rodríguez, María Belén c/Google Inc. s/daños y perjuicios", sentence of the 28th October 2014*

UN Council on Human Rights, "*Promotion, Protection and Enjoyment of Human Rights on Internet,*" June 29th 2012

Universal Declaration of Human Rights, article 12

[21]Basterra M, "*Definiendo el alcance y los límites de la responsabilidad de los buscadores de Internet.*"

The Right to Be Forgotten According to the Brazilian Precedents

Marcos Alberto Rocha Gonçalves

Abstract Within the various aspects of the online privacy fender, the discussions about "right to be forgotten", understood as the opposition to a public memory, of past events, outdated and dissonant characteristics in relation to their contemporary personality aspects. In the digital environment, it reaches the storage and public availability of decontextualized data and information, through news in a content server, mention in database, return in a search provider or even a post in social network, which represents a form of permanent and oppressive memory of the subject's current personality. In Brazil, the Superior Court of Justice (STJ) already had the opportunity to address the issue of the right to be forgotten on the Internet; the arguments demonstrate the preponderance of the right to information and communication, evoking the forgetfulness only when the subject can detach himself from the recounting of history and this does not prevent rememorating of socially relevant fact. The STJ removed the right to be forgotten cases of noncontractual liability for the conduct of search providers, but kept it in front of the application providers, adopting the so-called "notice and take down" regime. It is therefore perceived that the right to to be forgotten, especially in the face of the challenges of the Internet, imposes a critical reflection that aims the guarantee of effective protection of the constitutionally protected values.

1 Introduction: What One Wants to Forget

The digital age established the emergence of a new social understanding of space and time, leading social, economic, and legal relationships to a multidimensional platform. In the relational web established in what has been called cyberspace, information has become an asset of extreme value, since its collection, storage and use has direct influence on the formatting of the identity itself of the subjects from which they emerge.

M. A. R. Gonçalves (✉)
Universidade do Estado do Rio de Janeiro, Rio de Janeiro, Brazil

F. Werro (ed.), *The Right To Be Forgotten*, Ius Comparatum – Global Studies in Comparative Law 40, https://doi.org/10.1007/978-3-030-33512-0_13

This contemporary, invisible and multidimensional relational space referred to as the Internet, by its technical characteristics, causes the arising of its own gnosiology, which establishes new values, purposes and limits—or their absence. The human condition is transformed, online, into a multiple set of data, which handling is capable of producing deep impacts in the concrete life offline, affecting especially the privacy exercise spaces.

In such concern, there is a growing vulnerability of privacy in the digital environment. Not for other reason, after adopting, on July 16, 2012, Resolution 20/8 on the promotion, protection and exercise of Human Rights on the Internet, in which it "affirms that the same rights that people have offline must also be protected online, in particular freedom of expression",[1] the United Nations Human Rights Council also adopted, on April 1, 2015, Resolution 28/16,[2] declaring that equal rights offline and online include the right to privacy.

In the scope of the several aspects of online privacy protection, the growing aspect of legislative, doctrinal and precedents appreciation concerns the effects of the virtually endless digital data storage capacity and its almost instantaneous access—especially with technical and economic importance evolution of the search engines—, and the scenario is even more serious in relation to personal data, captured daily and intermittently by mobile applications, social networks and content providers. It speaks about the denaturation of the time lapse,[3] approaching and mixing past and present, preventing life from following its natural flow of constant clearance of subjective identities.

The discussions about the *right to be forgotten* have been developed in the context of such uneasiness, which is understood, in a general and common way, as the opposition to a public memory, of past events, that bind the subject to outdated and dissonant features in relation to their contemporary personality aspects. In the digital environment, it reaches the storage and public availability of decontextualized data and information, through news in a content server, mention in database, return in a search in a provider of such nature or even a post in social network, which represents a form of permanent and oppressive memory of the subject's current personality.

In the Brazilian scenario, in recent years, the theme has been subject to constant and in-depth appreciation of doctrine and precedents, which accordingly seek to establish their conceptual limits and practical applications, deserving an analytical approach to the recent course of the matter and the current state of art.

[1]https://documents-dds-ny.un.org/doc/RESOLUTION/GEN/G12/153/25/PDF/G1215325.pdf?OpenElement. Accessed 10 Feb 2017.

[2]https://documents-dds-ny.un.org/doc/UNDOC/GEN/G15/068/78/PDF/G1506878.pdf?OpenElement. Accessed 10 Feb 2017.

[3]On the theme, see: Prigogine (2011).

2 The Protection of Privacy as a Theoretical Foundation of the Right to Be Forgotten

The evolution, on a geometric scale, of the arrangement and storage of personal data in the network resulted in the inclusion in the legal discussions of the architecture of mechanisms of respect and protection of privacy in the virtual environment, among which the right to be forgotten—even that such first right has been presented, as a legal claim, compared to data arranged in previous media, such as television, as it will be seen in the exemplary cases below, such experience serving as base for application of the theme in the virtual environment.

Notwithstanding the importance of the theme, there is no express rule in the Brazilian legal system regarding the *right to be forgotten*. Thus, since it is agreed that the local legal system is based on the *Civil Law* matrix, with a strong connection to the legislative construction—although the open interpretation of the standards is conceived, especially from the understanding of ordering unity with the centrality of the Constitution, in a tradition inaugurated in the Italian doctrine of Pietro Perlingieri[4]—, such absence requires the normative implementation of this right from a mechanism of interpretation arising from the content of the constitutional rule, included in the general rule of privacy protection.

In such concern, the Brazilian Constitution of 1988 created the right to privacy, pursuant to article 5, subitems X, XI and XII,[5] as a fundamental right, legal category of rights that represents, in the domestic scope, the reach of the Human Rights in the international scope. Its implementation in private relations is provided, in addition to the direct application of the constitutional rule, by Article 21 of the Civil Code, which provides that "the private life of the individual is inviolable and the judge, upon request of the interested party, shall take the measures required to prevent or cause an act contrary to such rule to be ceased".

The insertion of the right to privacy as a fundamental right is fitted in the creation of a network of rights based on the protection and development of the human dignity,

---

[4]On the theme, *see:* Perlingieri (2008).

[5]*All persons are equal before the law, without any distinction whatsoever, Brazilians and foreigners residing in the country being ensured of inviolability of the right to life, to liberty, to equality, to security and to property, on the following terms:*

(...)

IX – the expression of intellectual, artistic, scientific, and communications activities is free, independently of censorship or license;

X – the privacy, private life, honour and image of persons are inviolable, and the right to compensation for property or moral damages resulting from their violation is ensured;

XI – the home is the inviolable refuge of the individual, and no one may enter therein without the consent of the dweller, except in the event of flagrante delicto or disaster, or to give help, or, during the day, by court order;

XII – the secrecy of correspondence and of telegraphic, data and telefone communications is inviolable, except, in the latter case, by court order, in the cases and in the manner prescribed by law for the purposes of criminal investigation or criminal procedural finding of facts.

established as one of the foundations of the Republic.[6] Therefore, in view of such multiple and wide protection of the human dignity, it must be read, in the Brazilian legal system, the sense and the scope of the privacy protection, without, of course, disregarding the insertion of such protection and such incentive to development in the contemporary social and historical context.

In the context of the information society[7]—which most proper name may be, as mentioned by José de Oliveira Ascenção, *communication society*,[8] since, if it is not capable of producing the Capability Approach established by Amartya Sen,[9] established the hyperconnectivity life—the right to privacy shall mean a function wider than the "right to be let alone", created by Samuel D. Warren and Louis D. Brandeis.[10]

As appointed by Anderson Schreiber, it is the right assigned to the individual to keep under his/her control his/her set of personal data and "transcends such domestic scope to reach any environment where personal data of its holder circulates",[11] including, of course and especially, the Internet.

In the same concern, Daniel Bucar says that the privacy is "the possibility of the person knowing, controlling, addressing and interrupting the flow of personal information related thereto, making possible to have an exact and previous awareness of the information space on which his/her personality will be developed",[12] being an original power to control his/her own information, which can be redeemed always when required to ensure the free construction of the identity.

It is precisely in such extent that one can affirm that the right to be forgotten is included, in the Brazilian legal system, in the scope of the legal discipline of right to privacy, since it "incorporates an expression of timely control of data, which fills with the chronological factor the current triad of privacy protection tools, complemented by the spatial and contextual controls".[13]

More specifically, the triad referred to in the aforementioned doctrine contemplates the control of flow of personal information (spatial), the assurance of accuracy

---

[6]**Article 1.** The Federative Republic of Brazil, formed by the indissoluble union of the states and municipalities and of the Federal District, is a legal democratic state and is founded on:

(. . .)

III - the dignity of the human person.

[7]"The expression *information society* appeared in Europe, in the international conference of 1980, where the European Community hold a meeting of intellectuals to assess the future of a new society called as such, in view of the regulation of the freedom of circulation of services and measures for implementation of access to the goods and services by the Member States" (Martins 2014, pp. 3–4).

[8] *"Information Society" is not a technical concept: is a slogan. It would be better to speak about a communication society, since what is intended to push is the communication, and only in a much wide sense all* messages can be qualified as information" (Ascensão 2002, p. 71).

[9]Sen (1999).

[10]Warren and Brandeis (1890), pp. 193–220.

[11]Schreiber (2013), pp. 135–137.

[12]Bucar (2013), p. 8.

[13]Bucar (2013), p. 8.

of the characteristics of the individual holder of the information reflected therein (contextual) and the protection of the natural occurrence of the voluntary changes in the exterior projection of the individual during the life, with a possible necessity of not being remembered anymore by past events, which characterized him/her differently from the present.[14]

It is precisely from this theoretical background that the assessment by court of the right to forgotten is being constructed, with the cases described below serving as an example.

3 Turning Off the Light of the Stars: Application of the Precedents to the Right to Be Forgotten

In the absence of a law that establishes the limits to the *right to be forgotten*, the jurisdictional activity—based on the doctrinal fabric built on the subject—has been seeking to give shape to the claims related to such right. Notwithstanding the lack of precedents on the issue, it is possible to infer the Brazilian precedent tendency from this framework.

In such concern, a search for decisions that deal with the right to be forgotten in the database of decisions of the Superior Court of Justice—the court responsible for

---

[14]Bucar (2013), p. 9.

It is also worth mentioning, as a possible basis for the *right to be forgotten*, the establishment, in Brazilian criminal law, of the right of the convicted person to dissociate himself/herself from past acts with a view to guarantee his/her resocialization. Finally, but not least, Law 12965, of April 23, 2014, known as the Civil Rights Framework for the Internet and its possible connection with the *right to be forgotten*. This happens because art. 7, sub. X of said law ensures to the Brazilian Internet user *"the definitive exclusion of the personal data he/she may have provided to a certain Internet application, upon its request, at the expiry of the relationship between the parties, except the hypotheses of obligatory custody of records provided for in this law"*.

In view of the legal provision, however, its interpretation is disputed in the doctrine as a modality of right to be forgotten, since it suggests being most proper a keeping of personal data than a circulation of right to intend the exclusion, deletion or even decoupling of past information out of context.

About the theme, see: "However, it was in the Criminal Law that the right to be forgotten mostly developed, when the convicted person had ensured the right to be released from the memory of the criminal act he/she practiced, in order to make his/her re-socialization effectively possible. In addition, such right has a constitutional status under the basis of the prohibition of penalty for life (art. 5, III, and XLVII, b of the Federal Constitution/1988); that is, the perennial remembrance of the conviction would be an effect of the penalty that would follow the convicted person *at eternum*. Therefore, the Sentence Execution Act (Law 7210/1984) provides, in its art. 202, that the criminal facts shall not appear in the police report of the convicted person, except in case he/she performs a new criminal infringement or other cases provided for in law. In other words, the convicted person, after having complied with the penalty, is entitled to have those facts forgotten. In addition to such provision, art. 93 of the Criminal Code and art. 748 of the Criminal Procedure Code bring such right to the convicted person, as a presupposition to his/her re-insertion in the society" (de Lima 2015, pp. 511–543).

the unification of the precedents in infraconstitutional matters and for ensuring the correct application of federal laws—shows that the Court dealt with the matter in collegiate decisions, fundamentally on three themes: (a) the claim of those convicted of past crimes of not having the records of the crimes permanently available for consultation or that the records of crimes committed long before are not considered in the sentencing of current crimes, since in the Brazilian legal system the social conduct of the defendant is a factor that increases the sentence; (b) application of the Amnesty Law (Law 6,683 of August 28, 1979) and the forgetting related to amnesty to crimes committed during the period of military dictatorship in Brazil,[15] and (c) protection of the right to privacy—scope that is interesting to this study.

Thus, in the framework of decisions related to the right to privacy, it is pointed out that the Court has chosen, firstly, to differentiate the incidence of *right to be forgotten* in the television media and in the Internet, creating specific limiting standards for each one of them—although to a certain extent, complementary.

In relation to the television media, two relevant judgments, with different results, had the application of the *right to be forgotten* and established limits for application of the rule.

In the first, Special Appeal 1,334,097, the convicted persons who had already spent some time in jail for a crime that was nationally known as the "Candacária massacre" in 1993, as well as others accused of the same crime who were judged not guilty, intended to prevent the transmission of a documentary that recalled the crime. In the report of the issued decision, it was said:

> In the case hereof, the main aspect of the dispute is the lack of contemporaneousness of the information about past facts, which reopened old wounds already overcame by the plaintiff and resumed the distrust of the society concerning his nature. The Plaintiff aims at the proclamation of his right to be forgotten, a right not be remembered against his will, especially concerning discrediting facts, of criminal nature, in which he was involved but in relation to which he was subsequently judged not guilty.[16]

Indeed, as mentioned before, the decision established the distinction between the application of the right to be forgotten in the television media and in the Internet, providing that the second "challenges solutions of technical character, with attention, for example, to the possibility of sharing of information and international circulation

[15]The Amnesty Law granted, pursuant to its art. 1, "amnesty to all those who, during the period between September 2, 1961 and August 15, 1979, committed political or similar crimes, electoral crimes, had suspended political rights and the employees of the Direct and Indirect Management, foundations linked to the government, the employees of the Legislative and Judiciary Branches, the Military and the trade union directors and representatives, punished based on Institutional and Complementary Acts." The period referred to in the law was marked by the occurrence of a Military Dictatorship in Brazil, which historical context may be better known in the series: Gaspari (2002a, b, 2003, 2004).

[16]REsp (Special Appeal) 1334097/RJ, Judge Rapporteur LUIS FELIPE SALOMÃO, Fourth Panel, judged on 05/28/2013. Available at: https://ww2.stj.jus.br/websecstj/cgi/revista/REJ.cgi/ITA?seq=1239004&tipo=0&nreg=201201449107&SeqCgrmaSessao=&CodOrgaoJgdr=&dt=20130910&formato=HTML&salvar=false.

of content, which can touch sensitive themes, such as the sovereignty of the states-nations".

It was also recognized that the cases of such nature, in which the right to be forgotten is attempted, are hypotheses of conflict between fundamental rights, mainly the privacy and the freedom of speech (materialized by the right to information).

Concerning the freedom of press (and, therefore, freedom of communication), the Superior Court of Justice, firstly, reaffirmed the consolidated position related to the limits of such right. It said in the decision: "the freedom of press, by not being absolute, has some limitations, some as '(I) the ethic commitment with the credible information; (II) the preservation of the so-called personality rights, including among it the rights to honor, to image, to privacy and to intimacy; and (III) the prohibition of broadcasting of journalism criticism aiming at defaming, slandering or insulting the person *(animus injuriandi vel diffamandi)*".

In compliance with those limits, the following criticism to the right to be forgotten was dialectically accepted:

> i) the acceptance of the so-called right to be forgotten is an attack to the freedom of speech and press; ii) the right to make disappear the information showing a person shall mean a loss of the history itself, which means that the right to be forgotten violates the right to memory of the whole society; iii) to think about a right to be forgotten is a sign that privacy is the censorship of our time; iv) the aforementioned right to be forgotten would be in conflict with the very idea of rights, because they have the ability to regulate the relationship between the individual and the society, while the first pretends that such relationship does not exist - a "delusion of modernity"; v) the right to be forgotten would have the power to make disappear records about perverse crimes and criminals, which entered to the social, police and judicial history, information of undeniable public interest; vi) a thing is, in its essence, whether lawful or unlawful, and it is not possible that a lawful information becomes unlawful only by the elapsing of time; vii) when someone is included in a fact of collective interest, the protection of intimacy and privacy is mitigated to the benefit of the public interest and, in addition, a second publication (the memory, that is in conflict with the forgetting) does nothing more than reaffirm a fact that is already of public knowledge; viii) and, finally, that police shows reporting past events, such as cruel crimes or famous killers are and have always been absolutely normal in Brazil and abroad, being inherent to the journalism activity itself.

Taking into consideration the possible criticism and the limits of the right to information, the decision supported the right to be forgotten, based on the need to protect the human dignity through the right to privacy, observing the timely and social relevance context of the news. In conclusion, it was declared that "an exception to the right to be forgotten includes the genuinely historical facts - which historical character shall be analyzed on a case-by-case basis -, which public and social interest shall survive to the elapsing of time, provided that the narrative unrelated to those involved becomes impossible".

In addition, the limit of the right to be forgotten is the qualification of the fact which remembering as historically and socially relevant is wanted to be prevented and the impossibility of disengagement of the involved subject in the story-telling.

On the same basis, the second relevant case involving television media had a different result. It is the Special Appeal 1,335,153, in which the siblings of Aída

Curi, a victim of homicide occurred in 1958, which at the time of the facts was also of national repercussion, claimed the prohibition of a journalism documentary film which intended to tell the story many years later. They argued that the right to be forgotten is also applicable to the victim and her relatives, whose privacy is affected when they are exposed to traumatic events happened in the past.

In this judgment, exactly by the argument of historical character and impossibility of disengagement of the involved parties (in such case, the sister of the plaintiffs) in the fats, the sustaining of the right to be forgotten would represent, in a weighted judgement, an infringement to the right of information and to the freedom of press. The following was declared in the judgment[17]:

> With effect, the right to be forgotten now recognized for all, offender and offended parties, is not applicable to the case hereof which showed again, decades after the crime, an event that became of public domain, in a way that would make impossible the press activity for purposes of showing the Aida Curi case, without Aida Curi.

Critically analyzing the judgments described above, Daniel Sarmento appoints that the solution for the cases should give even more emphasis to the criterion of public interest of the information, with possible superposition of the protection to the freedom of communication in relation to the right to privacy—especially in cases related to the rights to be forgotten and media actuation, containing such nature of interests.

As a point of support to his arguments, Sarmento states:

> The discipline of the question cannot threaten the freedom of press, speech, right of access to information of public interest or the cultivation of history and collective memory. Therefore, there is no way to extend the right to be forgotten to the information that is of public interest and such interest, as already emphasized before, does not disappear only by the elapsing of time.[18]

On the other hand, the Superior Court of Justice already had the opportunity to face the theme of the right to be forgotten related to an Internet content. It occurred, firstly, in Special Appeal 1,316,921, which aimed at the deindexing of certain contents in the results of the Google search engine. Such case is similar to the Google *versus* González case and is useful to demonstrate the divergence of the understanding of the Brazilian Court and that issued by the European Court.

In the Brazilian case, a famous host of a children's shows known as Xuxa (whose name is Maria da Graça Xuxa Meneghel) filed a demand against Google do Brasil, intending to compel the search engine to "remove from its Internet search website

[17]REsp 1335153/RJ, Judge Rapporteur Minister LUIS FELIPE SALOMÃO, FOURTH PANEL, judged on 05/28/2013. Available at: https://ww2.stj.jus.br/websecstj/cgi/revista/REJ.cgi/ITA?seq=1237428&tipo=0&nreg=201100574280&SeqCgrmaSessao=&CodOrgaoJgdr=&dt=20130910&formato=HTML&salvar=false. It is worth to be mentioned that the case involving the murder of Aida Curi was, after the judgement by the Superior Court of Justice, submitted to the Federal Supreme Court, which will analyze if the aforementioned decision infringed the Constitution. Such judgment has not yet begun.

[18]Sarmento (2016), p. 230.

named GOOD SEARCH the results related to the search of the expression 'xuxa pedophile' or, also, any other that may associate the name of the plaintiff, written partially or fully and regardless of wording, either correct of mistaken, to any criminal practice".[19]

When appreciating the merits of the issue, the Superior Court of Justice reached the understanding that:

> The search engines cannot be obliged to eliminate from their system the results arising from the search for a certain term or expression, and not even the results that appoint to a specific picture or text, regardless of indication of the URL of the website where it is included. It is not possible, under the pretext of making more difficult the broadcasting of unlawful or offensive content in the web, repress the collective right to information. Taking into consideration the involved rights and the potential risk of infringement of each of them, the scales should tilt to the warranty of the freedom of information ensured by art. 220, 1st § of the Federal Constitution/88, mainly taking into consideration that the Internet represents, today, an important vehicle of mass social communication.

Despite the existence of a certain technical lack of accuracy, since the deindexation is a possible technical activity and promoted by Google by its own standards and criteria since the decision made by the European Court of Justice, the used central arguments clearly demonstrate the definition of preponderance of the right to information and to communication, refusing, without any exception, the right to be forgotten.

The same understanding was consolidated at the time of judgment of the Special Appeal 1,593,873, in which the plaintiff SMS (the name was preserved by the Court as a result of the discussed fact) intended "the definitive blocking of its search system made through her name, since they could lead to pages that show her nude images".[20]

In the legal basis to the decision, the Brazilian Court made expression mention to the European precedent Google v. González, showing the parameters, fundamentals and limits of the decision of the Community Court. However, the Superior Court of Justice understood again that:

> The role of the search engines is restricted to the identification of Internet pages where certain data or information, even unlawful, is being freely broadcast. As stated above, the appellant does not store information and images appointed by the appellee, therefore there is no reason to consider that it is responsible for them.
>
> Therefore, even that its search engines facilitate the access and consequent publishing of pages which content is potentially illegal, the fact is that those pages are public and compose the worldwide web of computers and, therefore, are shown in the result of the search

[19]REsp 1316921/RJ, Judge Rapporteur Minister NANCY ANDRIGHI, THIRD PAINEL, judged on 26/06/2012. Available at: https://ww2.stj.jus.br/websecstj/cgi/revista/REJ.cgi/ITA?seq=1161904&tipo=0&nreg=201103079096&SeqCgrmaSessao=&CodOrgaoJgdr=&dt=20120629&formato=HTML&salvar=false.

[20]REsp 1593873/SP, Judge Rapporteur Minister NANCY ANDRIGHI, THIRD PANEL, judged on 11/10/2016. Available at: https://ww2.stj.jus.br/websecstj/cgi/revista/REJ.cgi/ITA?seq=1553533&tipo=0&nreg=201600796181&SeqCgrmaSessao=&CodOrgaoJgdr=&dt=20161117&formato=HTML&salvar=false.

> engines. If the page has unlawful content, the offended party shall take measures aiming their own suppression, which with they will be automatically excluded from the virtual search results of the search engines.

After those considerations, the decision established the following general parameters to be applied to the search engines:

> As a result of the characteristics of the providers of search applications in the Internet, as summarized above, this Superior Court of Justice reached the understanding that the search engines: (i) are not responsible for the content of the result of the searches made by their users; (ii) cannot be obliged to exercise a previous control of the content of the results of searches made by each user; and (iii) cannot be obliged to delete from their system the results arising from the search of a certain term or expression.[21]

Concerning the position shown in the appointed decisions, part of the Brazilian doctrine has a critical position, appointing, especially, the mistake mentioned above concerning the deindexation possibility. It is the case, for example, of the lesson taught of André Nery Costa, to whom, in view of the technical impossibility of a search engine to modify its own content of information resulting from the search made, the deindexation as a form to effectively perform the right to be forgotten is, at least, possible and useful. Cota provides that "it cannot be tolerated that the contents of the websites shown in the search cannot be monitored, the only solution being to question the sites that may have directly disclosed the information".[22]

Even more sharp is the criticism of Daniel Bucar, for whom "the recognition of supposed technical barriers to the application of the right to be forgotten in the virtual environment allows to understand that that cybernetic space is immune to the accrual of any legal discipline".[23] The author argues that it can not be admitted that the virtual environment represents a space immune to law, "STJ itself, in view of the legislative vacuum about the Internet existing in Brazil, has already had the opportunity to act as the creator of a true primary source of law and delimit parameters for the accountability of the service provider for extracontractual damages occurred on the web".[24]

---

[21]Interpreting the precedents, Carlos Affonso Souza and Chiara Spadaccini de Teffé declare: "According to the judgments on the subject in the STJ, it is possible to state that the following precedent has been established: a. the search engines are not liable for the content of the results of searches carried out by their users on their platform nor can they be required to exercise prior control over such results, since they only show contents available on the web that relate to the expressions selected by the users themselves and entered into the search; b. the person who feels damaged should file a claim against the person directly responsible for any damage caused, that is, the one who actually published the illegal content in his/her website and not the search engine that indexes the information freely found on the web; and c. taking into consideration the rights involved and the potential risk of infringement of each one of them, the assurance of freedom of information shall prevail, especially considering that the Internet represents important means of communication today" (Souza and Teffé 2018).

[22]Costa (2013), p. 204.

[23]Bucar (2013), p. 5.

[24]Bucar (2013), p. 5.

Despite the establishment, on one hand, of a jurisprudential tendency not to impose on the search engines the responsibility for the implementation of the *right to be forgotten*, nor even for the form of deindexation, there is, on the other hand, the consolidation of a court practice that gives effectiveness to Law 12.965 of April 23, 2014, known as the *Civil Rights Framework for the Internet*, in relation to other types of application servers that, in theory, have the same functional reason of the *right to be forgotten*.

Indeed, those criticisms have recently received some repercussion in court decisions, but have not consolidated an understanding different from the reported one. Nevertheless, it is worth mentioning the judgment of Special Appeal 1,660,168, which decision, by majority vote, was issued to ensure the removal of the name of the complainant from news that involved her in a case of fraud in public contest—from which unlawful act she was found not guilty. In this case, whenever the name of the plaintiff was searched, the first results were related to the news about the alleged fraud. Thus, the decision ordered that the search engines should stop showing such news when the search was made only by the name of the applicant, while maintaining the information of the websites when the search was made for the fact itself. This is the abstract of the decision: "SPECIAL APPEAL. CIVIL LAW. ACTION FOR OBLIGATION TO DO. 1. OMISSION, CONTRADICTION OR DIMNESS. ABSENCE. 2. *EXTRA PETITA* JUDGMENT. NOT CONFIGURED. 3. SEARCH ENGINE IN THE INTERNET. PROTECTION OF PERSONAL INFORMATION. LEGAL POSSIBILITY OF THE PETITION. NO LINK BETWEEN NAME AND RESULT OF SEARCH. FACT PECULIARITIES. CONCILIATION BETWEEN INDIVIDUAL RIGHT AND COLLECTIVE RIGHT TO INFORMATION. 4. DAILY FINE APPLIED. EXHORBITANT INITIAL VALUE. EXCEPTIONAL REVISION. 5. PARTIALLY GRANTED SPECIAL APPEAL. 1. It is discussed the possibility of determining the breach of the link established by search application providers in the Internet between the name of the damaged party, used as exclusive search criterion, and the news appointed in the results. 2. The Court of origin has faced all issues submitted by the parties, deciding on the strict limits of the demand and refusing, in an express and coherent way, all the bases that formed the free conviction of the Court. 3. The precedents of this Higher Court have a reiterated understanding in order to exclude the responsibility of Internet search engines for the presented search results, recognizing the impossibility of assigning to them the role of censor and imposing on the damaged party the directing of its claim against content providers, responsible for the provision of undue content on the internet. Precedents. 4. There are, however, very exceptional circumstances in which it is necessary the punctual intervention of the Judiciary Branch to cease the link created in the databases of search engines between personal data and search results that are not relevant for the public interest to information, either by the eminently private content or by the elapsing of time. 5. In these exceptional situations, the right to privacy and to be forgotten, as well as the protection of personal data, should prevail in order to allow the involved persons to continue their lives with reasonable anonymity, with the discrediting fact not being frequently recalled and perpetuated by automated search systems. 6. The breach of the said link without the exclusion of the news also makes compatible the individual interest of the holder of personal data and collective interest of access to information, insofar as it makes it possible to locate the news to those who direct their search by providing search arguments related to the fact reported, but not to those who exclusively seek the personal data of the protected individual. 7. In the present case, after more than a decade since the reported fact, when the name of the applicant is informed as the exclusive search criterion, the first presented result remained a link to news of his possible involvement in a discrediting, but unproven fact, despite of the existence of a lot of subsequent information about him available on the worldwide web. 8. The arbitration of daily fine should be reviewed whenever its initial value is clearly disproportional, is negligible or excessive, as is the case hereof. 9. Partially granted special appeals."

This is because, as shown above, the Superior Court of Justice, facing the issue of the relationship of the providers with the non-pecuniary damages incurred by the Internet user, established criteria to delimit the liability from concepts taken from the *Civil Rights Framework for the Internet.*

In order to better understand the issue, it should be noticed that the regulatory definition of providers has been established by art. 5 of the Civil Rights Framework for the Internet, which provides:

> Article 5. For the purposes of this law, the terms below shall have the following meaning:
> (...)
> V - internet connection: enabling a terminal to send and receive data packages over the Internet, by assigning or authenticating an IP address;
> (...)
> VII - internet applications: the set of functionalities that can be accessed through a terminal connected to the Internet.

To that extent, in view of the regulated activities, two types of providers are equally established, both intermediates in the relationship of the user with the Internet: the access or connection provider and the application provider—the last category includes the search engines, although, as observed, according to the case law, they are subject to the differentiated liability rule.

On one hand, the liability of the access (or connection) providers is defined by art. 18 of the Civil Rights Framework for the Internet, which provides: "The internet connection provider shall not be held civilly liable for damages arising from content generated by third parties."

On the other hand, the liability of the application providers is established by art. 19 of the same law, according to which:

> In order to ensure freedom of expression and to prevent censorship, the internet application provider may only be held civilly liable for damages arising from content generated by third parties if, after a specific court order, it does not take measure, within the scope and technical limits of its service and within the established period, to make unavailable the content appointed as infringing, except as otherwise provided for in law.

The removal of the content of nudity scenes or of sexual acts of a private nature does nor require a court order, it being enough a simple notification by the user, pursuant to art. 21[25] of the *Civil Rights Framework for the Internet.*

As taught by Carlos Afonso Souza, "here lies perhaps one of the most heated controversies of the Law: the Marco Civil provides that intermediaries are only held liable if they fail to fulfill a Court order requesting the removal of content".[26] However, the author continues:

---

[25] Article 21. The Internet application provider that makes available content generated by third parties will be held liable for the violation of the privacy resulting from the disclosure, without the permission of its participants, of images, videos or other material containing scenes of nudity or private sexual acts when , upon receipt of notification by the participant or his/her legal representative, it does not promote, in a diligent manner, within the scope and technical limits of its service, the unavailability of such content.

[26] Souza and Lemos (2017).

> What the Marco Civil sets forth is a safeguard for application providers in the sense that they will only be held liable if they do not comply with a Court order requesting the removal of the offensive material. This provision does not prevent intermediaries from determining their own requirements for removing content once notified by the alleged victims of damages arising out from materials made available through their platforms.[27]

Despite the express wording of art. 19 of the *Civil Rights Framework for the Internet*, the Superior Court of Justice appears to have reached an expanded understanding of art. 21 of the *Civil Rights Framework for the Internet*, since it has established that the liability of the application provider for damages caused by a user, by disclosing content that is offensive to honor and privacy (not necessarily containing nudity or sex scene) is established from the provider's refusal to exclude the content by simple notification from the alleged victim. This is what is read, for example, in the decision of Special Appeal 1,568,935[28]:

> Thus, if someone makes available offensive material on his/her personal page, whether on his/her social network (Facebook, Google+, Instagram, Linkedin, etc.) or on a blog type website (Blogger, Wordpress, Tumblr, etc.), in which there are no editorial controls of content, the provider can not be held liable for making the website available to the offender. It will be held liable only if, after notification for removal, it does not take any action.

It is possible to interpret, from the precedents of the Superior Court of Justice, the adoption, even in an interpretative and functional extension of the legal provisions, of the "notice and take down" regime in view of the recognition by the victim of a violation of honor and privacy.

From this premise, it can be taken the conclusion that, even if the Superior Court of Justice has dismissed the imposition of conduct that gives effectiveness to the *right to be forgotten* to the "search engines" (kind of application provider), it promotes normative openness for the implementation of such right to other types of application providers—such as social networks, blogs and streaming channels.

This happens because, as stated above, the right to privacy, protected by arts. 19 and 21 of the Civil Rights Framework for the Internet, is protected by the Constitution as a category of fundamental right.[29] In addition, assuming that the damage to honor or privacy can emerge from the historical decontextualization of a legal past fact, the *right to be forgotten* is imposed on the application servers, at least, from a court order to suppress the content (art. 19 of the Civil Rights Framework for the Internet) or, also, by notifying the victim (art 21 of the Civil Rights Framework

[27]Souza and Lemos (2017).

[28]REsp 1568935/RJ, Judge Rapporteur Minister RICARDO VILLAS BÔAS CUEVA, THIRD PANEL, judged on 04/05/2016. Available at: https://ww2.stj.jus.br/processo/revista/documento/mediado/?componente=ITA&sequencial=1501300&num_registro=201501011370&data=20160413&formato=HTML.

[29]Such understanding, moreover, is in line with Resolution 20/8, adopted on 16 July 2012, by the United Nations Human Rights Council about the promotion, protection and enjoyment of human rights on the Internet in which it "affirms that the same rights that people have offline must also be protected online, in particular freedom of expression" and Resolution 28/16, adopted on 1 April 2015, which provides that equal rights offline and online include the right to privacy.

for the Internet)—following the general criteria applicable to the implementation of the *right to be forgotten* already established, for example, for television media, especially in relation to the weighting of the fundamental right to information and communication.

To that extent, despite the Superior Court of Justice having established the understanding that application providers are not obliged to monitor or interfere with the content included on the web by the users (such as, for example, in the Special Appeal 1,308,830[30]), the effectiveness of the *right to be forgotten* is ensured by determining, based on the precedents, the duty of the application provider to remove content that recalls a fact of the past that is in itself an attack on privacy and intimate life, under penalty of being held liable for the damages caused by the non-removed content.

[30]CIVIL AND CONSUMER. INTERNET. CONSUMER RELATIONSHIP. APPLICATION OF THE CONSUMER DEFENSE CODE. SERVICE FREE OF CHARGE. INDIFFERENCE. CONTENT PROVIDER. PRIOR INSPECTION OF THE CONTENT OF THE INFORMATION POSTED ON THE WEBSITE BY THE USERS. UNNECESSITY. OFFENSIVE CONTENT MESSAGE. NONPECUNIARY DAMAGE. RISK INHERENT TO THE BUSINESS. INEXISTENCE. AWARENESS OF EXISTENCE OF UNLAWFUL CONTENT. IMEDIATE REMOVAL. DUTY. AVAILABILITY OF MEANS FOR IDENTIFICATION OF EACH USER.

DUTY. IP NUMBER REGISTRATION. SUFFICIENCY. 1. The commercial exploration of the Internet subjects the consumer relationships arising therefrom to Law No. 8.078/90. 2. The fact that the service provided by the internet service provider is free of charge does not eliminate the consumer relationship, since the meaning of remuneration, set forth in art. 3, 2nd paragraph of the Consumer Defense Code shall be wide, in order to include the supplier's indirect gain. 3. The prior inspection, by the content provider, of the content of the information posted on the web by each user is not an activity intrinsic to the provided service, therefore it can not be considered as in failure, according to art. 14 of the Consumer Defense Code, the website that does not examine and filter the data and images included therein. 4. The nonpecuniary damage resulting from messages with offensive content included in the website by the user does not constitute a risk inherent to the activity of the content providers, therefore the strict liability provided for in art. 927, sole paragraph, of the Civil Code/02 does not apply. 5. When informed that a text or image has unlawful content, the provider must act vigorously, removing the material immediately, under penalty of being jointly and severally liable with the direct author of the damage, due to the omission practiced. 6. By offering a service through which users are allowed to freely express their opinion, the content provider must be careful to provide means for identifying each of those users, preventing anonymity and assigning to each manifestation a certain and established author. From the point of view of the average diligence expected from the provider, it must take the measures that, according to the specific circumstances of each case, are within its reach for the individualization of the website users, under penalty of subjective liability for fault *in omittendo*. 7. The initiative of the content provider to maintain a channel for complaints in a virtual social network hosting website is commendable and consistent with the expected position in the provision of this type of service—to maintain means that allow identification of each user (and any abuse practiced by it)—but the mere provision of the tool is not enough. It is critical that there is effective adoption of measures to investigate and settle the complaints made, keeping the complainant informed of the measures taken, otherwise it will create only a false sense of security and control. 8. Special appeal not granted. (REsp 1308830/RS, Judge Rapporteur Minister NANCY ANDRIGHI, THIRD PANEL, judged on 05/08/2012, DJe 06/19/2012).

4 Conclusion Notes

It is undisputable the fact that legal science, now inside the multidimensional space of technological relations referred to as the Internet, has much to adapt in order to present effective solutions to the litigation that will come from this new architecture of relational mechanisms.

This adaptation is actually more complex in the absence of a normative system forged for this new relational environment, especially reaching the *right to be forgotten*, especially when taking into consideration the almost infinite capacity of data storage in the network.

Notwithstanding the difficulties, the Brazilian precedents begin to construct, even initially, by means of hermeneutical mechanisms applied to the constitutional rules, the interpretative paths that protect the data, especially in view of the decontextualized retrieval of private information. It was precisely the path taken until today that was sought to be demonstrated, establishing an analytical cut in the decisions issued by the Superior Court of Justice, which in a significant part of the judgments considered the right to be forgotten in view of the right to information and to freedom of expression.

It was observed, therefore, that even with little technical strictness, the Court decided for the prevalence of the right to information, evoking a right to be forgotten only when the subject can detach himself/herself from the re-telling of the story and this does not prevent the recalling of a socially relevant fact.

Subsequently, it has been observed that the Superior Court of Justice, in a more extensive interpretation of the Civil Rights Framework for the Internet (Law 12.965/2014), removed the *right to be forgotten* from cases of non-contractual liability for the conduct of search engines, but keeping it in relation to the application providers, adopting the so-called *notice and take down* regime, where the social network is notified, for example, so that it deletes something and, if it does not do so, is liable for damages to honor and privacy.

It is therefore perceived that the right to be forgotten, in the new technological context in which it is included, imposes a critical reflection that aims at the guarantee of effective protection of the constitutionally protected values.

References

Ascensão JO (2002) Direito da Internet e da sociedade da informação. Forense, Rio de Janeiro

Bucar D (2013) Controle temporal de dados: o direito ao esquecimento. Civilistica.com. http://civilistica.com/controle-temporal-de-dados-o-direito-ao-esquecimento/. Accessed 22 Mar 2017

Costa ABN (2013) Direito ao esquecimento na Internet: a *Scarlet letter* digital. In: Direito e mídia, Coord. Anderson Schreiber. Atlas, São Paulo, pp 184–206

de Lima CRP (2015) Direito ao esquecimento e Internet: o fundamento legal no direito comunitário europeu, no direito italiano e no direito brasileiro. In: Doutrinas Essenciais de Direito Constitucional. Editora Revista dos Tribunais, São Paulo, pp 511–543

Gaspari E (2002a) A Ditadura Envergonhada. Companhia das Letras, São Paulo

Gaspari E (2002b) A Ditadura Escancarada. Companhia das Letras, São Paulo
Gaspari E (2003) A Ditadura Derrotada. Companhia das Letras, São Paulo
Gaspari E (2004) A Ditadura Encurralada. Companhia das Letras, São Paulo
Martins GM (2014) O Direito ao esquecimento na Internet. In: Direito Privado e Internet, coord. Guilherme Magalhães Matins. Atlas, São Paulo, pp 3–28
Perlingieri P (2008) O Direito Civil na legalidade constitucional (trans: de Cicco MC). Renovar, Rio de Janeiro
Prigogine I (2011) O fim das certezas. Editora Unesp, São Paulo
Sarmento D (2016) Liberdades comunicativas e "direito ao esquecimento" na ordem constitucional brasileira. In: Revista Brasileira de Direito Civil. IBDC, São Paulo, pp 190–232
Schreiber A (2013) Direitos da Personalidade. Atlas, São Paulo
Sen A (1999) Development as freedom. Oxford University Press, New York
Souza CA, Lemos R (2017) Brazilian courts and the internet – rulings before and after the Marco Civil on intermediary liability. https://publixphere.net/i/noc/page/OI_Case_Study_Brazilian_Courts_and_the_Internet. Accessed 18 May 2017
Souza CA, Teffé CS (2018) O STJ e o direito ao esquecimento. https://www.jota.info/opiniao-e-analise/artigos/o-stj-e-o-direito-ao-esquecimento-05042018. Accessed 07 May 2018
Warren SD, Brandeis LD (1890) The right to privacy. Harv Law Rev 4:193–220

Analyzed Decisions

REsp 1308830/RS, Judge Rapporteur Minister NANCY ANDRIGHI, THIRD PANEL, judged on 05/08/2012. Available at: https://ww2.stj.jus.br/websecstj/cgi/revista/REJ.cgi/ITA?seq=1142916&tipo=0&nreg=201102574345&SeqCgrmaSessao=&CodOrgaoJgdr=&dt=20120619&formato=PDF&salvar=false
REsp 1316921/RJ, Judge Rapporteur Minister NANCY ANDRIGHI, THIRD PANEL, judged on 06/26/2012. Available at: https://ww2.stj.jus.br/websecstj/cgi/revista/REJ.cgi/ITA?seq=1161904&tipo=0&nreg=201103079096&SeqCgrmaSessao=&CodOrgaoJgdr=&dt=20120629&formato=HTML&salvar=false
REsp (Special Appeal) 1334097/RJ, Judge Rapporteur LUIS FELIPE SALOMÃO, FOURTH PANEL, judged on 05/28//2013. Available at: https://ww2.stj.jus.br/websecstj/cgi/revista/REJ.cgi/ITA?seq=1239004&tipo=0&nreg=201201449107&SeqCgrmaSessao=&CodOrgaoJgdr=&dt=20130910&formato=HTML&salvar=false
REsp 1335153/RJ, Judge Rapporteur Minister LUIS FELIPE SALOMÃO, FOURTH PANEL, judged on 05/28/2013. Available at: https://ww2.stj.jus.br/websecstj/cgi/revista/REJ.cgi/ITA?seq=1237428&tipo=0&nreg=201100574280&SeqCgrmaSessao=&CodOrgaoJgdr=&dt=20130910&formato=HTML&salvar=false
REsp 1568935/RJ, Judge Rapporteur Minister RICARDO VILLAS BÔAS CUEVA, THIRD PANEL, judged on 04/05/2016. Available at: https://ww2.stj.jus.br/processo/revista/documento/mediado/?componente=ITA&sequencial=1501300&num_registro=201501011370&data=20160413&formato=HTML
REsp 1593873/SP, Judge Rapporteur Minister NANCY ANDRIGHI, THIRD PANEL, judged on 11/10/2016. Available at: https://ww2.stj.jus.br/websecstj/cgi/revista/REJ.cgi/ITA?seq=1553533&tipo=0&nreg=201600796181&SeqCgrmaSessao=&CodOrgaoJgdr=&dt=20161117&formato=HTML&salvar=false
Resp 1660168/RJ, Judge Rapporteur. for Appellate Decision Minister MARCO AURÉLIO BELIZZE, THIRD PANEL, judged on 05/08/2018. Available at: https://ww2.stj.jus.br/websecstj/cgi/revista/REJ.cgi/ITA?seq=1628798&tipo=0&nreg=201402917771&SeqCgrmaSessao=&CodOrgaoJgdr=&dt=20180605&formato=PDF&salvar=false

Rapport Canadien: Le déréférencement à l'ère numérique – une approche hybride pour faire le pont entre la vision européenne et américaine du « droit à l'oubli »

Karen Eltis and Pierre Trudel

> *"The one thing [the victorious plaintiff] Costeja did not want us to know about him is now the only thing the entire world knows about him" (Commentaires du comédien John Oliver au sujet des conséquences inattendues de la l'arrêt* Costeja *de 2014, en ligne,* You Tube, < *https://www.youtube.com/watch?v=r-ERajkMXw0* > *à la cinquième miniute de la séquence.)*
>
> *« Dis-moi ce que tu oublies, je te dirai qui tu es » (Augé 2001 : 26).*

Résumé Ce rapport analyse la notion « hybride', du déréférencement en droit canadien. Un droit certes en émergence mais qui représente un réel potentiel de procurer les équilibres qui semblent nécessaires afin de garantir à la fois la libre circulation de l'information et la dignité des personnes. Il est bien établi au Canada que les tribunaux peuvent, après avoir constaté qu'un document se trouvant sur Internet contrevient aux lois, ordonner sa suppression. Les juges ordonnent aussi la suppression des liens hypertexte menant à des documents contraires aux lois. Il y a donc un équilibre entre le droit de rechercher librement des informations et le droit de protéger la vie privée et la réputation. Dès lors qu'il est démontré qu'un document viole la réputation d'une personne ou sa vie privée, un juge peut ordonner sa suppression de même que l'effacement des hyperliens qui y conduisent. Le « droit à l'oubli » tel que décrit ici-haut doit être distingué du droit de faire effacer les liens générés par un moteur de recherche crée par la décision de la Cour de justice européenne. En droit canadien, on ne peut postuler que le droit de la protection des renseignements personnels procure un droit au déréférencement des résultats de

K. Eltis (✉)
Faculté de droit, Université d'Ottawa, Ottawa, ON, Canada

Princeton University CITP (2016–2018), Princeton, NJ, USA
e-mail: Karen.Eltis@uOttawa.ca

P. Trudel
Faculté de droit, Centre de recherche en droit public, Université de Montréal, Montreal, QC, Canada
https://www.pierretrudel.net

F. Werro (ed.), *The Right To Be Forgotten*, Ius Comparatum – Global Studies in Comparative Law 40, https://doi.org/10.1007/978-3-030-33512-0_14

recherche. La garantie constitutionnelle de la liberté d'expression, entendue comme protégeant la liberté de rechercher des informations ne contrevenant pas à la loi, s'oppose à une application d'un « droit au déréférencement » qui prétendrait se réclamer du droit à la protection des renseignements personnels.

Abstract This report analyses the "hybrid" notion of dereferencing in Canadian law. An emerging right representing a real potential to provide the balances that seem necessary to ensure both the free flow of information and the dignity of people. It is well established in Canada that courts may, after finding that a document on the Internet is in violation of the law, order its removal. Judges also order the removal of hypertext links to documents that are contrary to the law. There is therefore a balance between the right to freely seek information and the right to protect privacy and reputation. Once it is shown that a document violates a person's reputation or privacy, a judge may order its removal as well as the deletion of the hyperlinks that lead to it. The "right to be forgotten" as described above must be distinguished from the right created by the decision of the European Court of Justice to erase links generated by a search engine. In Canadian law, it can not be assumed that privacy law provides a right to erase search results. The constitutional guarantee of freedom of expression, understood as protecting the freedom to search for information that does not contravene the law, precludes the application of a "right to dereference" which would be linked to the right to the protection of personal information.

1 Introduction

La technologie joue un rôle incontestablement crucial dans nos récits personnels, et ce au delà des frontières envisagées par les traditions juridiques, ancrées fermement dans les logiques territoriales. Compte tenu du rôle ubiquitaire du réseau, il est apparu très tôt que l'usage d'Internet n'est en pratique possible que moyennant la disponibilité des moteurs de recherche ces outils capables d'identifier rapidement l'information qui est susceptible de répondre aux besoins de l'internaute.

Les moteurs de recherche agglomèrent des informations sur les personnes ou sur les diverses entités à propos desquelles on peut formuler une requête de recherche exprimée habituellement sous forme de mots-clés. Ils sont essentiels aux internautes qui ont à trouver de l'information. Ils délivrent à l'utilisateur des documents ou des liens à des documents doivent être le plus pertinents possibles eu égard à sa requête. Les informations sont repérées dans des espaces virtuels; elles sont en principe publiques. La plupat du temps, les moteurs de recherche tiennent compte du contexte et de l'historique de recherche de l'utilisateur. Il est donc pratiquement impossible de postuler qu'une requête de recherche sur un nom spécifique va forcément générer les mêmes résultats.

Par leur efficacité, les moteurs de recherche contribuent puissamment à réduire les phénomènes d'«obscurité pratique » qui rendent souvent difficiles l'accès et la compilation d'un ensemble de documents portant sur une personne ou sur un sujet déterminé. C'est probablement ce qui explique qu'ils ont été ciblés comme des ressources susceptibles de mettre à mal le « droit à l'oubli ».

L'émergence des technologies de l'information s'inscrit dans un contexte socioculturel appelant un cadre normatif vivant et évolutif. C'est pourquoi ce rapport propose de contribuer à l'amélioration de la protection des droits de la personne par la promotion d'une meilleure cohérence des principes fondamentaux dans les environnements désormais sans frontières engendrés par les logiques du numérique. Pour rendre compte du droit en émergence du fait de la « révolution technologique » qui ne connaît pas de frontières, le rapport met de l'avant la perspective bi juridique canadienne et la méthode comparatiste[1].

En cohérence avec son système de droit civil comme de Common Law, le Canada est particulièrement bien placé pour « faire le pont » entre traditions européennes et américaines et apporter un éclairage sur les grandeurs et faiblesses, les tenants et aboutissants du « droit à l'oubli »[2]. Le champ de vision bi juridique et bilingue procure des opportunités de transcender « les conceptions orthodoxes » du droit[3].

Ce rapport se penche donc sur la notion « hybride', du déréférencement en droit canadien. Un droit certes en émergence mais qui représente un réel potentiel de procurer les équilibres qui semblent nécessaires afin de garantir à la fois la libre circulation de l'information et la dignité des personnes.

Dans un premier temps, nous abordons les particularités du cadre normatif canadien et québécois (en réponse aux questions 1–5). Dans un deuxième temps, nous proposons quelques observations d'ordre général et contextuel pour mieux caractériser l'accueil réservé au Canada à l'arrêt Costeja (Q6-11) de la Cour de justice de l'Union européenne. Enfin, nous proposons en conclusion certaines pistes à envisager afin de répondre aux défis de la protection des droits dans l'espace virtuel (Q 12).

---

[1]Voir pour des analyses de l'état du droit canadien sur ce sujet: Gratton and Polonetsky (2017); Saint-Laurent (2015), pp. 185–197; Pierre Trudel, « Moteurs de recherche, déréférencement, oubli et vie privée en droit québécois », (2016) 21 *Lex electronica* 89. En ligne : http://www.lex-electronica.org/s/1535.

[2]Eltis (2011), pp. 69–95; Eltis (2016a), pp. 355–380.

[3]Selon les mots du professeur Robert Leckey, doyen de la Faculté de droit de l'Université McGill, voir à : < https://www.mcgill.ca/law/about/deans-welcome >.

2 La protection canadienne du droit à l'oubli : Une protection assurée par le droit commun (Questions 1-5)

Il est bien établi au Canada que les tribunaux peuvent, après avoir constaté qu'un document se trouvant sur Internet contrevient aux lois, ordonner sa suppression. Les juges ordonnent aussi la suppression des liens hypertexte menant à des documents contraires aux lois. Il y a donc un équilibre entre le droit de rechercher librement des informations et le droit de protéger la vie privée et la réputation. Dès lors qu'il est démontré qu'un document viole la réputation d'une personne ou sa vie privée, un juge peut ordonner sa suppression de même que l'effacement des hyperliens qui y conduisent.

En droit civil québécois les tribunaux appliquent les règles générales de la responsabilité civile. Dans ce cadre, ils ont considéré qu'en certaines circonstances, on peut commettre une faute de violation d'oubli. Certains auteurs en ont déduit l'existence d'un droit à l'oubli. Ainsi, le rappel d'évènements survenus dans le passé a été jugé fautif lorsqu'il y a absence de démonstration d'un intérêt public de la part de la personne qui a fait la divulgation. Les exemples suivants sont tirés de la jurisprudence québécoise.

Dès 1889, la Cour supérieure du Québec a estimé qu'un journal *Le Violon* avait eu tort de faire revivre certaines «accusations depuis longtemps oubliées» concernant le demandeur. Cette décision fut confirmée par la Cour de révision[4]. Plus récemment, dans l'affaire *Lévesque*[5], la Cour Supérieure a dû trancher sur une revendication du droit à l'oubli. Le requérant, Lévesque, poursuivait le *Journal de Québec* pour avoir rappelé le crime qu'il avait commis deux ans auparavant dans une «piquerie» de la ville de Montréal. Lévesque avait alors été impliqué dans une bagarre entre groupes criminels. La juge a conclu que le *Journal* n'avait pas commis de faute étant donné que l'information dévoilée était accessible au public. De plus, puisque l'objet de l'article portait sur l'incendie de la « piquerie » où Levesque avait jadis commis son crime, l'information divulguée demeurait d'intérêt public.

Par contre, dans une affaire semblable[6], Gilbert Ouellet a poursuivi le journal *Photo-Police* pour avoir publié un article relatant le crime commis par sa défunte épouse dix ans auparavant. Cette dernière avait tué leurs quatre enfants pour ensuite s'enlever la vie. Le juge de la Cour du Québec a conclu que l'article publié était «sensationnaliste» et qu'il ne pouvait être justifié par l'intérêt du public à l'information. Dans une autre affaire de la Cour du Québec[7], le juge Barbe rappelle qu'il est difficile pour celui qui participe à des « activités publiques de nature politique » d'invoquer un droit à l'oubli. Se référant aux propos de la juge Piché

[4]*Goyette* c. *Rodier* (1889) 20 R.L. 108,110 (C. Rév).

[5]*Lévesque* c. *Communications Quebecor inc.* (C.S., 1999-06-21), SOQUIJ AZ-99021730, J.E. 99-1527, [1999] R.R.A. 681.

[6]*Ouellet* c. *Pigeon*, REJB 1997-03106, 1997 (C.Q.).

[7]*Mathieu* c. *Presse ltée (La)*, (C.Q., 1998-11-24), SOQUIJ AZ-99036093, B.E. 99BE-169.

dans *Szabo c. Morissette*[8], le juge de la Cour du Québec mentionne que « celui qui est à l'origine de l'histoire ne peut blâmer d'autres que lui-même s'il n'a pas aimé qu'on parle de lui ».

La faute de violation de l'oubli telle que reconnue en droit québécois découle de la diffusion d'une information autrefois connue mais en conférant à celle-ci une portée temporelle et spatiale différente de celle découlant de la diffusion initiale. Ce qui est jugé fautif et sanctionné est la redivulgation considérée injustifiable dans le contexte où elle se produit. À ce titre, la légitimité juridique de la prétention à l'oubli se structure par l'appréciation du contexte de la diffusion de l'information. L'oubli est donc un droit pour la personne concernée lorsqu'il est jugé déraisonnable de diffuser l'information. Alors, la diffusion est jugée fautive, c'est à dire qui n'aurait pas été faite par une personne raisonnable oeuvrant dans des circonstances analogues. Le contexte dans lequel s'effectue la diffusion est un facteur très important dans ce processus de détermination de son caractère fautif

2.1 Les limites du droit à l'oubli

Le « droit à l'oubli » tel que décrit ici-haut doit être distingué du droit de faire effacer les liens générés par un moteur de recherche crée par la décision de la Cour de justice européenne. En droit canadien, on ne peut postuler que le droit de la protection des renseignements personnels procure un droit au déréférencement des résultats de recherche[9].

La garantie constitutionnelle de la liberté d'expression, entendue comme protégeant la liberté de rechercher des informations ne contrevenant pas à la loi, s'oppose à une application d'un « droit au déréférencement » qui prétendrait se réclamer du droit à la protection des renseignements personnels. Dans *Alberta (Information and Privacy Commissioner)* c. *Travailleurs et travailleuses unis de l'alimentation et du commerce, section locale 401*[10], la Cour suprême du Canada, dans une décision unanime, a invalidé la loi albertaine sur la protection des renseignements personnels en ce qu'elle prohibait la prise d'images dans un lieu public.

La Loi attaquée interdisait de recueillir des renseignements, ici des images de personnes franchissant une ligne de piquetage et ne comportant aucune information intime, sans le consentement de celles-ci. Dans la situation présentée à la Cour, aucun détail concernant le mode de vie ou les choix personnels des intéressés n'avait été dévoilé. Or, la loi, à l'instar des autres lois sur la protection des renseignements personnels en vigueur au Canada ne fait aucune distinction : tout renseignement

[8](1993) R.R.A. 554, J.E. 93-1385 (C.S.).

[9]*C.L.* c. *BCF Avocats d'affaires*, 2016 QCCAI 114 (CanLII), 14 avril 2016, en ligne : <http://canlii.ca/t/gr5q0>.

[10][2013] 3 RCS 733.

personnel y est traité de la même façon, même ceux qui ne relèvent pas de la vie privée. Il est interdit de le collecter et de le diffuser sans consentement sauf pour des motifs définis de manière très étroite. C'est cette absence de possibilité de laisser un espace à l'exercice des autres droits fondamentaux qui rend excessive la loi sur la protection des renseignements personnels. En somme, la Cour fait échos à une évidence : il existe des renseignements portant sur les personnes qui ne relèvent pas de la vie privée de celles-ci. La Cour rappelle la nécessité de baliser les interdictions se trouvant dans les lois sur la protection des renseignements personnels. Telles que rédigées, ces lois interdisent de capter, conserver et diffuser toute information relative à une personne identifiable sans sa permission et cela même lorsqu'elle se trouve dans des lieux publics. La Cour juge que de tels interdits limitent la liberté d'expression de façon déraisonnable.

La Cour explique que ces lois doivent comporter des balises afin de permettre l'exercice des activités expressives ne portant pas sur des matières relevant de l'intimité des personnes. La Cour a jugé que la loi albertaine sur la protection des renseignements personnels empêchait de recueillir des renseignements personnels, telles que des prises d'images ou des vidéos lors d'une manifestation au cours de laquelle le public pouvait facilement observer les personnes qui y prennent part.

Cette décision de la Cour suprême invalide l'approche qui a prévalu au Canada depuis plus de trois décennies en matière de protection des renseignements personnels. Portées par un mouvement qui semble postuler que la vie privée est le seul droit fondamental à devoir être protégé, ces lois ignorent pratiquement les impératifs de la libre circulation de l'information dans les espaces publics. En invalidant la loi albertaine, la Cour met fin à ce déséquilibre.

Bien sûr la Cour reconnaît la légitimité de protéger le droit à la vie privée et d'assurer que la collecte et la communication de renseignements personnels soient encadrées. Mais elle vient rappeler que tout renseignement personnel n'est pas automatiquement un renseignement sur la vie privée d'une personne, surtout s'il s'agit d'un renseignement se trouvant légitimement dans l'espace public. Il est donc excessif de considérer tout renseignement personnel comme étant assujetti a son bon vouloir. Les libertés expressives imposent de baliser la faculté de l'individu à l'égard des informations le concernant en tenant compte des droits des autres et du public en général.

Pour l'heure, bien qu'il porte sur un phénomène passablement distinct de ceux qui sont concernés par les moteurs de recherche, ce prononcé de la Cour suprême laisse planer d'importants doutes sur la possibilité, en droit canadien, d'un droit au déréférencement qui se fonderait sur les principes issus des lois sur la protection des renseignements personnels.

Par contre, les moteurs de recherche génèrent des liens hypertextes. Dans *Crookes c. Newton*[11], la Cour suprême du Canada a examiné la question de savoir si l'incorporation dans un texte d'hyperliens menant à des propos prétendument diffamatoires équivaut à la « diffusion » de ces derniers. Selon les six juges

[11][2011] 3 RCS 269.

majoritaires de la Cour, une personne ne peut en diffamer une autre simplement en publiant un hyperlien menant au site Web ou à un document d'un tiers qui contient des propos diffamatoires : la juge en chef McLachlin et le juge Fish expliquent qu'« […] un hyperlien, en lui-même, ne devrait jamais être assimilé à la "diffusion" du contenu auquel il renvoie »[12]. Ils souscrivent à l'analyse de la Juge Abella qui écrit que : « Le fait de mentionner l'existence d'un contenu et/ou l'endroit où il se trouve par le biais d'un hyperlien ou de toute autre façon, sans plus, ne revient pas à le diffuser »[13].

Les juges majoritaires estiment que les hyperliens s'apparentent aux notes de bas de page d'un document papier. Si l'hyperlien offre, contrairement aux notes de bas de page, un accès immédiat au site Web d'un tiers auquel il renvoie, les lecteurs n'en savent pas moins que ce lien les mènera à une source différente.

2.2 *Les recours, la mise en œuvre et l'efficacité*

Les recours en responsabilité civile sont les voies privilégiées pour sanctionner les diffusions intempestives et fautives de faits passés qui ne correspondent pas à un intérêt public démontrable. Quant au droit au déréférencement, il est possible en vertu des principes du droit commun dès lors qu'il est démontré que le document vers lequel pointe un hyperlien contrevient à la loi.

Dès lors qu'il est judiciairement établi qu'un document fautif est en ligne, les tribunaux peuvent rendre les décisions et ordonnances nécessaires afin de protéger les droits des personnes. Au Canada, le droit à l'oubli s'entend du droit de faire supprimer l'information fautive ramenant dans le présent des faits passés. Il est rarement invoqué sauf dans les situations ou les rappels de faits passés ne sont pas d'intérêt public.

3 Une vue contextuelle du droit à l'oubli : l'accueil réservé à l'arrêt Costeja au Canada (Questions 6-11)

L'exemple le plus fragrant des changements qualitatifs entrainés par l'ère numérique semblerait-il est celui d'un phénomène que le quotidien Britannique *The Guardian'* surnomme « the Internet's "law of unintended consequences" »[14].

Le cas de M Costeja (précité) est maintenant bien connu. Moins connu est celui de son prédécesseur, qui a en réalité déclenché le débat entourant le « droit à l'oubli ».

[12] Au paragraphe 47 de la décision.

[13] Au paragraphe 42 de la décision.

[14] http://www.theguardian.com/world/blog/2014/may/14/mario-costeja-gonzalez-fight-right-forgotten.

Il s'agit d'un médecin espagnol nommé, Hugo Guidotti Russo. Les faits de cette cause, qui ne s'est pas rendue devant les tribunaux, sont pourtant très saillants et semblent résumer la problématique bien mieux que ceux de M Costeja[15].

Il se résument comme suit. Il y a plus de vingt cinq ans, une des patientes du docteur Rossi a allégué la négligence professionnelle[16]. La dispute, alors documentée dans un journal local, fut réglée et le docteur Rossi pratiqua depuis sans incident. Néanmoins, et compte tenu de la nature même d'Internet, toute recherche à son sujet produisait sans faute les allégations publiées dans ce reportage choquant, cru et décontextualisé. Inévitablement, cela avait pour effet d'éclipser la résolution amiable du dossier (hors cour) et d'entacher vingt ans de pratique professionnelle respectable.

Les moteurs de recherche génèrent des liens hypertextes en appliquant des algorithmes. Ces algorithmes fonctionnent de façon automatisée, en des temps s'exprimant en fractions de secondes. Pour y arriver, ces algorithmes[17] analysent des masses considérables de données incluant notamment des données sur le contexte dans lequel se trouve la personne qui introduit la requête de recherche, son historique de recherche, sa position géographique etc.

Par contre, en raison du caractère contextuel de ce traitement algorithmique, il paraît impossible de postuler que l'introduction des mêmes mots dans des requêtes de recherche lancées en différents points du réseau par des personnes ayant forcément un historique de recherche différent va forcément générer les mêmes résultats.

On peut même se demander si le type de résultats qu'obtient un individu en introduisant son propre nom dans une requête de recherche sera identique à celui obtenu par une autre personne se trouvant dans un contexte donnant à penser qu'elle a des intérêts éloignés de ceux de la première personne.

Mais la Cour de justice de l'Union européenne semble avoir ignoré cet aspect pourtant fondamental. Elle semble avoir été préoccupée par une perception au sujet d'un autre type de décontextualisation. Sa décision a ultimement exigé que les moteurs de recherche (qualifiées de contrôleurs de données) déréférencent les résultats que l'on prétend être décontextualisés[18] en vertu du « droit à l'oubli » qu'elle déduit de l'article 12 de la Directive européenne 95/46/EC.

Comme exposé ailleurs dans ce texte, les tribunaux canadiens sont soucieux de respecter la liberté d'expression, valeur constitutionnelle enchassée dans la Charte

---

[15]Eltis (2016b).

[16]Paul Sonne, Max Colchester, & David Roman, "Plastic Surgeon and Net's Memory Figure in Google Face-Off in Spain" *The Wall Street Journal* (7 March 2011).

[17]Un algorithme est un ensemble d'instructions donné à un ordinateur permettant d'obtenir un résultat en effectuant des calculs. Voir : Ghatnekar (2012–2013), p. 171.

[18]*Ibid.* As Bernal explains, only search results arising from a search under a particular name are removed. Neither the underlying source material itself, nor the same (contentious) search results obtained when searched for in any other way are required to be removed. See e.g. Paul Bernal, "Is Google Undermining the « right to be forgotten'?", *CNN Opinion* (7 July 2014), online: CNN <http://www.cnn.com/2014/07/07/opinion/bernal-google-undermining-privacy-ruling/>.

canadienne (et québécoise). La plupart des groupes ayant pris part à une consultation du Commissaire à la vie privée sur le droit au déréférencement ont estimé qu'un tel droit est difficilement conciliable avec les impératifs de la liberté d'expression telle qu'elle est comprise en droit canadien.

Au nom des mêmes préoccupations, la décision fut aussi attaquée aux États-Unis et plus tard au Royaume Uni[19] et dénoncée comme : [TRADUCTION] « la plus importante menace à la liberté d'expression ».[20] Cette critique trahit une divergence importante entre la vision continentale de la vie privée, axée sur la protection procédurale des données et l'approche absolutiste des États-Unis qui pour sa part repose sur le Premier Amendement. Cette division tel que noté s'étend au delà des États-Unis et semble caractériser l'approche de la common law qui pour toutes fins pratiques rejeté un droit au déréférencement basé sur la protection des données nonobstant les particularités de Internet[21].

Ainsi, le ministre britannique de la Justice, Simon Hughes a exprimé son opposition farouche à l'arrêt *Costeja* qu'il caractérise de [TRADUCTION] « censure [22]inexécutable qui met en cause la liberté d'expression et d'information »[23].

---

[19]European Union Committee - Second Report EU Data Protection law: a 'right to be forgotten'?, < https://publications.parliament.uk/pa/ld201415/ldselect/ldeucom/40/4002.htm >.

[20]See Rosen (2012), p. 88, online: Stanford Law Review <http://www.stanfordlawreview.org/online/privacy-paradox/right-to-be-forgotten>. Many American scholars view as this topic as the biggest threat to free speech on the Internet in the coming decade.

[21]Walker (November 20, 2012), December 2012. https://ssrn.com/abstract=2017967 or https://doi.org/10.2139/ssrn.2017967.

[22]"We have criticized the government of China... for closing down people's right to information. There are other countries with strict information access. It is not a good position for the EU to be in to look as if it is countenancing restrictions in the access of the citizen to access to information because it could be a very bad precedent". Stuart Lauchlan, "Britain pledges to fight Europe's Right to be Forgotten bad law" *diginomica* (10 July 2014), online: Diginomica <http://diginomica.com/2014/07/10/britain-pledges-fight-europes-forgotten-bad-law/>.

[23]Hughes said: "If politicians think they can delete findings about their expenses, that's not going to happen. If people think they can delete their criminal history, it won't occur. It looks to me as if it may be an unmanageable task. It will be a phenomenal task. It's not technically possible to remove all traces of data loaded on to the internet from other sources. You can't exercise the right to be forgotten. The information system could not be made to do it".

Owen Bowcott, "EU 'right to be forgotten' law unenforceable, says justice minister", The Guardian (9 July 2014), online: The Guardian.< http://www.theguardian.com/technology/2014/jul/09/eu-right-to-be-forgotten-law-unenforceable-justice-minister-simon-hughes>.

4 Conférer un pouvoir de limiter l'expression à des acteurs privés?

Le mécanisme institué par la Cour de justice européenne est ainsi conçu qu'il est plus facile pour un moteur de recherche de donner suite aux demandes de déréférencement que d'y résister. Pour chaque moteur de recherche, faire l'effort d'analyser et ensuite d'expliquer que certains liens conduisent à des documents qui présentent un réel intérêt pour le public représente un coût que certains d'entre eux pourraient ne pas souhaiter encourir[24].

Mais au delà de la question de censure comme telle, se pose la question de savoir s'il est compatible avec le respect de l'État de droit de confier à une entreprise privée, américaine, un pouvoir judiciaire normalement réservé à une institution étatique (conseil constitutionnel ou analogue) ; le pouvoir de trancher les limites raisonnables de la liberté d'expression? Et ce dans l'absence de balises claires et connues ?

En effet Peter Barron de Google semble lui-même s'être confessé aux nouvelles BBC News que la compagnie « apprend en faisant » (« we're "learning as wc go" ») adoptant inévitablement une approche ad hoc à l'exécution du jugement. Une autre remarque édifiante semble confirmer cette incertitude : « no one really knows what the criteria is . . . So far, we're getting a lot of noes... It's a complete no man's land »[25].

Car la transparence et la reddition de comptes ne sont pas faciles à cultiver lorsqu'on laisse la pondération de valeurs constitutionnelles délicates à des acteurs privés étrangers et possiblement réticents à se metre à juger d l'opportunité du maintien en ligne d'un lien hypertexte vers un document. Le cadre juridique garantissant la protection des des droits de la personne énoncé au Canda dans Charte des droits et libertés, vise exclusivement les mesures gouvernementales. Les droits fondamentaux sont constitutionnalisés - en tant que rempart contre les abus du gouvernement[26]. Mais à l'ère numérique il va sans dire que le pouvoir d'enfreindre - - même par mégarde - les valeurs constitutionnelles - y compris, la liberté d'expression ne réside pas exclusivement – dans les instances de l'Etat. Au contraire, dans un monde post-Costeja, les parties privées, à savoir les «contrôleurs de données » ayant une influence globale, utilisant l'intelligence artificielle, sont devenues des arbitres involontaires du discours public à l'échelle globale », et ce dans l'absence de balises précises ou de transparence. C'est la réalité que la Cour suprême du Canada semble reconnaître (même si indirectement) dans les arrêts clé *Douez* et *Equustek*[27].

---

[24]Wechsler (2015), pp. 135–165.

[25]Mr. Wadsworth of the U.K. ORM firm Igniyte, referring to the ECJ's recent decision (Mark Scott, "European Companies See Opportunity in the 'Right to Be Forgotten', *The New York Times* (8 July 2014), online: The New York Times < http://www.nytimes.com/2014/07/09/technology/european-companies-see-opportunity-in-the-right-to-be-forgotten.html?_r=0)>.

[26]Art 32.

[27]*Douez* c. *Facebook Inc.*, 2017 SCC 33 et *Google Inc.* c. *Equustek Solutions*, 2017 SCC 34.

En effet, Facebook, Twitter et d'autres plates-formes se voient assigner la discrétion de déterminer quelle expression supprimer, ce qui conduit inévitablement à des approches ad hoc de ces entreprises, tâche normalement réservé au Législateur et aux tribunaux.La difficulté augmente lorsque les « équilibreurs » sont des acteurs corporatifs inexpérimentés et ayant peu d'incitatif économique à agir dans l'arbitraige des demandes de déréférencement. De plus, ces acteurs sont pour la plupart situés aux Etats-Unis, là où le cadre juridique demeure très protecteur pour les intermédiaires d'Internet. Enfin, les entreprises doiminantes du web maîtrisent des algorithmes et des solutions fondées sur de l'intelligence artificielle qu'ils déploient de plus en plus à cette fin[28]. C'est dire l'ampleur des enjeux que pose la solution mise de l'avant par la Cour européenne lorsqu'on a le moindre souci de garantir le respect de l'État de droit.

5 Prochaines étapes (Q12) : Réviser la notion d'action gouvernementale et la responsabilité des plateformes au delà de Costeja

5.1 L'action gouvernementale et la Charte

Pour applanir les contradictions aux discordes entre les logiques issues du droit civil et celles découlant du common law en ce qui concerne « droit au déréférencement » sur le plan transfrontalier, la notion d'« action gouvernementale » doit être soigneusement revue. Autrement, les ultimes arbitres des limites appropriées aux droits fondamentaux pourraient être des algorithmes ou d'autres formes d'intelligence artificielle déployées par des plates-formes qui, ne possèdent pas la légitimité démocratique que l'on reconnaît habituellement aux juges dans les sociétés démocratiques.

Alors que nous luttons pour définir les limites du discours dans un monde post-Charlottesville[29], reconnaissons l'importance de maintenir la surveillance des tribunaux sur les valeurs constitutionnelles et les limites appropriées de l'expression, plutôt que de les laisser devenir des domaines de l'inconnu (tranchés par des « bots » d'intelligence artificielle).

Qui plus est et dans un deuxième temps, les intermédiaires doivent être responsabilisés selon le modèle du « **contrôle contextuel** » décrit ici bas, et ce encore sur le plan transfrontalier.

---

[28]http://csrcl.huji.ac.il/people/inadvertently-appointing-digital-judges-canadian-perspective-restricting-speech-and.

[29]http://csrcl.huji.ac.il/people/inadvertently-appointing-digital-judges-canadian-perspective-restricting-speech-and?ref_tid=3718.

5.2 *La responsabilisation des intermédiaires*

Il y a quelques années, le professeur Jonathan Zittrain a baptisé le droit à l'oubli tel qu'élaboré dans *Costeja* de "bad solution to a very real problem"[30] [une mauvaise solution à un problème réel] ». Reconnaissant donc que les renseignements déroutants ou decontextualisés représentent un problème réel mais rejetant néanmoins la solution retenue en Europe, ce dernier s'est plus récemment penché sur la nécessité de repenser le U.S. *Communications Decency Act's* et l'immunité totale que cette loi octroie aux intermédiaires aux Etats-Unis par le biais de son article §230[31] :

> Section 230 nearly entirely eliminated the liability of Internet content platforms under state common law for bad acts, such as defamation, occasioned by their users. The platforms were free to structure their moderation and editing of comments as they pleased, without a traditional newspaper's framework in which to undertake editing was to bear responsibility for what was published. If the New York Times included a letter to the editor that defamed someone, the Times would be vulnerable to a lawsuit (to be sure, so would the letter's author, whose wallet size would likely make for a less tempting target). Not so for online content portals that welcome comments from anywhere—including the online version of the New York Times.[32]

Mais la rationalité sous-jacente de cette immunité ne convient plus à une industrie qui n'est plus dans son enfance (selon le professeur Zittrain « an infant industry has grown up »)[33]. Ainsi pour nos fins, même des commentateurs proéminents américains, toujours résistants à la désindexation comme réponse aux maux de la decontextualisation, semblent maintenant envisager une certaine responsabilisation des « contrôleurs de données » compte tenu des pouvoirs de ces derniers qui croissent de manière exponentielle. Il s'agit donc d'un moment charnière pour peaufiner le droit au déréférencement par le biais de la responsabilisation contextuelle des moteurs de recherche puisqu'on constate dores et déjà des fissures dans l'armure de l'immunité de ces derniers.

Si le déréférencement dans son état actuel ne s'accorde pas avec la vision américaine (ou britannique) de la liberté d'expression, le Canada semble particulièrement bien situé pour proposer une approche mesurée au déréférencement. Selon cette approche hybride, l'accent n'est pas sur la protection des données mais sur la protection des individus. Autrement dit, l'idée est de

---

[30]Jonathan Zittrain "Don't Force Google to Forget", *New York Times* (14 May 2014), online: The New York Times http://www.nytimes.com/2014/05/15/opinion/dont-force-google-to-forget.html. See also: David Streitfeld, "European Court Lets Users Erase Records on Web", The New York Times (13 May 2014), online: The New York Times<http://www.nytimes.com/2014/05/14/technology/google-should-erase-web-links-to-some-personal-data-europes-highest-court-says.html?_r=0>.

[31]https://www.law.com/therecorder/sites/therecorder/2017/11/10/cda-230-then-and-now-does-intermediary-immunity-keep-the-rest-of-us-healthy/.

[32]*Id.*

[33]*Id.*

permettre aux protections existantes (enchâssées dans la Charte québécoise et canadienne ainsi que dans le Code civil du Québec entre autres) de s'« étendre » aux interactions des canadiens dans le « cyberespace », suivant la logique de *Douez c. Facebook* inter alia qui vise à pallier les inéquités inhérentes au monde numérique et son « caractère automatique ».

Pour que le droit au déréférencement soit opérable sur le plan transfrontalier, il serait prudent de s'éloigner graduellement de l'emphase formelle sur les logiques du droit de la protection des données sur lequel on tente de fonder le « droit à l'oubli »[34]. Le raisonnement sur lequel s'appuie la Cour européenne pour fonder une exigence de déréférencement procède d'une importation dans le droit commun, des notions issues du droit de la protection des données personnelles. Négligeant les différences qui existent entre les données personnelles à caractère privé en fonction desquelles furent développés les lois sur la protection des données personnelles dans les années '70 et les informations du domaine public, qui n'ont jamais été destinées au régime juridique des données personnelles lors de la conception des lois sur la protection des données, la Cour a choisi d'importer et de plaquer aux données ayant un caractère public le modèle concu pour assurer la protection des données intimes. Le glissement conceptuel n'est pas expliqué par la Cour européenne et il est difficile à concilier avec l'impératif d'équilibre entre les droits et libertés qui caractérise les raisonnements des tribunaux canadiens.

Par contre, lorsqu'il est démontré que le propos contrevient à une loi ou viole un droit fondamental, les tribunaux canadiens n'ont aucune hésitation à ordonner le déréférencement. Dans *Corriveau* c. *Canoe*,[35] une affaire où l'intermédiaire avait concédé être responsable de tous et chacun des propos diffusés sur un blogue, une cour québécoise a condamné un intermédiaire de retirer des propos diffamatoires. Suivant cette logique fondée sur les droits de la personne (contrairement à la protection des données), la Cour fédérale du Canada a pour sa part et pour toutes fins pratiques reconnu un droit au déréférencement « Canadien » dans A.*T. c. Globe24H.com*[36] et ce au delà des fontières canadiennes[37]. Mais dans l'une et l'autre des situations, le caractère diffamatoire ou contraire aux lois applicables avait été constaté par la Cour.

Selon l'article 22 de la *Loi concernant le cadre juridique des technologies de l'information*[38] du Québec, « le prestataire qui agit à titre d'intermédiaire pour offrir des services de référence à des documents technologiques, dont un index, des hyperliens, des répertoires ou des outils de recherche » n'est pas responsable des

---

[34](surtout puisque la définition des données personnelles est très problématique).

[35]2010 QCCS 3396.

[36]https://www.canlii.org/fr/ca/cfpi/doc/2017/2017cf114/2017cf114.html.

[37]Notons que cette cause impliquait un site roumain qui avait republié/ ré-indexé des décisions judiciaires canadiennes dans le but d'exiger un paiement en échange du retrait de ces renseignements.

[38]*Loi concernant le cadre juridique des technologies de l'information,* RLRQ, c. C.1.1. Voir (2012), pp. 189 et ss.

activités accomplies au moyen de ces services. » La possibilité d'engager sa responsabilité peut découler notamment de sa connaissance de fait que les services qu'il fournit servent à la réalisation d'une activité à caractère illicite et s'il ne cesse promptement de fournir ses services aux personnes qu'il sait être engagées dans cette activité.

En vertu de la législation québécoise, étant donné une disposition législative explicite[39] selon laquelle ils n'ont pas d'obligation de surveiller, il est logiquement impossible de considérer que les moteurs de recherche effectuent un « traitement » de données personnelles qui pourrait découler de l'introduction de mots qui s'avèreraient correspondre au nom d'une personne. Tout au plus, ils pourraient avoir une responsabilité suite à l'introduction de tels mots s'ils ont connaissance préalable du caractère illicite du document référencé.

D'ailleurs, contrairement au droit européen, les lois canadiennes sur la protection des renseignements personnels ne retiennent pas la notion de « traitement ». Les lois canadiennes régissent la collecte, l'utilisation et la communication de renseignements personnels. Pour constituer une collecte au sens de ces lois, il faut que l'entité ait au minimum une connaissance de la teneur et du sens des renseignements sur lesquels il acquiert ainsi le contrôle[40].

6 Le contrôle contextuel : entre le droit civil et la common law

Enfin, cette convergence des deux grandes traditions juridiques réunies en droit canadien se traduit dans l'approche privilégiée par une Cour australienne qui dans l'arrêt *Duffy c. Google* s'est inspirée d'une décision de la Cour suprême du Canada (obiter) pour responsabiliser le moteur de recherche. Dans cette cause qui emprunte explicitement la logique de la Cour suprême du Canada dans Crookes c. Newton (obiter) - et ouvre la possibilité de responsabilité pour les intermédiaires lorsque ceux-ci vont au-delà de fournir des hyperliens - le tribunal australien tient le moteur de recherche responsable pour ne pas avoir déréférencé des sites diffamatoires ainsi que d'avoir proposé ces sources par le biais de sa fonction « autocomplete » entre autres. Cette approche qui s'inspire du droit Canadien concorde fort mieux avec la logique qui anime le déréférencement dans l'ère numérique et atteint un meilleur équilibre entre liberté d'expression et droit au respect de la vie privée et à la réputation, valeurs sur lesquelles on prétend fonder le droit à l'oubli.

Qui plus est, cette vision contextuelle ancrée dans les droits de la personne plutôt que dans la protection des données, permet à des tribunaux compétents, plutôt qu'à

[39] *Loi concernant le cadre juridique des technologies de l'information*, art. 27.

[40] Pour pouvoir collecter et conserver des documents contenant des renseignements personnels, il faut au minimum acquérir le contrôle sur ceux-ci. Sur cette question voir : Gautrais and Trudel (2010), p. 59 et s.

des acteurs corporatifs inexpérimentés, de juger de ce qui est réellement decontextualisé voir même diffamatoire (plutôt que simplement «non pertinent» selon Costeja) et doit donc être désindexé[41]. Le test du « contrôle contextuel »[42] illusre les difficultés sérieuses précédemment décrites et caractéristiques de l'ère numérique.

7 Le retour de *LICRA c. Yahoo*? - l'avenir de la juridiction extraterritoriale sur le déréférencement

Les systèmes juridiques contemporains postulent que chaque état applique ses normes dans les limites de ses propres frontières[43]. De toute évidence, cette supposition qui sous-tend nos cadres normatifs ne correspond plus à la réalité de l'ère numérique.

Comme l'explique si éloquemment le Professeur Yuval Shany, Directeur du nouveau Centre sur la cyber sécurité à Jérusalem « les interactions actuelles ne se produisent pas principalement sur le territoire physique, mais dans un cyberespace. La réalité est très différente de celle dans laquelle nos lois ont été créées et sont appliquées »[44].

Dans un arrêt très récent, présageant de l'avenir, le tribunal du district de la Californie, district du Nord a émis une ordonnance enjoignant à une compagnie canadienne (Equustek) d'exécuter une ordonnance mondiale de désindexation que la Cour suprême du Canada avait rendue contre Google quelques mois avant. L'injonction accordée dans le cadre d'un différend de propriété intellectuelle/secret de commerce obligeait Google à déréférencer tous les sites Web qui vendaient des produits qui contrevenaient à la marque de commerce de Equustek, la plus haute juridiction du Canada ayant déclaré comme suit: «Internet n'a pas de frontières – son habitat naturel est mondial »[45].

Malgré cette réalité, les présupposés traditionnels du droit restent intimement liés au territoire. Le «renversement» impensable de la Cour suprême du Canada devant un tribunal inférieur de l'autre côté de sa frontière illustre habilement cette dissonance de plus en plus dérangeante entre les racines intransigeantes du droit ancrés

---

[41]BCF avocats C.L. c. BCF Avocats d'affaires 2016 QCCAI 114, droit de rectification en vertu de la Loi sur le secteur privé,) « l'entreprise doit prendre tous les moyens raisonnables pour rectifier les renseignements de la demanderesse à l'interne (sur son site Internet), ce qui n'équivaut toutefois pas à un devoir de déréférencement (à l'externe, sur le reste de la Toile) » . La décision ne traite pas du devoir d'un intermédiaire.

[42]"Duffy v Google: is this the end of the Internet as we know it?" *Defamation e-bulletin* (30 October 2015). Online: http://www.landers.com.au/publications/dispute-resolution/duffy-v-google-is-this-the-end-of-the-internet-as-we-know-it/.

[43]Voir par exemple Yuval Shany, Jerusalem.

[44]*Id.*

[45]*Google Inc. c. Equustek Solutions Inc*, 2017 CSC 34, au par. 41 (J.Abella).

fermement dans le territoire et le caractère sans frontières du cyberespace. Reflétant le changement important de circonstances évoqué par l'ère numérique en ce qui concerne la portée extraterritoriale des jugements, Google s'est tournée vers un tribunal dans son état de la Californie, qui, comme indiqué, a bloque l'injonction accordée par la décision canadienne. Elle a jugé (sur la base de l'article 23 de la Communications Decency Act – précité, bien que Google ait plaide le Premier Amendement) que l'injonction mondiale n'aurait aucun effet au-delà des frontières étroites du Canada, autorisant ainsi Google à «remettre en vente» les résultats de recherche contestés. Google.ca, malgré la décision canadienne.

Inutile de dire qu'une telle décision fait non seulement violence à la plus haute cour du Canada, mais rend sa décision sans effet. Car, dans un monde où le commerce électronique ne connaît pas de frontières, quel est l'effet pratique de désindexer des résultats uniquement à l'intérieur de« frontières données? Il confirme également l'appréhension évidente de la Cour canadienne dans une affaire antérieure, Douez c. Facebook (précité), que les choix d'ordre public consacrés par la loi canadienne finiraient par être dénués de sens (ou à tout le moins inapplicables) à l'ère d'Internet. Cette difficulté est aggravée lorsque les normes en jeu sont des valeurs constitutionnelles, comme dans la liberté d'expression, qui attire une attention particulière dans *Equustek.*

Equustek souligne donc l'incapacité de l'Etat à règlementer la conduite qui défie les frontières traditionnelles de brique et de mortier efficacement. Cela crée inévitablement un vide juridique où la forme de conduite la plus omniprésente (soit le cybercommerce) est de manière absurde gouvernée par des normes dépassées inadaptées aux interactions dans le cyberespace.

8 Conclusion

Les controverses au sujet du « droit à l'oubli » envisagé comme procurant un droit de forcer à la suppression des liens vers des documents qui ne contreviennent pas aux lois sont emblématiques de la désuétude de lois protection des données personnelles fondées sur la fiction du « consentement ». Reflétant l'informatique centralisée des années 1970, ces législations ne procurent plus les équilibres appropriés pour garantir que l'utilisation des données générées par la collectivité se fera dans le respect des droits fondamentaux et des valeurs démocratiques. Cela apparaît de plus en plus comme une régulation d'autrefois pour encadrer les pratiques du futur.

Faute d'innover sur les mécanismes de régulation, on s'épuise à tenter d'appliquer des lois persistant à postuler que les données relatives à une personne ne peuvent être utilisées que moyennant son consentement et uniquement pour des finalités précises. Cette vision individualiste est fondée sur la fiction d'un « consentement » que les internautes et tous les usagers d'objets connectés donneraient en pleine connaissance de cause.

Avec ce type de loi, la régulation qui compte vraiment est celle qu'imposent les plateformes de ce monde!

Désormais, les données massives essentielles à la création de valeur dans les environnements connectés sont laissées à la disposition des entreprises sans réelles obligations d'en garantir la protection. Avec le droit au déréférencement tel que créé en Europe, on s'enfonce plus profondément dans une vision formaliste fondée sur le »consentement » et l'hypothèses selon laquelle il est encore possible de déterminer les finalités d'une information qui circule dans le cyberespace.

Or, les traitements de données massives procèdent de logiques qui font fi des finalités au nom desquelles elles ont été initialement collectées. Devenues « Big Data », ce ne sont plus des données « appartenant » aux individus. Massivement utilisées, les données sont une ressource commune à tous, comme l'air et l'eau que nous utilisons. Comme l'eau, l'air ou les fréquences radioélectriques, les données sont au cœur de la création de valeur fondée sur l'IA. Leur utilisation doit se concevoir comme un privilège régi par des balises que les États doivent avoir le courage d'imposer et de faire respecter. Mais pour arriver à une telle refondation du cardre juridique de la protection des libertés, il faut oser remttre en cause les certitudes et les dogmes qui tiennent trop souvent lieu de fondement au droit de la protection des données personnelles.

ANNEXE - Réponses au questionnaire

The Right to Be Forgotten / *Le droit à l'oubli – Questionnaire Franz Werro*

Question 1

- Comment votre droit protège-t-il le droit à l'oubli ? Le droit à l'oubli est-il consacré demanière spécifique dans une loi ou découle-t-il de dispositions générales ?

How is the right to be forgotten protected under your law? Does your law specifically grant a right to be forgotten or does this right derive from a more general framework?

- En droit civil québécois les tribunaux appliquent les règles générales de la responsabilité civile. Dans ce cadre, ils ont considéré qu'en certaines circonstances, on peut commettre une faute de violation d'oubli. Certains auteurs en ont déduit l'existence d'un droit à l'oubli. Ainsi, le rappel d'évènements survenus dans le passé a été jugé fautif lorsqu'il y a absence de démonstration d'un intérêt public de la part de la personne qui a fait la divulgation. Les exemples suivants sont tirés de la jurisprudence québécoise.
- Dès 1889, la Cour supérieure du Québec a estimé qu'un journal *Le Violon* avait eu tort de faire revivre certaines «accusations depuis longtemps oubliées»

concernant le demandeur. Cette décision fut confirmée par la Cour de révision[46]. Plus récemment, dans l'affaire *Lévesque*[47], la Cour Supérieure a dû trancher sur une revendication du droit à l'oubli. Le requérant, Lévesque, poursuivait le *Journal de Québec* pour avoir rappelé le crime qu'il avait commis deux ans auparavant dans une «piquerie» de la ville de Montréal. Lévesque avait alors été impliqué dans une bagarre entre groupes criminels. La juge a conclu que le *Journal* n'avait pas commis de faute étant donné que l'information dévoilée était accessible au public. De plus, puisque l'objet de l'article portait sur l'incendie de la «piquerie» où Levesque avait jadis commis son crime, l'information divulguée demeurait d'intérêt public.

- Par contre, dans une affaire semblable[48], Gilbert Ouellet a poursuivi le journal *Photo-Police* pour avoir publié un article relatant le crime commis par sa défunte épouse dix ans auparavant. Cette dernière avait tué leurs quatre enfants pour ensuite s'enlever la vie. Le juge de la Cour du Québec a conclu que l'article publié était «sensationnaliste» et qu'il ne pouvait être justifié par l'intérêt du public à l'information. Dans une autre affaire de la Cour du Québec[49], le juge Barbe rappelle qu'il est difficile pour celui qui participe à des «activités publiques de nature politique» d'invoquer un droit à l'oubli. Se référant aux propos de la juge Piché dans *Szabo c. Morissette*[50], le juge de la Cour du Québec mentionne que « celui qui est à l'origine de l'histoire ne peut blâmer d'autres que lui-même s'il n'a pas aimé qu'on parle de lui ».
- La faute de violation de l'oubli telle que reconnue en droit québécois découle de la diffusion d'une information autrefois connue mais en conférant à celle-ci une portée temporelle et spatiale différente de celle découlant de la diffusion initiale. Ce qui est jugé fautif et sanctionné est la redivulgation considérée injustifiable dans le contexte où elle se produit. À ce titre, la légitimité juridique de la prétention à l'oubli se structure par l'appréciation du contexte de la diffusion de l'information. L'oubli est donc un droit pour la personne concernée lorsqu'il est jugé déraisonnable de diffuser l'information. Alors, la diffusion est jugée fautive, c'est à dire qui n'aurait pas été faite par une personne raisonnable oeuvrant dans des circonstances analogues. Le contexte dans lequel s'effectue la diffusion est un facteur très important dans ce processus de détermination de son caractère fautif

Question 2

- Quelles sont les limites au droit à l'oubli selon votre droit ?

  What are the limits to the right to be forgotten under your law?

---

[46]*Goyette c. Rodier* (1889) 20 R.L. 108,110 (C. Rév).

[47]*Lévesque c. Communications Quebecor inc.* (C.S., 1999-06-21), SOQUIJ AZ-99021730, J.E. 99-1527, [1999] R.R.A. 681.

[48]*Ouellet c. Pigeon*, REJB 1997-03106, 1997 (C.Q.).

[49]*Mathieu c. Presse ltée (La),* (C.Q., 1998-11-24), SOQUIJ AZ-99036093, B.E. 99BE-169.

[50](1993) R.R.A. 554, J.E. 93-1385 (C.S.).

- Le « droit à l'oubli » doit être distingué du droit de faire effacer les liens générés par un moteur de recherche crée par la décision de la Cour de justice européenne. En droit canadien, on ne peut postuler que le droit de la protection des renseignements personnels procure un droit au déréférencement des résultats de recherche[51].
- La garantie constitutionnelle de la liberté d'expression, entendue comme protégeant la liberté de rechercher des informations ne contrevenant pas à la loi, s'oppose à une application d'un « droit au déréférencement » qui prétendrait se réclamer du droit à la protection des renseignements personnels. Dans *Alberta (Information and Privacy Commissioner) c. Travailleurs et travailleuses unis de l'alimentation et du commerce, section locale 401*[52], la Cour suprême du Canada a invalidé la loi albertaine sur la protection des renseignements personnels en ce qu'elle prohibait la prise d'images dans un lieu public.
- La Loi attaquée interdisait de recueillir des renseignements, ici des images de personnes franchissant une ligne de piquetage et ne comportant aucune information intime, sans le consentement de celles-ci. Dans la situation présentée à la Cour, aucun détail concernant le mode de vie ou les choix personnels des intéressés n'avait été dévoilé. Or, la loi, à l'instar des autres lois sur la protection des renseignements personnels en vigueur au Canada ne fait aucune distinction : tout renseignement personnel y est traité de la même façon, même ceux qui ne relèvent pas de la vie privée. Il est interdit de le collecter et de le diffuser sans consentement sauf pour des motifs définis de manière très étroite. C'est cette absence de possibilité de laisser un espace à l'exercice des autres droits fondamentaux qui rend excessive la loi sur la protection des renseignements personnels. En somme, la Cour fait échos à une évidence : il existe des renseignements portant sur les personnes qui ne relèvent pas de la vie privée de celles-ci. La Cour rappelle la nécessité de baliser les interdictions se trouvant dans les lois sur la protection des renseignements personnels. Telles que rédigées, ces lois interdisent de capter, conserver et diffuser toute information relative à une personne identifiable sans sa permission et cela même lorsqu'elle se trouve dans des lieux publics. La Cour juge que de tels interdits limitent la liberté d'expression de façon déraisonnable.
- La Cour explique que ces lois doivent comporter des balises afin de permettre l'exercice des activités expressives ne portant pas sur des matières relevant de l'intimité des personnes. La Cour a jugé que la loi albertaine sur la protection des renseignements personnels empêchait de recueillir des renseignements personnels, telles que des prises d'images ou des vidéos lors d'une manifestation au cours de laquelle le public pouvait facilement observer les personnes qui y prennent part.

---

[51] *C.L. c. BCF Avocats d'affaires*, 2016 QCCAI 114 (CanLII), 14 avril 2016, en ligne : <http://canlii.ca/t/gr5q0>.

[52] [2013] 3 RCS 733.

- Cette décision de la Cour suprême invalide l'approche qui a prévalu au Canada depuis plus de trois décennies en matière de protection des renseignements personnels. Portées par un mouvement qui semble postuler que la vie privée est le seul droit fondamental à devoir être protégé, ces lois ignorent pratiquement les impératifs de la libre circulation de l'information dans les espaces publics. En invalidant la loi albertaine, la Cour met fin à ce déséquilibre.
- Bien sûr la Cour reconnaît la légitimité de protéger le droit à la vie privée et d'assurer que la collecte et la communication de renseignements personnels soient encadrées. Mais elle vient rappeler que tout renseignement personnel n'est pas automatiquement un renseignement sur la vie privée d'une personne, surtout s'il s'agit d'un renseignement se trouvant légitimement dans l'espace public. Il est donc excessif de considérer tout renseignement personnel comme étant assujetti au bon vouloir du sujet. Les libertés expressives imposent de baliser la faculté de l'individu à l'égard des informations le concernant en tenant compte des droits des autres et du public en général.
- Pour l'heure, bien qu'il porte sur un phénomène passablement distinct de ceux qui sont concernés par les moteurs de recherche, ce prononcé de la Cour suprême laisse planer d'importants doutes sur la possibilité, en droit canadien, d'un droit au déréférencement qui se fonderait sur les principes issus des lois sur la protection des renseignements personnels.

Question 3

- Quels sont, dans votre droit, les moyens de droit pour mettre en oeuvre son droit à l'oubli ?

What are, in your law, the legal remedies available to enforce the right to be forgotten?

- Les recours en responsabilité civile sont les voies privilégiées pour sanctionner les diffusions intempestives et fautives de faits passés qui ne correspondent pas à un intérêt public démontrable. Quant au droit au déréférencement, il est possible en vertu des principes du droit commun dès lors qu'il est démontré que le document vers lequel pointe un hyperlien est contraire à la loi.

Question 4

- Dans le prolongement de la question précédente, est-ce que votre droit permet à une personne qui s'estime lésée par une information sur internet d'obtenir une réparation de son dommage ou de son tort moral ? Si oui, est-ce que la mise en oeuvre d'une telle action en responsabilité est réalisable en pratique ?

As a follow-up to the previous question, does your law allow the plaintiff to receive material or immaterial damages? If yes, is such remedy realistic in practice?

- Dès lors qu'il est judiciairement établi qu'un document fautif est en ligne, les tribunaux peuvent rendre les décisions et ordonnances nécessaires afin de protéger les droits des personnes.

Question 5

- De manière générale, comment évaluez-vous la mise en oeuvre du droit à l'oubli dans votre droit ? Est-elle efficace ? Le droit à l'oubli est-il souvent utilisé en pratique ? Existe-t-il des obstacles particuliers à sa mise en oeuvre ?

In general, how do you assess the implementation of the right to be forgotten in your law? Is it effective? Is it used in practice? Are there particular obstacles in the implementation of this right?

- Au Canada, le droit à l'oubli s'entend du droit de faire supprimer l'information fautive ramenant dans le présent des faits passés. Il est rarement invoqué sauf dans les situations ou les rappels de faits passés ne sont pas d'intérêt public.

Question 6

- Comment les tribunaux et les auteurs de doctrine ont-ils accueilli la décision *Google* c. *González* de la CJUE dans votre État ?

How did courts and commentators in your country welcome the ECJ ruling on Google v González?

- Un grand nombre d'auteurs ont souligné le caractère déséquilibré du prononcé européen qui d'ailleurs ne dit rien du droit des internautes d'accéder à l'information licite qui se trouve en ligne.

Question 7

- Pour les ressortissants d'un État qui ne fait pas partie de l'Union européenne, est-ce que les tribunaux de votre État ont suivi la décision de la CJUE ? Pensez-vous qu'ils vont le faire ?

For those who are from a country that is not part of the European Union, did your courts follow the ECJ ruling on the right to be forgotten? Is it likely Do that they will follow it?

- Il faut souhaiter que les tribunaux canadiens se tiendront loin de l'approche représentée par la décision de la CJUE.

Question 8

- Est-ce que votre droit accordait déjà un droit à l'oubli sur internet similaire à celui consacré par la CJUE ?

Did your law already grant a similar right to be forgotten than the one stated in the ECJ ruling?

- En droit canadien, on ne peut postuler que le droit de la protection des renseignements personnels procure un droit au déréférencement des résultats de recherche[53]. Un droit au déréférencement qui se fonderait sur les principes issus des lois sur la protection des renseignements personnels apparaît poser d'importants problèmes de compatibilité avec la liberté d'expression.

Question 9

- Pour mettre en oeuvre la décision de la CJUE, Google a mis en place un formulaire permettant à toute personne intéressée de déposer une requête pour déréférencer une information qui la concerne. Sur la base de cette demande, Google doit faire une pesée des intérêts entre l'intérêt privé de la personne à déréférencer son information et l'intérêt public à ce que l'information soit publique. Google ne rend toutefois pas publique la manière dont il traite les requêtes de déréférencement. En particulier, Google n'informe pas le public du nombre de demandes qu'il reçoit, du type de demande, du cercle des personnes concernées, du nombre d'acceptation et de refus et des raisons des refus. Pensez-vous que Google doive améliorer la transparence dans la mise en oeuvre du droit à l'oubli ?

To implement the ECJ ruling, Google has created a form in which anyone interested can submit a request to have information about him-or herself be delisted. Based on this request, Google will weigh between the private interest of the petitioner and the public interest to be informed. Google does not disclose the ways in which it deals with requests. In particular, Google does fully not disclose, the category of requests that are excluded or accepted, the proportion of requests and successful de-listings and, among others, the reason for the denial of delisting. Do you think that Google should be more transparent about the ways it uses to implement the right to be forgotten?

- Dans beaucoup de pays démocratiques, on commencerait par s'étonner qu'un tribunal supposément attaché au respect de l'État de droit ne trouve aucun problème à confier à une entreprise, le soin de décider des conflits entre le droit du public d'accéder à des documents licitement en ligne et les revendications de ceux qui aimeraient mieux que leurs faits publics passés soient rendus introuvables. À cet égard, la question de transparence ne serait pertinente que dans la mesure ou l'on trouve conforme à l'État de droit, le fait de confier à une société privée le soin d'arbitrer entre les droits fondamentaux.

Question 10

- Est-ce que les citoyens de votre État font usage du formulaire de Google pour mettre en oeuvre le droit à l'oubli sur internet ?

  Is the procedure prepared by Google used in your country?

[53] *C.L. c. BCF Avocats d'affaires*, 2016 QCCAI 114 (CanLII), 14 avril 2016, en ligne : <http://canlii.ca/t/gr5q0>.

- Le droit au déréférencement n'est possible en droit canadien qu'une fois constaté par une instance judiciaire que le document vers lequel pointe l'hyperlien est contraire aux lois.

Question 11

- Des réformes sont-elles prévues au niveau législatif pour renforcer ou modifier la protection du droit à l'oubli dans votre droit ?

Is there any upcoming legal reform in your country whose purpose is to reinforce or modify the right to be forgotten?

- On considère généralement qu'il y a des priorités plus urgentes au plan de la protection des renseignements personnels que celles qui consiste à censurer l'information du domaine public.

Question 12

- Quelle devrait être à votre avis la prochaine étape dans la protection du droit à l'oubli ? Pensez-vous que les États devraient protéger davantage la personnalité des utilisateurs sur internet ? Pensez-vous que l'Union européenne devrait modifier ou adapter ses normes qui protègent le droit à l'oubli ?

In your opinion, what should be the next step in the protection of the right to be forgotten? Do you think that one must go further and strengthen the right to be forgotten? Do you think that the European Union should modify or adapt its legislation on the right to be forgotten?

- Les lois actuelles de protection des données personnelles fondées sur la fiction du « consentement » ne procurent plus les cadres appropriés pour garantir que l'utilisation des données générées par la collectivité se fera dans le respect des droits fondamentaux et des valeurs démocratiques. En Europe, là où la réflexion est parfois plus avancée sur ces questions, on continue d'envisager la régulation des données associées aux personnes comme on le faisant dans le dernier quart du 20e siècle. Une pareille approche donne une régulation d'autrefois pour encadrer les pratiques du futur.
- Faute d'innover sur les mécanismes de régulation, on s'épuise à tenter d'appliquer des lois persistant à postuler que les données relatives à une personne ne peuvent être utilisées que moyennant son consentement et uniquement pour des finalités précises. Cette vision individualiste s'accroche à la fiction d'un « consentement » que les internautes et tous les usagers d'objets connectés donneraient en pleine connaissance de cause. Cette approche produit un résultat déficient: pratiquement la plupart des utilisations des données par les géants du web, même celles qui soulèvent beaucoup d'inquiétudes, sont autorisées. Nous avons tous cliqué le rituel « J'accepte » dès lors que nous avons décidé d'utiliser une application, un site ou un objet de ce monde connecté !
- Les lois actuelles sur les données personnelles ne font que gérer l'abandon de nos libertés aux conditions définies par les géants du web. À cet égard, le droit au déréférencement est un mécanisme qui confère à ceux ui en ont les moyens, les

capacités de compliquer la vie de ceux qui recherchent une certaine transparence à l'égard des gens de pouvoir comme les professionnels ou les personnalités publiques.

- Désormais, les données massives sont essentielles à la création de valeur dans les environnements connectés. Pour l'heure, les lois persistent à imposer que ces masses d'information ne soient utilisées que pour des « finalités » définies. Avec le droit au déréférencement tel que créé en Europe, on s'enfonce plus profondément dans une vision formaliste fondée sur le »consentement » et l'hypothèses selon laquelle il est encore possible de déterminer les finalités d'une information qui circule dans le cyberespace.
- Or, les traitements de données massives procèdent de logiques qui font fi des finalités au nom desquelles elles ont été initialement collectées. Devenues « Big Data », ce ne sont plus des données « appartenant » aux individus. Massivement utilisées, les données sont une ressource commune à tous, comme l'air et l'eau que nous utilisons. Comme l'eau, l'air ou les fréquences radioélectriques, les données sont au cœur de la création de valeur fondée sur l'IA. Leur utilisation doit se concevoir comme un privilège régi par des balises que les États doivent avoir le courage d'imposer et de faire respecter.

References

Eltis K (2011) Breaking through the 'Tower of Babel': a 'Right to be Forgotten' and how trans-systemic thinking can help re-conceptualize privacy harm in the age of analytics. Fordham Intellect Prop Media Entertain Law J 22:69–95

Eltis K (2016a) The Anglo-American/Continental privacy divide? How civilian personality rights can help reconceptualize the 'Right to be Forgotten' toward greater transnational interoperability. Canadian Bar Rev 94:355–380

Eltis K (2016b) Courts, litigants and the Digital age, 2nd edn. Irwin Law, Toronto

Gautrais V, Trudel P (2010) Circulation des renseignements personnels et Web 2.0. Éditions Thémis, Montréal, p 59 et s

Ghatnekar S (2012–2013) Injury by algorithm. Loyola Los Angeles Entertain Law Rev 33:171

Gratton E, Polonetsky J (2017) Droit À L'Oubli: Canadian perspective on the global 'Right to Be Forgotten' debate. Colorado Technol Law J 15(2):337

Rosen J (2012) The right to be forgotten. Stanf Law Rev Online 64:88

Saint-Laurent G (2015) Vie privée et « droit à l'oubli » : que fait le Canada? Revue de droit de l'Université du. Nouveau-Brunswick 66:185–197

Voir PT (2012) Introduction à la Loi concernant le cadre juridique des technologies de l'information. Éditions Yvon Blais, Cowansville, p 189 et ss

Walker R (2012) The Right to be Forgotten. Hastings Law J 64:257

Wechsler S (2015) The right to remember : the European Convention on human rights and the right to be forgotten. Colum J Law Soc Probs 49:135–165

Part III
Asia

A Japanese Equivalent of the "Right to Be Forgotten": Unveiling Judicial Proactiveness to Curb Algorithmic Determinism

Itsuko Yamaguchi

Abstract Despite the absence of any explicit basis for the so-called "right to be forgotten" in Japanese data protection statutes, there is a basis in Japanese privacy case law that provides part of the substance of such a right. This was elucidated in a landmark decision of the Supreme Court of Japan on January 31, 2017, regarding the issue of search engine liability. The Supreme Court held that, if certain substantive requirements are met, injunctive relief can be granted against a search engine operator to remove search results containing private facts. The level of protection provided for such a Japanese equivalent of the right to be forgotten, being subject to a heavily fact-specific balancing test formulated by this Supreme Court decision, can be roughly characterized as somewhat eclectically in-between the two ends of the spectrum represented by EU and US law respectively, in terms of how it seeks to strike a balance among the multiple competing interests including but not limited to privacy, the freedom of expression, access to information, and online platform business. This paper highlights the importance of such recent proactive judicial moves in Japan to curb algorithmic determinism, and also emphasizes the need to prepare for the mixed blessings of the next generation of self-learning algorithmic decision-making enabled by artificial intelligence (AI) and the latest smart information technologies.

I. Yamaguchi (✉)
Interfaculty Initiative in Information Studies, Graduate School of Interdisciplinary Information Studies, The University of Tokyo, Tokyo, Japan
e-mail: itsuko@iii.u-tokyo.ac.jp; http://www.iii.u-tokyo.ac.jp/faculty/yamaguchi_itsuko

F. Werro (ed.), *The Right To Be Forgotten*, Ius Comparatum – Global Studies in Comparative Law 40, https://doi.org/10.1007/978-3-030-33512-0_15

1 Introduction

What kind of stance does Japanese law take about the so-called "right to be forgotten", and how should it be assessed from a comparative law perspective?[1] To be sure, Japanese law and its actual practice may appear obscure or esoteric in the eyes of the rest of the World. That might be especially true for the right to be forgotten, whose scope and limits would be hard to delineate in any jurisdiction, because how this right is conceptualized inevitably impacts not only the fundamental rights and freedoms of individuals but also the future direction of innovation and the sharing economy in emerging cyber-physical systems. Moreover, Japanese law sometimes uses different labels or concepts of rights from those in other jurisdictions, even while these different labels and concepts may subtly overlap with those in other jurisdictions in substance. Nevertheless, this report aims to explicate the trajectory of the present status of Japanese law relating to the right to be forgotten, and particularly, attempts to unveil an intriguingly proactive role played there by the Supreme Court of Japan.

In answering the above question, subsequent sections of this report clarify the following four points. First, Japan has no explicit statutory basis to grant the right to be forgotten in the sense construed in 2014 by the Court of Justice of the European Union (CJEU) in *Google Spain SL and Google Inc. v AEPD and Mario Costeja Gonzále.*[2] That has continued to be so, even under a set of Japanese data protection statutes which were thoroughly revised in 2015 and fully enforced in 2017 (Sect. 2).

Second, despite the absence of any explicit basis for the right to be forgotten in Japanese data protection law, there is a basis in Japanese privacy law that provides part of the substance of the right to be forgotten. This was elucidated in a landmark

[1]This report is partly based on the following works: Yamaguchi (2015a, b); presentations delivered at the University of Michigan Law School on Oct. 6, 2016, the University of British Columbia on Mar. 29, 2017, the Fudan University on May 28, 2017. English translation of Japanese text in this report is provided by the author, unless otherwise specified. Website information cited in the report was last visited on November 17, 2019. This work was partly supported by JSPS KAKENHI Grant Number 17K03501. The author hopes to express gratitude to Professor Junichi Hamada, Professor Franz Werro, and Ms. Claudia Hasbun for useful comments on the draft of this report, and to Mr. David Buist for carefully proofreading the draft. An earlier version of this report was published in "Japanese Reports for the XXth International Congress of Comparative Law (ICCLP Publications No.14)" in 2019, by the International Center for Comparative Law and Politics, Graduate School of Law and Politics, The University of Tokyo, Japan.

[2]CJEU, Case C-131/12 (May 13, 2014), http://eur-lex.europa.eu/legal-content/EN/TXT/?uri=CELEX:62012CJ0131; Official Journal of the European Union, C212, Vol. 57 (July 7, 2014), at 4–5. Regarding the long-debated issue of the territorial scope of the right of de-referencing, the CJEU finally held in Google LLC v. CNIL on September 24, 2019 that the operator of a search engine is not required to carry out a de-referencing on all versions of its search engine, but only on the versions of that search engine corresponding to all the Member States (Case C-507/17, ECLI:EU:C:2019:772). On the same day, with respect to de-referencing of sensitive data, the CJEU held in GC and Others v CNIL that the prohibition or restrictions relating to the processing of special categories of personal data should be applied also to the operator of a search engine, under certain conditions (Case C-136/17, ECLI:EU:C:2019:773).

decision of the Supreme Court of Japan on January 31, 2017.[3] In this eagerly awaited decision, the Court settled the issue of search engine liability in injunction cases, which had been divided in the lower courts. The Supreme Court held that, if certain substantive requirements are met, injunctive relief can be granted against a search engine operator to remove search results containing private facts (Sect. 3).

Third, from a comparative perspective, the level of protection provided for this Japanese equivalent of the right to be forgotten, being subject to a heavily fact-specific balancing test formulated by the Supreme Court in the above decision, can be roughly characterized as somewhat eclectically in-between the two ends of the spectrum represented by EU and US law respectively, in terms of how it seeks to strike a balance among the multiple competing interests including but not limited to privacy, the freedom of expression, access to information, and online platform business (Sect. 4).

Lastly, this paper concludes by highlighting the importance of such recent proactive judicial moves in Japan to curb algorithmic determinism. Even without a legislative mandate, the Japanese Supreme Court stepped forward to resolve the issue of search engine liability, by reasoning that "the provision of search results" should be regarded as not just "automatic" data processing by computer programs, but rather as a sort of "expressive conduct by the search service provider itself", because the programs were made in a way to achieve results in accordance with "the search service provider's policy". In doing so, the Court seems to be sending a message to society that it is high time to prepare for the mixed blessings of the next generation of self-learning algorithmic decision-making enabled by artificial intelligence (AI) and the latest smart information technologies (Sect. 5).

2 The Lack of Any Explicit Basis for the Right to Be Forgotten in Japanese Law

In exploring whether and to what extent Japan protects the so-called "right to be forgotten", we need to start from the simple fact that there is neither an explicit text nor a specific basis to grant such a right under the current Japanese Constitution and statutes, in the sense conceived by the CJEU in *Google Spain SL v AEPD* on the basis of the EU data protection Directive (95/46/EC), and particularly of Articles 7 and 8 of the Charter of Fundamental Rights of the European Union (2000/C 364/01). The post-World War II Constitution of Japan, which was promulgated in 1946 and has never been amended, does not explicitly stipulate "privacy" or "data protection" in the text. It was not until the 1960s that the Japanese courts became

[3]Case No. 2016 (Kyo) 45, 71 MINSHU 63 (Sup. Ct., Jan. 31, 2017). An unofficial translation of this decision in English is available at http://www.courts.go.jp/app/hanrei_en/detail?id=1511. For an analysis of this decision written by a research judge of the Supreme Court of Japan, *see* Takahara (2017).

rather active in protecting a certain kind of individual interest, later often labeled as "privacy", through statutory and constitutional interpretation.[4] More specifically, let's take a brief overview of the following two neighboring areas of the right to be forgotten: privacy and data protection. It is better to begin with the latter which is equipped with complicated but specific legislation, because it is easier to explain than the former which has been evolved mainly through case law in Japan.

2.1 *The Recently Revised Data Protection Legislation in Japan*

First, in the area of data protection law, unlike the General Data Protection Regulation (GDPR) of the European Union which contains an explicit textual mention of the "right to be forgotten" as part of the right to erasure in Article 17,[5] there is no such term in the list of rights of data subjects in Japanese data protection statutes as of today. It is worth mentioning that the absence of this term itself does not necessarily connote a reluctance in Japan to comply with the global standards of data protection. In fact, Japan has developed a multi-layered, complex set of data protection or information privacy statutes, whose legislative moves were synchronized with international initiatives such as the OECD Guidelines in 1980[6] and the EU data protection Directive (95/46/EC). The first Japanese national legislation, "Act on Protection of Computer Processed Personal Information possessed by Administrative Organs" (Act No. 95 of 1988), was enacted in 1988 but only covered the public sector.

---

[4]For a succinct overview of the post-World War II Japanese case law relating to privacy, *see, e.g.*, Ashibe (Rev. by Takahashi) (2015) and Hasebe (2014); *see also* Yamaguchi (2006). The first judicial recognition of "privacy" as a private right came from a district court judgment in 1964. In the reasoning of the Tokyo District Court on Sept. 28, 1964 ("Utage-no-Ato" [After the Banquet] case), 385 HANREI JIHO 12, privacy was defined as the "legal guarantee or right that a private life shall not be unduly disclosed". The Supreme Court judgments on Dec. 24, 1969 (Kyotofugakuren case), 23 KEISHU 1625, and on Apr. 14, 1981 (Referral of criminal record case), 35 MINSHU 620, have been later understood as recognizing privacy to be protected at the level of constitutional guarantee, but the exact term "privacy" appeared neither in the opinion of the Court in the former nor in the majority opinion of the latter. Article 13 of the Constitution of Japan provides that "[a]ll of the people shall be respected as individuals. Their right to life, liberty, and the pursuit of happiness shall, to the extent that it does not interfere with the public welfare, be the supreme consideration in legislation and in other governmental affairs", http://www.japaneselawtranslation.go.jp/law/detail_main?id=174 [unofficial English translation].

[5]Regulation (EU) 2016/679 of the European Parliament and of the Council of 27 April 2016 on the protection of natural persons with regard to the processing of personal data and on the free movement of such data, and repealing Directive 95/46/EC (General Data Protection Regulation), Official Journal of the European Union, L 119, Vol. 59 (May 4, 2016), at 43–44.

[6]The OECD Guidelines were updated in 2013. *See* the original version in 1980, OECD Guidelines on the Protection of Privacy and Transborder Flows of Personal Data (paras.7–14), http://www.oecd.org/sti/ieconomy/oecdguidelinesontheprotectionofprivacyandtransborderflowsofpersonaldata.htm.

The centerpiece of the current Japanese legal system for data protection, "Act on the Protection of Personal Information" (Act No. 57 of 2003),[7] has an eclectically hybrid structure combining the EU type of "omnibus" and baseline-protection approach to be applied to both the public and private sectors, and the US type of "sectoral" and market-oriented approach which prescribes specific duties of business entities to be applied in the private sector.[8] For example, a basic principle for both public and private sectors is laid down in Article 3 of the centerpiece Act, which says that "personal information should be processed prudently, based on the principle of respecting the personality of individuals" and thus "its proper processing" should be striven for. A list of specific duties of business entities in the private sector under this centerpiece Act corresponds roughly to eight basic principles in the OECD Guidelines. For example, the obligation of "proper acquisition" in Article 17 of this Act corresponds to the OECD's Collection Limitation principle, and the obligation to ensure the "accuracy of the content of data" in Article 19 of this Act corresponds to the OECD's Data Quality principle. Furthermore, specific duties of business entities in the public sector are more rigorously governed by a separate body of statutes, such as the "Act on the Protection of Personal Information Held by Administrative Organs" (Act No. 58 of 2003).

This centerpiece Act and related Japanese data protection statutes were thoroughly revised in 2015 and fully enforced in 2017. Prior to the recent revision, these statutes were often severely criticized not only within the country but also overseas in comparative studies, largely due to the lack of any independent enforcement body resembling the Data Protection Authority (DPA) under the EU data protection Directive and very limited availability of evidence to show the effectiveness of enforcement in Japan.[9] After revision, a new independent supervisory body called the "Personal Information Protection Commission" was established. Also, the revised centerpiece Act has clarified that the new Articles 28 to 30 are to grant the judicially enforceable right to data subjects to request "disclosure", "rectification", "addition", "erasure", and "cessation of use" of personal data in the private sector, in the same way as Articles 12, 27, and 36 of the Act No. 58 of 2003 do so in the public sector.[10] This revision is to foreclose a kind of interpretation employed in lower courts which placed emphasis on the self-regulatory complaint-processing mechanism established by the centerpiece Act and denied the judicial enforceability of these Articles.[11]

---

[7]A tentative, unofficial English translation of this centerpiece Act of 2003 is available at, http://www.japaneselawtranslation.go.jp/law/detail/?id=2781&vm=04&re=01&new=1.

[8]Concerning a comparative analysis of the "sectoral" approach to regulate consumer data privacy in the United States, *see, e.g.*, Solove and Schwarts (2018).

[9]*See, e.g.*, Greenleaf (2012, 2014).

[10]For example, for the public sector, Article 12 of the Act No. 58 of 2003 explicitly grants a "right" to request a disclosure of one's personal information against the administrative organs of the Japanese national government to "anyone" regardless of one's nationality or residency. For analysis of the possibility that this right could be used as an effective tool to check a mishandling of one's personal information by the Japanese government even from anyone overseas, *see, e.g.*, Yamaguchi (2014a).

[11]*See, e.g.*, the judgment of the Tokyo District Court (June 27, 2007), 1978 HANREI JIHO 27. For an analysis of the legislative intent and judicial enforcement of these Articles of the centerpiece Act, *see, e.g.*, UGA (2018).

Even under the revised Japanese data protection statutes, it is hard to find a clear legislative mandate for the obligation to remove a search result whose content was originally published by a third party, in the way that the CJEU in *Google Spain SL v AEPD* did, based on the interpretation that Google Inc. is the "controller" in respect of the processing of personal data under the EU data protection Directive (95/46/EC). Of course, it is yet to be seen how the level of protection provided under the revised Japanese data protection law would be assessed in comparison with the EU data protection law. At least, the European Commission started negotiations with key trading partners, including Japan, equipped with revised or updated legislations, regarding the Commission's so-called "adequacy decision" under the EU law for international data transfers. The Commission launched the procedure to ensure an adequate level of data protection in Japan in September 2018, and finally adopted its adequacy decision with respect to Japan on January 23, 2019.[12]

2.2 Privacy and the "Right to Personality" Under Japanese Case Law

Second, in the area of privacy law, the Constitution of Japan (1946) does not explicitly stipulate "privacy" in the text, as mentioned above. In the closely-related area of civil defamation tort law, for example, there are explicit statutory bases in Articles of 709, 710, 723 of the Civil Code (Act No. 89 of 1896), which grant a petitioner a set of remedies, particularly, to seek for compensation of damages for material and the non-material losses. Article 709 of the Civil Code provides a general principle of tort liability for damages, and Article 710 further provides that "non-property damage" such as infringement of the body, liberty, and "reputation" should be compensated.[13] Usually, the lack of such explicit text in the Constitution or statutes would imply a rather reserved role that the courts are expected to play in Japan. Such a judicial role might also be explained historically in the sense that the modern Japanese legal system is said to have developed through adopting mainly the continental European civil law tradition, which was only gradually supplemented by

[12]European Commission (2017, 2018); *see* Commission Implementing Decision (EU) 2019/419 of 23 January 2019 pursuant to Regulation (EU) 2016/679 of the European Parliament and of the Council on the adequate protection of personal data by Japan under the Act on the Protection of Personal Information (Text with EEA relevance), Official Journal of the European Union, L 76, Vol. 59 (March 19, 2019), at 1–58.

[13]Article 709 of the Japanese Civil Code provides that "[a] person who has intentionally or negligently infringed any right of others, or legally protected interest of others, shall be liable to compensate any damages resulting in consequence." Article 710 stipulates that "[p]ersons liable for damages under the provisions of the preceding Article must also compensate for damages other than those to property, regardless of whether the body, liberty or reputation of others have been infringed, or property rights of others have been infringed." This unofficial English translation is available at, http://www.japaneselawtranslation.go.jp/law/detail/?id=2057&vm=04&re=01&new=1.

the influence of Anglo-American common law under the current, post-World War II Constitution of Japan.[14]

Still, however, it would be fair to say that the post-World War II Japanese courts have been rather active in protecting rights and interests of individuals in specific cases concerning civil defamation and privacy invasion, although the flip side of such activeness might be seen as a lack of robustness in protecting conflicting interests such as the freedom of expression in specific cases.[15] Let's take a quick look at how privacy is protected under Japanese law, which laid the basis for the later judicial development relating to the Japanese equivalent of the right to be forgotten.

Simply put, it is generally said that since the 1960s, the Japanese courts have recognized the right to privacy not only as a private right protected under civil tort law, but also as a constitutional right which is construed to be guaranteed under the "pursuit of happiness" clause of Article of 13 of the post-World War II Constitution.[16] As a remedy for infringement of privacy, compensation for damages can be granted as a kind of "non-property damage" under Articles 709 and 710 of the Civil Code. However, there are no specific statutory provisions of injunctive relief for such non-property loss, and thus the legal basis of such injunctions was not so obvious, even in the case of infringement of reputation. This ambiguity had generated divisive interpretations of civil law, and gave rise to contention over the constitutionality of such injunctions, especially when they touched upon the freedom of expression.

The Supreme Court of Japan clarified this point in a judgment on June 11, 1986.[17] This was a case in which the Court upheld the constitutionality of a preliminary injunction to prevent the publication of a defamatory article in a monthly journal. Concerning the legal basis of the injunction, the Court explained in the reasoning that a person whose reputation was injured illegally may seek for an injunction on the basis of the right to reputation as "a right to personality" ("*jinkaku-ken*" in Japanese),[18] because reputation is a "highly important legally-protected interest" along with life and body, and such a right to reputation as a right to personality has "exclusivity" in the same way as a property right. Still, according to the Court, a "prior restraint" on the expressive conduct should be allowed "only under strict and definite requirements", as an "exception" to the purpose of Article 21 of the Constitution which guarantees the freedom of expression and prohibits censorship.[19]

---

[14]Regarding a comparative analysis of Japanese law in terms of a distinction of civil law and common law countries, *see, e.g.*, Beer and Ito (1996).

[15]For a quick overview of the post-World War II Japanese case law relating to the freedom of expression, *see, e.g.*, Yamaguchi (2002).

[16]For the development of privacy law in Japan, *see supra*note 4.

[17]*See* the judgment of the Supreme Court on June 11, 1986 (Hoppo Journal case), 40 MINSHU 872, http://www.courts.go.jp/app/hanrei_en/detail?id=82 [unofficial English translation].

[18]This report translates "*jinkaku-ken*" as the "right to personality", with the aim of conveying the nuance of the generic and elastic nature of this right, and to avoid leaving the impression that this right is exactly the same as the so-called "general right of personality" developed by German case law.

[19]Article 21 of the Constitution of Japan stipulates that "[f]reedom of assembly and association as well as speech, press and all other forms of expression are guaranteed" (paragraph 1), and that "[n]o

Unsurprisingly, the Supreme Court's answer of setting the "right to personality" as the solid basis for injunctive relief in civil defamation cases, in turn, poses yet another question of the nature and scope of this right. At least from a historical and comparative law perspective, the old Civil Code of Japan (1890), which was primarily based on the French Civil Code (1804), did not have a provision for the right to personality. This old Civil Code was promulgated, but met with criticisms and was eventually not enforced. The current Civil Code of Japan (1896) was influenced heavily by then-ongoing drafts of the German Civil Code (Bürgerliches Gesetzbuch: BGB). Article 710 of the current Japanese Civil Code, which covers not only body and liberty but also "reputation" as non-property damage to be compensated, was based on the first draft of BGB, although any mention of the term "reputation" itself was removed in the later-codified, current Section 823(1) of BGB.[20]

German influence remains strong in Japanese civil law interpretation. Still, a research judge of the Japanese Supreme Court in the defamation injunction case above commented that this judgment of the Supreme Court in 1986 was not meant to grant "the 'general right of personality', like West Germany".[21] According to this comment, this judgment took the same stance as common academic theories in Japan that injunctive relief can be granted not as the direct effect of tort law, but based on the right to reputation as a right to personality which has exclusivity like a property right, and it should be an issue in the future "what kind of personality interests other than life, body, and reputation" injunctive relief can be granted for.[22] Indeed, it seems that this "right to personality" as the legal basis of injunctive relief has a quite generic and elastic nature, in the sense in which it is later employed in privacy cases.[23] This is then increasingly invoked in the recent claims for preliminary injunctions to remove search results.

censorship shall be maintained, nor shall the secrecy of any means of communication be violated" (paragraph 2). This translation is available at the website, *supra* note 4. The courts may issue a "preliminary" injunction, as an "order of provisional disposition" ("*kari-shobun-meirei*" in Japanese) under the paragraph 2, Article 23 of "Civil Provisional Remedies Act" (Act No. 91 of 1989).

[20] *See, e.g.*, Igarashi (2003) and Uchida (2008).

[21] Kato (1989).

[22] Kato (1989), pp. 2630–2632.

[23] For an example of contentious cases in which injunctive relief was sought on the basis of the "right to personality" covering a mixture of multiple claims including defamation and privacy invasion, *see* the judgment of Supreme Court on Sept. 24, 2002 ("Ishi-ni-oyogu Sakana" [Fish swimming in Stone] case), 1802 HANREI JIHO 60.

3 The Winding Trajectory of Preliminary Injunctions to Remove Search Results

Roughly since the European Commission proposed a draft of the GDPR[24] in 2012, and undoubtedly since the CJEU ruling in *Google Spain SL v AEPD* construed that the search engine operator must be regarded as a "controller" under the EU data protection Directive and thus obliged to remove the search results, issues of what the right to be forgotten means and how it can be enforced in transnational contexts have been contentious even in Japan.[25] Commentators in Japan are divided as to whether, and if so, how such EU law influenced Japanese law about search engine liability. Still, it seems that the CJEU ruling motivated attorneys at law on the side of petitioners in Japan to bring a motion for a preliminary injunction in Japanese courts against not a subsidiary in Japan but the headquarters of Google Inc. located in the United States,[26] resulting in a series of Japanese court rulings described below. At least, it would be fair to say that, part of the substance of the right to be forgotten can be found in Japan, not straightforwardly by following the CJEU ruling to recognize this right as such, but rather by developing good old Japanese civil defamation and privacy case law to allow a preliminary injunction to remove a certain kind of search result, as elucidated by the Supreme Court's decision on January 31, 2017.

3.1 The First Preliminary Injunction to Remove Search Results

Before we look at specific cases, we need to note that there has been a long-debated "open justice" problem concerning preliminary injunctions under Japanese law. The above-mentioned judgment of the Supreme Court of Japan in 1986 clarified that, if certain substantive and procedural requirements are met, a preliminary injunction against defamatory expressive material can be granted. Although it is often said that the number of motions for such injunctions has significantly increased, the full facts and reasoning of preliminary injunctions in defamation and especially privacy-related cases are not usually published. They occasionally appear in print publications of court reporters on a selected basis, and thus it is not easy to grasp a whole picture of what exactly is going on in relevant cases.[27]

---

[24]In the lengthy legislative process of the GDPR, the term "right to be forgotten" in Article 17 appeared in the European Commission's draft on January 25, 2012, but was omitted in the European Parliament's legislative resolution on March 12, 2014. After the CJEU ruling on May 13, 2014, this term reappeared in Trialogue negotiations, and was finally included as part of the right to erasure in Article 17 of the GDPR. The GDPR applies to EU member states from May 25, 2018.

[25]*See, e.g.*, Ishii et al. (2015) and Shishido et al. (2015).

[26]*See, e.g.*, Kanda (2015); Ishii et al. (2015), pp. 8–16.

[27]Concerning the "open justice" problem relating to preliminary injunctions, *see, e.g.*, Shishido et al. (2015), pp. 76–77; *see also* Yamaguchi (2012). According to the statistics, there has been a

The first court ruling to allow a preliminary injunction against a search engine operator to remove search results was said to be the Tokyo District Court's decision on October 9, 2014.[28] It was made only after a series of rulings of the lower courts dismissing the petitioner's defamation and/or privacy claims for damages and/or injunctive relief for various reasons.[29]

In this case, a petitioner filed a motion for a preliminary injunction based on his right to privacy against a respondent, the headquarters of Google, Inc. located in the United States (not the Japanese subsidiary thereof). The litigated articles displayed in search results referred to the petitioner's prior association with a group of reckless motor-cycle riders, who had previously committed a serious crime. In declining to immunize the search engine operator, the Tokyo District Court acknowledged that the search results of the URLs, titles, and excerpts of the relevant websites are displayed "automatically and mechanically" based on a set of algorithms, and not by the artificial manipulation by the respondent. Nevertheless, the district court found that the petitioner's "right to personality" was clearly infringed by part of the said search results' titles and excerpts themselves. This decision of the Tokyo District Court granted a preliminary injunction for the removal of 122 articles, out of 237 articles the petitioner sought for, although removal of only a small part of these articles was sustained on appeal.[30] Once the gist of this district court's decision

continuing increase in Internet-related cases filed in the Tokyo District Court seeking preliminary injunctions. The number of such Internet-related cases was 607 (63.2%), out of the total number of 960 cases seeking orders of provisional disposition under the paragraph 2, Article 23 of the Civil Provisional Remedies Act (*see supra* note 19) in the Fiscal Year 2014, and it was 680 (64.9%) out of 1048 cases in the FY 2015. These Internet-related cases include not only motions for removal of articles, but also motions for disclosure of information to identify a sender under "Act on the Limitation of Liability for Damages of Specified Telecommunications Service Providers and the Right to Demand Disclosure of Identification Information of the Senders" (Act No. 137 of 2001; *see infra* note 44), etc. For detailed analysis of statistics and general trends, *see* Seki (2016).

[28]Some unpublished cases relating to the issue of search engine liability are gradually being included in electronic databases. This decision of the Tokyo District Court on Oct. 9, 2014, Case No. 2014 (Yo) 2002, is now available in D1-law.com database, ID No. 28252702.

[29]*See, e.g.*, a judgment of the Tokyo District Court on Nov. 6, 2009, Case No. 2008 (Wa) 11998, available at 2009WLJPCA11068002 (claims for defamation, etc. to seek for injunctive relief and damages against Google Japan Inc.); a judgment of the Tokyo District Court on Feb. 18, 2010, Case No. 2009 (Wa) 25234, available at 2010WLJPCA02188010 (defamation claims for injunctive relief and damages against Yahoo Japan Corporation); a judgment of the Tokyo District Court on Dec. 21, 2011, Case No. 2011 (Wa) 25033, available at LEX/DB25490833 (defamation claims for damages and injunctive relief against Google Japan Inc. and Yahoo Japan Corporation); a judgment of the Kyoto District Court on Aug. 7, 2014, Case No. 2013 (Wa) 2893, available at LEX/DB25504803 (defamation and privacy infringement claims for damages and injunctive relief against Yahoo Japan Corporation). For an overview of such development of case law and self-regulatory regimes relating to obligations of search engine providers to remove search results in Japan, *see, e.g.*, Uga (2016); Shishido et al. (2015), pp. 72–79; *see also* Yamaguchi (2015a), pp. 195–196.

[30]The decision of the Tokyo District Court on October 9, 2014 was affirmed in part and reversed in part by the decision of the same court in the proceeding of the objection on July 14, 2016 (Case No. 2015 (Mo) 53974, available at D1-law ID No. 28252703), which sustained the removal of only

was reported by mass media, it spurred debate in Japan, especially over possible influence from the CJEU ruling in May 2014 on the Japanese courts.

3.2 *The First District Court Ruling for the "Right to Be Forgotten" and Reversal on Appeal*

Furthermore, the first Japanese court ruling which explicitly mentioned the "right to be forgotten" in its reasoning was said to be a decision of the Saitama District Court on December 22, 2015.[31] This ruling, however, was later reversed by the Tokyo High Court. The case was then taken up by the Supreme Court, which handed down the above-mentioned decision in January 2017.

The case came to the district court with a motion for a preliminary injunction sought by a petitioner against a respondent, Google, Inc., to remove search results based on the petitioner's "right to personality" relating to "rehabilitation". The said search result contained web-page information referring to the petitioner's prior history of arrest in 2011 on suspicion of breaching the child prostitution clause of "Act on Regulation and Punishment of Acts Relating to Child Prostitution and Child Pornography, and the Protection of Children" (Act No. 52 of 1999).[32] These search results were displayed when the petitioner's address and full name were entered into the search engine.

In the reasoning of this ruling in December 2015, the Saitama District Court said as follows: [e]ven if a criminal's prior arrest history has been once reported and known to society, he/she has the right to have his/her private life respected as "a right to personality", and has "an interest in not having his/her rehabilitation hindered", and therefore, there should be a "right to be forgotten" from society with respect to a prior crime after a certain period of time has passed, depending on the nature of the crime, among other factors.[33] Undoubtedly, this ruling became contentious in Japan,

---

3 articles which falsely indicated that the petitioner committed blackmail, out of 66 litigated articles; this was mainly because the court took into considerations that the petitioner published the relevant facts by himself. On appeal, this Tokyo District Court decision in 2016 was affirmed by the Tokyo High Court on January 12, 2017 (Case No. 2016 (La) 1295, available at D1-law ID No. 28252704), and by the Supreme Court on July 19, 2017 (Case No. 2017 (Ku) 141 & Case No. 2017 (Kyo) 2, available at D1-law ID No. 28252705).

[31]Case No. 2015 (Mo) 25159, 2282 HANREI JIHO 78; this Saitama District Court decision was handed down in the proceeding of the objection, and sustained the initial decision of the same court on June 25, 2015 (Case No. 2015 (Yo) 17).

[32]An unofficial translation of this Act is available at, http://www.japaneselawtranslation.go.jp/law/detail/?id=2592&vm=04&re=01&new=1. According to the fact of this district court's decision, the petitioner had paid a fine of 500,000 yen [approximately 4500 US dollars] by a summary court's order under summary procedure.

[33]Concerning the interest of not having rehabilitation hindered for those who with past criminal convictions, *see* the judgment of the Supreme Court on February 8, 1994 (Non-fiction "Gyakuten" [Reversal] case), 48 MINSHU 149, http://www.courts.go.jp/app/hanrei_en/detail?id=1300

not only because of its mentioning of this novel concept of the right to be forgotten, but also because of the way in which an existing multi-factor balancing test was applied to this specific case, particularly in relation to how much weight should be given to the public interest in publishing the fact of arrest for a crime of child prostitution 3 years ago.

On appeal, the Tokyo High Court, as an intermediate court of appeals, delivered a decision on July 12, 2016, reversing the decisions below, and dismissing the petitioner/appellee's motion for a preliminary injunction.[34]

First, the Tokyo High Court said in the reasoning that the "right to be forgotten" argued by the appellee has "no explicit statutory basis" in Japan and its "requirements and effect" are not clear. It should be said that, according to the high court, the appellee's claim for injunctive relief based on the "right to be forgotten" as part of a right to personality is not different in substance from the existing claims for injunctive relief based on the right to reputation or privacy as part of a right to personality, and thus "there is no need for independent consideration".

Second, about the appellant's claim to be "a mere intermediary" and therefore entitled to immunity from liability, the Tokyo High Court took a negative position similar to the district court mentioned below. The high court said that "even if the search results were generated automatically and mechanically, this is done through a program equipped with an algorithm determined by the appellant" and "the title and snippet can be said to function usually as an independent expression" if we look at the way in which the appellant's service is used practically. In applying the existing balancing criteria in defamation and privacy case law, the high court seems to have tilted slightly more toward the freedom of speech and right to know than the Saitama District Court, and thus denied the need for injunctive relief in this case.[35]

In the final appeal procedure, the Supreme Court affirmed this Tokyo High Court decision, and dismissed the petitioner/intermediate appellee/final appellant's appeal. In an uncertain climate where the lower courts were divided over the issue of search engine liability, the Supreme Court held that, if certain substantive requirements are met, injunctive relief can be granted against a search engine operator to remove search results containing private facts. In the next section, we will take a close look at the holding and reasoning of the Supreme Court decision, whose six-factor balancing test would prescribe a level of protection to be granted in subsequent cases with similar facts and issues in Japan.

[unofficial English translation]. In this tort case, the Supreme Court of Japan took the interest of so-called "passage of the time" into consideration, and formulated the multiple-factor balancing test to decide the availability of compensatory damage. The term "privacy" had never appeared in this judgment in 1994, but it was cited by the Supreme Court as its own privacy precedent in the judgment on March 14, 2003 (Nagara-gawa incident news-reporting case), 57 MINSHU 229, http://www.courts.go.jp/app/hanrei_en/detail?id=628 [unofficial English translation], etc.

[34]Case No. 2016 (La) 192, 71 MINSHU 82.

[35]*See, e.g.*, Uga (2016), p. 31.

4 The Supreme Court's Six-Factor Balancing Test Under Privacy Law

4.1 The Substantive Requirements of Injunctive Relief to Remove Search Results

The judgment of the Supreme Court of Japan on January 31, 2017 set up the following substantive requirements of injunctive relief for removal of certain search results. The Court's original sentence in Japanese is too lengthy to be translated into English, so it is broken up into a few paragraphs.

> [T]he illegality of the conduct of providing website information including URLs containing articles with private facts of the said person as part of search results in response to search request terms about the person, should be decided by weighing up various circumstances concerning the following "legal interest of not having the said facts published" and "reasons to provide information including the said URLs as search results":
>
> [①] the nature and content of the said facts, [②] the extent to which private facts of the said person were distributed as a result of the provision of information including the said URLs, and the extent of the damage specifically suffered by the said person, [③] the social status and influence of the said person, [④] the purpose and significance of the said articles, [⑤] the social circumstances at the time of the said articles' publication, and the subsequent changes, and [⑥] the need to mention the said facts in the said articles, etc.;
>
> and as a result, if it is "clear" that the "legal interest of not having the said facts published" is overriding, then it should be reasonably interpreted that it is possible to request the search service operator to remove information including the said URLs from search results.[36]

In order to understand the importance and limit of the holding of this judgment, there are a few points to be noted. First, the basis of injunctive relief here is not the right to be forgotten as such but "privacy." Based on its own precedent, the Supreme Court of Japan in this case confirmed that the interest of not having facts relating to the individual's private life published unduly, shall be entitled for "legal protection."[37] The Court does not say anything about the right to be forgotten or defamation, so it remains to be seen how the availability or requirements of injunctive relief for the removal of search results on the basis of the right to be forgotten or right to reputation will be decided.

Second, regarding the issue of search engine liability, the Supreme Court puts an end to the mere-intermediary argument that search engine operators should be categorically immunized from liability for third party content.[38] The Supreme

[36]Case No. 2016 (Kyo) 45, 71 MINSHU 63, 66 (Sup. Ct., Jan. 31, 2017).

[37]Case No. 2016 (Kyo) 45, 71 MINSHU 63, 65 (Sup. Ct., Jan. 31, 2017) (citing the judgments of the Supreme Court of Japan on Apr. 14, 1981 (Referral of criminal record case), 35 MINSHU 620; on Feb. 8, 1994 (Non-fiction "Gyakuten" [Reversal] case), 48 MINSHU 149; on Sept. 24, 2002 ("Ishi-ni-oyogu Sakana" [Fish swimming in Stone] case), 1802 HANREI JIHO 60; on Mar. 14, 2003 (Nagara-gawa incident news-reporting case), 57 MINSHU 229; and on Sept. 12, 2003 (University students list case), 57 MINSHU 973).

[38]71 MINSHU at 65–66. *See also* Takahara (2017), pp. 120–121.

Court explained this point in the reasoning, stating that, although the collection, arrangement and provision of the said information was “conducted automatically by programs”, the programs were made in a way to achieve the results in accordance with the “search service operator’s policy”, and thus the provision of the search results has “an aspect of expressive conduct by the search service provider itself”. Moreover, the Court rightfully emphasizes the “significant role” that search engine operators have been playing through the provision of search results as a platform of information distribution on the Internet in modern society.

Third, in applying the requirements of injunctive relief, the Supreme Court found in the end that it was not “clear” in this specific case.[39] In comparison with the Tokyo High Court’s balancing test which took the factors of “seriousness” and “irreparability” of the petitioner’s damage into consideration, the Supreme Court seems to take a middle ground in the following two senses. Firstly, the Supreme Court did not endorse the high court’s approach in favor of the freedom of expression. Secondly, the Supreme Court did not just simply maintain the *ad hoc* balancing approach in precedents of privacy cases but added a further substantive requirement of “clarity”.[40]

The Supreme Court’s holding of the availability of injunctive relief to remove the search results accompanied by such substantive requirements with six-factor balancing will have a significant impact not only on subsequent court rulings, but also on the self-regulatory regimes of search engine operators who are more susceptible to voluntarily remove a search result which is adjudicated to be “clearly” illegal within each jurisdiction.[41]

---

[39]71 MINSHU at 66-67.

[40]*See, e.g.*, the judgments of the Supreme Court of Japan, Feb. 8, 1994, and on Mar. 14, 2003, *supra* note 33. *See also* Takahara (2017), p. 121.

[41]For example, Yahoo Japan Corporation published a report of the Advisory Committee relating the search results and privacy, and a policy to respond the requests to remove the search results voluntarily on March 30, 2015, available at, https://publicpolicy.yahoo.co.jp/2015/03/3016.html [in Japanese]*; see infra* notes 45–48. As a subsequent development of case law, there has been a series of rulings of the lower courts which cited the balancing test by the Supreme Court decision on January 31, 2017 and dismissed the petitioner’s privacy claims for the removal of search results based on the right to personality; *see, e.g.*, a decision of Nagoya High Court on Mar. 31, 2017, Case No. 2016 (La) 284, 2349 HANREI JIHO 28 (privacy claims against Google Inc.); a decision of Takamatsu High Court on July 21, 2017, Case No. 2017 (La) 11, 2354 HANREI JIHO 40 (privacy claims against Google Inc.); a judgment of Tokyo High Court on July 2, 2018, Case No. 2017 (Ne) 5296, available at D1-Law ID No. 28263456 (privacy claims against Yahoo Japan Corporation); *see also* a judgment of Tokyo High Court on Aug. 23, 2018, Case No. 2018 (Ne) 1104, 2391 HANREI JIHO 14 (citing the Supreme Court judgment on June 11, 1986 (Hoppo Journal case), 40 MINSHU 872, and the Supreme Court decision on Jan. 31, 2017, to dismiss defamation claims for removal of search results against Google LLC).

4.2 A Variable and Modest Level of Protection Under Japanese Law?

While this test formulated by the Supreme Court has the advantage of being heavily fact-specific and flexible enough to let the lower courts grapple with a vexing balancing job, it needs further clarification in a more transparent and foreseeable way, especially for gray-zone cases. To the extent that the Japanese equivalent of the right to be forgotten is subjected to the Supreme Court's balancing test, its level of protection may become quite variable, and thus might end up offering only a modest level of protection, in comparison with its counterpart's higher level of protection under the CJEU ruling in 2014 and Article 17 of the GDPR.

Still, no jurisdiction can escape from the same difficulty of clarifying criteria in order to strike a fair balance among equally important competing values such as privacy and the freedom of expression. In particular, given that the CJEU ruling relating to the right to be forgotten in 2014 held that "Article 12(b) and subparagraph (a) of the first paragraph of Article 14 of Directive 95/46 are to be interpreted as meaning that, when appraising the conditions for the application of those provisions, it should inter alia be examined whether the data subject has a right that the information in question relating to him personally should, at this point in time, no longer be linked to his name by a list of results displayed following a search made on the basis of his name, *without it being necessary* in order to find such a right that the inclusion of the information in question in that list *causes prejudice* to the data subject",[42] it seems to be quite challenging to elucidate what specific value or interest such right aims to serve, and what criterion shall be applied to reconcile relevant competing values and interests.

Furthermore, it can be also said that the level of the Japanese equivalent of the right to be forgotten is substantially more than anything that could be recognized under US law, mainly because the US Supreme Court has granted full First Amendment protection to Internet communication,[43] and additionally because the US Congress has made an explicit policy choice to grant federal immunity to Internet service providers under the Communications Decency Act of 1996, Section 230 (47 U.S.C. § 230(c)), this immunity having later been held to be broad enough by the courts to cover search engines as well.[44]

---

[42]CJEU, Case C-131/12, *Official Journal of the European Union*, at 5 (*see* supra note 2; emphasis added).

[43]*See, e.g.*, Reno v. ACLU, 521 U.S. 844, 869-870 (1997).

[44]*See, e.g.*, Zeran v. America Online, Inc., 129 F.3d 327 (4th Cir. 1997); Blumenthal v. Drudge, 992 F. Supp. 44, 51-52 (D.D.C. 1998); Garcia v. Google, 786 F.3d 733, 745-746 (9th Cir. 2015); Manchanda v. Google, No. 16-CV-3350 (JPO), 2016 WL 6806250, at *3 (S.D.N.Y. Nov. 16, 2016); *see also* Marshall's Locksmith Serv. Inc. v. Google, LLC, 925 F.3d 1263, 1267-1272 (D.C. Cir. 2019); Force v. Facebook, Inc., 934 F.3d 53, 64 (2d Cir. 2019). Regarding the issue of liability of online intermediaries in Japan, the scope of a statutory immunity for Internet service providers granted under "Act on the Limitation of Liability for Damages of Specified Telecommunications Service Providers and the Right to Demand Disclosure of Identification Information of the Senders"(*see* supra note 27) is much more limited, compared to the Communications Decency Act, Section 230, in the United States. Unofficial English translation of the former Act is available at, http://www.japaneselawtranslation.go.jp/law/detail/?id=2088&vm=04&re=01&new=1.

5 Conclusion: Preparing for What Comes Next

In the aftermath of the CJEU ruling in *Google Spain SL v AEPD* on May 13, 2014, it is well known that Google quickly set up a webform for processing removal requests to comply with the ruling, but this specific webform is offered for requests based on data protection law in Europe and thus covers only limited relevant countries.[45] Those who do not have some connection to such relevant countries, including most people in Japan, are offered instead a general webform to request the removal of certain kinds of personal information to be processed according to Google's removal policies.[46] As with the basic framework of Japanese data protection statutes applicable to the private sector mentioned in Section 2 of this report, Japanese government discussions on the future policy of the right to be forgotten tend to put more emphasis on improving the self-regulatory, market-based approach in complaint processing procedures,[47] in contrast to the EU data protection reforms that have codified this right explicitly in the GDPR. Still, given that the Japanese Supreme Court's decision on January 31, 2017 has taken a step forward toward taking a clear stance on the issue of search engine liability and foreclosed the mere-intermediary argument for categorical immunity, search engine providers are expected to respond to search removal requests voluntarily and in a more transparent manner.[48]

Indeed, ongoing highly-divisive debates about the "right to be forgotten" in many jurisdictions seem to symbolize the growing need for a new conception of rights in times of rapid change and for a rebalancing among competing values at stake in today's globally networked communities. Interestingly, there was a similar factual change in the latter half of the twentieth century, when the "right to know" emerged in many countries against the background of one-way mass communication, in which big government and traditional types of mass media played a dominant or asymmetric role and ordinary individuals were mere passive users.[49] Nowadays in the 2010s, the formerly-divided roles of speaker, user, and intermediary have become interchangeable, and individual users can actively engage in sharing and collaborative creation via networking platforms, while our identities may well be defined digitally and often misconfigured invisibly.

What comes next are automated or even autonomous algorithmic decision-making systems with minimal human intervention, empowered by the latest smart technologies such as artificial intelligence (AI), big data analytics, the Internet of

---

[45] *See, e.g.*, Google France SARL (2014).

[46] *See* Google (2018).

[47] *See, e.g.*, Soumusho [Ministry of Internal Affairs and Communications], the Japanese Government (2015).

[48] Regarding the recent self-regulatory measures by the search engine operators, *see e.g.*, Sogabe (2018).

[49] In Japan, it was a decision of the Supreme Court on Nov. 26, 1969 (Hakata-station TV film subpoena case), 23 KEISHU 1490, that explained the important role of news-reporting and news-gathering by mass media to serve the people's "right to know" in democratic society.

Things (IoT), and robotics, of which search engine algorithms are an early example of a specific application. If such systems could automatically learn and improve by themselves for more general application, then we need to call for a comprehensive investigation into how such technologies impact our daily lives, society, and legal systems, and what should be done in response.[50] More specifically, search engine algorithms in the early days were said to be conceptually quite simple, in the sense that they worked to determine the rank of a page based on the number of links given to that page.[51] So long as algorithms function in the way they were originally programmed, the traditional legal concept of foreseeability and causal responsibility would work well. Certainly, these concepts were invoked in the reasoning for the decision of the Japanese Supreme Court on January 31, 2017, where it was stated that the "provision of search results" has an aspect of "expressive conduct by the search service provider itself" because their computer programs were made in a way to achieve the results in accordance with "the search service provider's policy". However, as algorithms are increasingly programmed in ways that enable them to learn, decide, and constantly update by themselves, we ought to consider closely who exactly should be held liable for harms caused by algorithmic decision-making, among complexly-related multiple entities who are involved in the design and practical implementation of such algorithms.[52]

There is a whole list of tasks that must be accomplished in order to prepare for the mixed blessings of next-generation algorithmic decision-making. Among these is the need to clarify what set of values we should aim to achieve and how to combine multiple means to realize them, including legal regulation/protection, self-regulation, technological measures, education and information ethics. Indeed, besides the issues relating to search engines, there are also other specific practical examples, such as autonomous driving, medical diagnosis, criminal justice, and weapons, in which algorithmic decision-making enabled by the latest smart technologies such as AI, IoT, and robotics are quickly coming into play, and therefore, policy choices as well as legal questions are constantly recurring about how to deal with liability and immunity for relevant multiple entities so as to ensure justice, fairness, and accountability.[53] At least for the time being, however smart or complex, an AI computer program is just a creation of humans; it can be a powerful means, but it is not yet good at setting normative ends by itself.[54]

Japanese law's recent proactive judicial move to curb algorithmic determinism, through allowing preliminary injunctions to remove search results under privacy

---

[50]For more detailed analysis to tackle with such questions, *see, e.g.*, Yamaguchi (2014b, 2015b, 2018a).

[51]*See* Tutt (2017).

[52]*See, e.g.*, Tutt (2017), pp. 94–96, 109; Balkin (2015, 2018); Scherer (2016).

[53]*See, e.g.*, U.S. Executive Office of the President (2016); *see also* Recent Cases (2017); U.K. Parliament (2016). For an overview of legal concepts and policy issues in Japan relating to emerging information and communications technologies, *see, e.g.*, Hamada (2017).

[54]*See, e.g.*, Nishigaki (2016); Kimura (2018); *see also* Yamaguchi (2018b, c).

case law, even without an explicit legislative mandate, might provide yet another workable example to strive for the rebalancing of fundamentally important values. Although the eclectic stance taken by Japanese law is different from the relevant EU or US laws, it is based on the shared aspiration to realize the fundamental rights of individuals in substance or essence, without stifling innovation, irrespective of differences of terminology.

References

Ashibe N (2015) (Rev. by Takahashi K), KENPO [Constitutional Law], 6th edn., pp 119–127 [in Japanese]

Balkin JM (2015) The path of robotics law. Calif Law Rev Circuit 6:45, 46, 51–55

Balkin JM (2018) Free speech in the algorithmic society: big data, private governance, and new school speech regulation. U.C. Davis Law Rev 51:1149

Beer LW, Ito H (1996) The Constitutional Case Law in Japan, 1979 Through 1990, pp 5–6, 18–20

European Commission (2017) Communication from the Commission to the European Parliament and the Council: exchanging and protecting personal data in a globalised World (Brussels, Jan. 10, 2017. COM(2017) 7 final), pp 4, 8, 10, 16, https://ec.europa.eu/newsroom/document.cfm?doc_id=41157

European Commission (2018) International data flows: Commission launches the adoption of its adequacy decision on Japan (Press release)(Brussels, Sept. 5, 2018), http://europa.eu/rapid/press-release_IP-18-5433_en.pdf

Google (2018) "Google kara no Joho no Sakujyo" [Removal of information from Google] (Japanese language version) (2018), https://support.google.com/websearch/troubleshooter/3111061?hl=ja

Google France SARL (2014) Google's letter on "Questionnaire addressed to Search Engines by the Article 29 Working Party regarding the implementation of the CJEU judgment on the "right to be forgotten"" (July 31, 2014), https://docs.google.com/file/d/0B8syaai6SSfiT0EwRUFyOENqR3M/edit?pli=1

Greenleaf G (2012) Independence of data privacy authorities (Part II): Asia-Pacific experience. Comp Law Secur Rev 28:121, 126

Greenleaf G (2014) Asian data privacy laws: trade and human rights perspectives. Oxford University Press, Oxford, pp 264–265

Hamada J (2017) Joho-tsushin Seisaku no Aratana Dankai ni Mukete [Toward a new stage of information and communications policy research]. J Inf Commun Policy 1(1):3–8. [in Japanese]

Hasebe Y (2014) Privacy in the Age of Ubiquitous Computing. Percorsi Costituzionali 1:133

Igarashi K (2003) JINKAKUKEN GAISETSU [Law of personality], pp 2–13. [in Japanese]

Ishii K, Kanda T, Mori R (2015) Kensakukekka-sakujyo no Karishobunkettei no Toraekata to Kigyo wo fukumu Net-Joho no Sakujyogimu [Understanding of Preliminary Injunction to Remove the Search Results and Practice of the Removal of Information on the Net including Company Practice]. NBL 1044:7–22. [in Japanese]

Kanda T (2015) NET KENSAKU GA KOWAI [Frightend by the net search], pp 43–54. [in Japanese]

Kato K (1989) Saikosaibansho Hanrei Kaisetsu [Sup. Ct. Case Review] (Sup. Ct., June 11, 1986). HOSO JIHO 41:2621, 2632. [in Japanese]

Kimura M (2018) AI to Keiyaku [AI and Contract]. In: Yanaga M, Shishido G (eds) ROBOT, AI TO HO [The laws of robots and artificial intelligence], pp 157–158. [in Japanese]

Nishigaki T (2016) BIG DATA TO JINKO CHINO [Big data and artificial intelligence], pp 153–161. [in Japanese]

Recent Cases, 130 HARV. L. REV. 1530 (2017)
Scherer MU (2016) Regulating artificial intelligence systems: risks, challenges, competencies, and strategies. Harv J Law Technol 29:353, 356–357, 397–398
Seki N (2016) Heisei 27 Nendo no Tokyo Chiho Sibansho Minji dai 9 bu ni okeru Minjihozenjiken no Gaikyo [General Trends in Cases for Civil Provisional Remedies in Subdivision No. 9, Civil Division, the Tokyo District Court]. KINYU HOUMU JIJO 2044:30, 31–32, 34–35. [in Japanese]
Shishido G, Monguchi M, Yamaguchi I (2015) Internet ni okeru Hyogen no Jiyu to Privacy: Kensaku Engine wo Chushin toshite, [The Freedom of Expression and Privacy: With a Focus on the Search Engines]. JURIST 1483:ii–v, 68–80. [in Japanese]
Sogabe M (2018) "Internet jono Johoryutsu no Kiban" toshite no Kensaku-service [Search Service as "Infrastructure of Distribution of information on the Internet"]. Q Jurist 25:51–53. [in Japanese]
Solove DJ, Schwarts PM (2018) Information privacy law, 6th edn., pp 786–792
Soumusho [Ministry of Internal Affairs and Communications], the Japanese Government (July 17, 2015) Houkokusho: Internet jo no kojinjoho riyoushajohotou no ryutsu heno taiou ni tsuite [Report: Concerning the Measures for Distribution of Personal Information and Users Information on the Internet], http://www.soumu.go.jp/menu_news/s-news/01kiban08_02000184.html [in Japanese]
Takahara T (2017) Saiko-sai Toki no Hanrei [Recent Cases of the Supreme Court]. Jurist 1507:119. [in Japanese]
Tutt A (2017) An FDA for algorithms. Adm Law Rev 69:83, 92–94
U.K. Parliament, House of Commons, Science & Technology Committee (Oct. 2016) Robotics and artificial intelligence, pp 16–26, https://www.publications.parliament.uk/pa/cm201617/cmselect/cmsctech/145/145.pdf
U.S. Executive Office of the President, National Science & Technology Council (Oct. 2016) Preparing for the future of artificial intelligence, pp 1–4, 13–14, 30, 37–38, https://obamawhitehouse.archives.gov/sites/default/files/whitehouse_files/microsites/ostp/NSTC/preparing_for_the_future_of_ai.pdf
Uchida T (2008) MINPO I [Civil Law I], 4th edn., pp 24–26. [in Japanese]
Uga K (2016) "Wasurerareru-kenri" nitsuite [Concerning the "Right to be Forgotten"]. Q Jurist 18:24–33. [in Japanese]
Uga K (2018) KOJINJOHOHOGO NO CHIKUJO KAISETSU [Commentary of personal information protection laws], 6th edn., pp 204–223. [in Japanese]
Yamaguchi I (2002) Beyond De Facto freedom: digital transformation of free speech theory in Japan. Stanf J Int Law 38:109, 114–119
Yamaguchi I (2006) Mass media and privacy in Japan: current issues, recent trends, and future challenges toward the "Ubiquitous Network Society". J Korea Inf Law 10(1):171–181
Yamaguchi I (2012) Net Jidai no Meiyo Kison, Privacy Shingai to "Jizen Yokusei": Hoppo Journal Jiken Hanketsu [Defamation, Privacy Invasion, and "Prior Restraint" in the Age of the Net: Hoppo Journal Case Judgment]. Q Jurist 1:50–58. [in Japanese]
Yamaguchi I (2014a) The checking value in "Information Privacy" concept: what lesson can be drawn from Japanese Law's Eclectic approach in-between the U.S. and EU law in the smart media and IT environment?. Paper presented at the 9th World Congress of the International Association of Constitutional Law, University of Oslo, Oslo, Norway, June 19, 2014
Yamaguchi I (2014b) Protecting privacy against emerging "Smart" big data surveillance: what can be learned from Japanese Law? PERCORSI COSTITUZIONALI 1:193
Yamaguchi I (2015a) EU Ho ni okeru "Wasurerareru Kenri" to Kensaku Engine Jigyosha no Kojin Data Sakujyo Gimu (CJEU, Preliminary Ruling in Google Spain SL Case, May 13, 2014) ["Right to be Forgotten" in EU Law and the Search Engine Operators' Obligation to Remove Personal Data]. In: Horibe M (ed) JOHOTSUSHIN HOSEI NO RONTEN BUNSEKI [Analysys of the Issues of information communication laws], Special Issue of NBL, No. 153, pp 181–196. [in Japanese]

Yamaguchi I (2015b) Internet ni okeru Hyogen no Jiyu [Freedom of Expression on the Internet]. In: Matsui S, Suzuki H, Yamaguchi I (eds) INTERNET HO [Intenet Law], pp 25–52. [in Japanese]
Yamaguchi I (2018a) Hyogen no Jiyu to Chosakuken: AI Jidai no "User Rights" Gainen to sono Check Kino [Freedom of speech and copyright: a concept of "User Rights" and its checking function in the age of AI]. Q Jurist 25:61. [in Japanese]
Yamaguchi I (2018b) Free speech, National Security, and privacy under stress of algorithmic surveillance: a comparative study of Japan and the United States. Paper presented at the 10th World Congress of the International Association of Constitutional Law, SungKyunKwan University, Seoul, South Korea, June 19, 2018
Yamaguchi I (2018c) Kokka Anzenhosho niokeru Algorithm niyoru Kanshi: Kenpo jo no Genron no Jiyu, Privacy to Platform Jigyosha no Yakuwari [Algorithmic Surveillance in National Security: Constitutional Free Speech and Privacy, and the Role of Platform Business Operators]. KENPO KENKYU 3:47. [in Japanese]

Limits and Prospects of *the Right to Be Forgotten* in Taiwan

Wen-Tsong Chiou

Abstract The multifaceted nature of the right to be forgotten suggests that the concept actually has many different roots and serves different interests. Four kinds of the right to be forgotten are explored and discussed in Taiwan's legal contexts. Both the right to request deletion of personal data and more specifically the right to de-indexing on internet are commonly understood as honoring individual will or choice. It could be easily outweighed by more compelling interests of public's right to know and the freedom of press. However, the right of oblivion for the purpose of social rehabilitation and the right as independence from power in an era of big data deepen the meaning of the right to be forgotten and relate the concept to more democratic values.

1 Introduction

Sharing a similar appearance with the right to erasure, the right to be forgotten (RTBF) is very often considered to be new wine in old bottles. Not only as a liberty right but also a claim right to impose obligations on others, RTBF is yet another example of the right to control one's personal data, which is commonly thought to be the core of a more fundamental right to informational privacy. Its multifaceted nature, however, suggests that RTBF actually has many different roots and serves different interests. As the concept is gradually explored and developed in Taiwan, I will suggest that seeing RTBF as simply an expression of individual will or choice that conveys an agent's self-interests or preferences wrongly dilutes its real significance. Because RTBF, as such, is no more than a permission or even an order to rewrite or erase history merely for personal gains, it could easily be outweighed by competing interests such as public's right to know or the freedom of press that opposes obscurity and contests any form of censorship. Instead, RTBF could and

W.-T. Chiou (✉)
Institutum Iurisprudentiae, Academia Sinica, Taipei, Taiwan
e-mail: wentsong@gate.sinica.edu.tw

F. Werro (ed.), *The Right To Be Forgotten*, Ius Comparatum – Global Studies in Comparative Law 40, https://doi.org/10.1007/978-3-030-33512-0_16

should be understood as a tool for social rehabilitation as well as resistance to the threat to personal independence in an era of big data and algorithms. Such a revised version of RTBF pushes us to think more deeply about the democratic values to which RTBF can contribute and to delineate more properly the extension and intension of the concept.

2 RTBF as an Omnibus Right to Erasure

The term "right to be forgotten" is not explicitly used in any statutory law in Taiwan. To the extent that the concept is understood as related to the erasure of personal data and the prohibition of its further dissemination, a more conventional right to request deletion of personal data, which is clearly stated in the Personal Data Protection Act (PDPA), is the long-existed basis of the relatively new term. Article 11 of the PDPA provides that data controller shall, on its own initiative or upon the request of the Party, delete personal information collected when the specific collection purpose no longer exists, or when the authorization of data use expires, or in cases when the data collection, processing, or data use is in violation of the provisions of this Act. However, the PDPA does not explicitly mention about the withdrawal of data subject's consent as a legal basis for requesting the deletion of personal information. The Ministry of Justice, which was the authority in charge of interpretation of the law before National Development Council took over its role in 2018, expounded only that the data shall not be *further processed* once the consent is withdrawn (MOJ 2015). The authority stopped short of explaining whether the data also needs to be deleted. It remains an untested theory that the withdrawal of consent entails the mandatory deletion of personal information when the data is still processed in a way compatible with its specific collection purpose.

Article 13 of the PDPA further provides that data controller shall make decisions upon receiving the deletion request within 30 days or at latest with an extension of another 30 days. Administrative fine no less than NT$20,000 but no more than NT $200,000 may be imposed by the government authority on a private data controller who fails to honor the deletion request in violation of the provisions of the PDPA. Article 28 of the PDPA also affords a legal ground for the compensation for non-material damages as a result of infringement on the rights of data subject, including the right to request deletion of personal data.

However, when data was collected on grounds other than individual consent, a wide array of secondary data uses may easily trump individual's right to request deletion of personal data. The point is most clearly demonstrated in a case involving secondary use of health insurance data.

In 2012, a number of individuals jointly filed a petition and later a lawsuit requesting the National Health Insurance Administration (NHIA), which is authorized by law to collect personal medical data for the purpose of health insurance payment, to cease processing their data for secondary medical research uses and to delete such data from a research database. As a single-payer mandatory social insurance run by NHIA, Taiwan's National Health Insurance (NHI) provides a universal coverage for its 23 million people. Collection of personal data for each and every NHI service is required by law with a specific purpose of approving insurance claims.[1] The tremendous volume and the centralized nature have made NHI data invaluable treasures for those who are eager to make the best use of large-scale health information and big data analytic tools. Since 2011, the NHIA has begun to provide a copy of its complete NHI data to Ministry of Health and Welfare (MHW) to create a NHI research database in company with other databases held by MHW. While the NHI research data released by MHW for public access is without direct identifiers, data belonging to a specific individual within the NHI database is still linkable among different datasets and to other databases using a universally unique identifier, that is, the scrambled national identity number. Although users of the database are required to take away only the aggregate results of their analyses, the possibility of indirect identification is never completely ruled out during or after on-site data analysis. After a 5-year lawsuit, the Supreme Administrative Court ruled that individual plaintiffs do not have a right to request deletion. The court reasoned that since the collection is authorized by law rather than individual consents, it is beyond the control of data subjects regarding the immediate and subsequent data use in the first place. Also, as long as the data is further used for medical research and is at least pseudonymized (no need to be entirely anonymized), a standalone opt-out right is not to be granted regardless of the importance of each specific research (Supreme Administrative Court 2017). Conceiving simply as the expression of individual will or choice, individual preferences against data use were nonchalantly outweighed by the ostensible public interest of medical research.

The plaintiffs later brought the case for constitutional review on the basis, among other things, that depriving individual's right to request deletion would amount to forced participation in research and thus encroaching on individual's constitutional right to privacy as independence. While the result of constitutional review is still pending thus far, the case poses an important question of whether there is some more critical values undergirding the recognition of the RTBF other than the respect for individual will or choice.

3 RTBF as the Right to De-indexing on the Internet

When the RTBF refers more specifically to the de-indexing of personal data from the results of search engines on the web, PDPA's right to request deletion of personal data may provide only a tenuous ground. Indexed information is very often collected and processed from and retained in the public domain where the right to erasure under PDPA is usually not applicable unless the correctness of the information is

[1]National Health Insurance Act, Art. 79 & 80.

contested. Alternatively, causes of action based on the Civil Codes may provide more appropriate legal grounds against internet service providers (ISPs) when the disclosure or dissemination of the true information is claimed to infringe on the reputation of a data subject.

For example, a plaintiff in a tort case requested the Taiwan Branch of Google International LLC to remove the search results containing a game-fixing scandal allegedly involving the plaintiff. As the CEO of a professional baseball team, the plaintiff of the tort case was prosecuted earlier for conspiring to fix the game and thus breach the trust. The result of the criminal trial cleared the CEO of the charge. News about his involvement in the scandal and other dishonest behaviors remains popular results in google search with the plaintiff's name as the query. People unaware of the criminal judgement continue to take in what they learned from those search results and make adverse comments about the plaintiff accordingly. The court that heard the tort case, without ruling directly on the substantive issues of the RTBF, found against the plaintiff. The court determined that Google Inc., which is considered to be the real search engine operator yet without local presence in Taiwan (Taiwan High Court 2016), is a different entity from Google International LLC. Suing Google International LLC for delisting is of no merit.

Courts in other cases involving the erasure request directed at ISPs, however, were more willing to resort to a balancing test, with which interests between data subjects and the opposing party or even the general public are weighed. In a tort case in which Yahoo Taiwan was requested to remove an allegedly defamatory article posted on a social media forum run by Yahoo Taiwan, the High Court rejected the plaintiff's damage claim for fear that the freedom of speech may be overly suppressed if an ISP is to play the role of a police censoring speech on the internet (Taiwan High Court 2013). Yet in another tort case, the defendant convicted in a criminal slander case was asked by the plaintiff to be responsible for having Google removed the search results containing the defamatory information. The district court was content with the effort to remove all search results from the domains of google.tw without holding that the defendant be further responsible for removing the same search results processed by an extraterritorial data processing server that is not a Taiwanese company (Taoyuan District Court 2015).

In short, the right to request deletion of personal data under the PDPA is implemented but is limited to data collected on the basis of individual consents. It provides only a shaky ground for requesting removal of search results containing public information from a search engine. Tort laws instead lay an alternative basis for requesting deletion of reputation infringing information on the internet. However, whether the information at issue infringes on reputation is never a plain fact and always needs to be settled by court before the removal request would be mandatorily taken. Weighing against the all too important freedom of press and the public's right to know, the interests of personal reputation are seldom viewed substantial enough to justify the request for delisting. Thus, court decisions that uphold ISP's obligation to remove contents or search result links with personal information about an individual are limited to situations where the ISP understands that the contents or search results may infringe upon reputations of others or fails to know as a result of gross negligence. Besides, the territorial rule that limits the jurisdiction only to tortious conducts or their effect occurring within the territorial boundaries of the state poses another challenge to the implementation of tort laws against extraterritorial reputation infringing acts.

4 RTBF as the Right of Oblivion

When the modern concept of RTBF was first introduced by Mayer-Schönberger in 2009 (Mayer-Schönberger 2009), it was closely linked to the right of oblivion. The effort to inquire into the institutional root of the legal concept helps to reposition the RTBF as a tool for social rehabilitation, which in turn is crucial for building a more tolerant society as a necessary condition for a democracy.

Forgetting is a very important function of human memory mechanisms. By reorganizing the content of what humans remember, forgetting allows humans to filter and delete part of the brain's stored information, keeping each person's limited memory for the most important and the most relevant matters. In addition to the reason of efficiency, the limit of memory also keeps our mental health in good shape by selectively remembering happy experience and forgetting negative one.

Although the society as a whole needs to keep its collective memory, through writing, narrative, or the preservation of space or architecture, for those things that should not be forgotten, it also depends on the physiological limits of personal memory in order to function properly. For example, a person does not have to worry that every bit of daily life is remembered by people whom one meets only occasionally, so that a tranquil life is possible. Neither will stupid things that one did in the far past will be kept replayed and reminded. Being forgotten means a state of not being marked by others and a possibility of restart a life all over again after time passes and situation changes.

Now, thanks to the almost unlimited data collection and storage technologies, a sea change of the memory capacity that human can control and deploy is happening. 24 hours of uninterrupted street surveillance video, driving recorders installed in many cars, mobile phone cameras that everyone carries with them. All these devices incessantly record all kinds of events in daily life. Etag, GPS, and mobile phone constantly send out location signals that can be received and logged by system operators. Other information is also continuously created as footprints of human activities. For instance, preference tags secretly attached to users who are browsing websites, profiles of a user's reading preference, "like and share" patterns collected by social networking sites, electricity use habits recorded by smart meters provided by the power company, consumers' purchase records that the credit card companies retain, telecommunication metadata created and maintained by the telecommunications companies, medical information collected by the national health insurance administration and private companies. The above data sets are just the tip of an iceberg of a vast amount of very detailed information that one can imagine about every aspect of a persons' life and that is created and collected by present-day ICT.

These latest development of information technologies have not only made everything leave traces but also made those traces literally unforgettable.

However, if to forget were to delete every bit of information created along human activities, it would have swamped the legal system and diluted the very meaning and function of RTBF. RTBF as the right of oblivion focuses instead on removing information that is detrimental to the legitimate goal of social rehabilitation. As the need for social rehabilitation arises only after the wrongdoing has been sanctioned and retributive justice has been realized, time is always a prerequisite for granting RTBF. The scope of RTBF as the right of oblivion is also limited to situations where the information at issue becomes hindrances for a meaningful return to the society. Such a right of oblivion is, however, inadequately protected in Taiwan.

Like other kinds of sensitive data, criminal records are protected under the PDPA. Article 6 of the PDPA provides that the collection, processing, or use of criminal records needs to be, among others, based on written consent of the data subject, specific statutory authorization, the necessity for compliance with a legal obligation to which the private controller is subject, or the necessity for the performance of a task carried out in the exercise of official authority vested in the public controller. The erasure of past criminal data for the reason of social rehabilitation is nonetheless not specifically recognized as a legal right under the PDPA.

The Juvenile Delinquency Act is the only law that explicitly requires removal of juvenile criminal records certain years after the execution of sentence or the completion of juvenile protective measures. Under the Juvenile Delinquency Act, the legal obligation of removal is imposed only on criminal justice agencies and those authorized private entities that retain the official juvenile criminal records. While the obligation to delete criminal records of a juvenile delinquent for the purpose of social rehabilitation is without exception, it has no legal force on a private entity that obtains juvenile criminal information from its own sources. On the other hand, the Protection of Children and Youths Welfare and Rights Act (PCYWRA) provides additional protection for juvenile delinquent from media exposure. The PCYWRA prohibits the media from reporting personal identifiable information of the minor in a juvenile delinquency case and authorizes the competent authority to order the removal of contents containing such information. The media prohibition can be lifted when it is determined, through a reviewing process involving both the child protection agency and the representatives of the press association, that the report of the identifiable information is for the benefits of the subject minor or is necessary to protect public interests. The PCYWRA's provision is, however, less about the right of oblivion arising after a lapse of certain time than a right not to be socially labeled in the first place.

If RTBF as the right of oblivion is to ensure a "clean slate" for individuals after a certain period of time thus allowing them to come back to a more solidary society, current statutory laws in Taiwan provide insufficient grounds. More specific clean slate legislations may be necessary to complement a general RTBF to achieve the goal of building a more tolerant society.

5 RTBF as Independence from Power

As individual behaviors, preferences, physical conditions, and psychological states are all under the scrutiny and study of data technologies through profiling and machine learning, the threat to democracy is lurking behind subjecting one's independence to the algorithmic power. It is in this light that RTBF should not be simply regarded as a permission or even an order to rewrite or erase history merely for personal preferences, which could easily be outweighed by competing interests of public's right to know or the freedom of press that opposes obscurity and contests any form of censorship. Instead, RTBF should be viewed as a weapon of resistance to the threat to personal independence in an era of big data and algorithms. RTBF as independence from power helps to preserve a minimum breathing space in which more flexible self-identity can be formed without the domination of all kinds of knowledge about human produced from a vast amount of varied data. The petitioners who argued for a standalone opt-out right and brought the National Health Insurance Act for constitutional review asked the constitutional court to recognize exactly RTBF as independence from power. Such a revised version of RTBF pushes us to think more deeply about the democratic values to which RTBF can contribute and to delineate more properly the extension and intension of the concept.

6 Conclusion

The right to be forgotten gained world-wide attention after ECJ's ruling in 2014. However, the right is recognized only indirectly through varied legal mechanisms in Taiwan. A more traditional right to request deletion of personal data under the Personal Data Protection Act is granted but is limited to data collected on the basis of individual consents and is subject to relatively broad secondary uses. It is not a proper basis for requesting removal of search results containing personal information from a search engine. While tort laws provide a more appropriate basis for requesting deletion of reputation infringing information on the internet, whether the information at issue infringes on reputation is always a question that needs to be settled in advance by courts. The right to be forgotten so understood is easily outweighed by competing interests of the public's right to know or the freedom of press. However, the multifaceted nature of the legal concept suggests that the right to be forgotten actually serves interests other than honoring individual will or choice. It has institutional root that is more essential to the value of democracy. First, the right to be forgotten was closely related to the concept of a "clean slate" for individuals after a certain period of time thus ensure social rehabilitation. Such a right of oblivion is inadequately protected now in Taiwan. More specific clean slate legislations may be necessary. Second, the right to be forgotten could also be contrived as a weapon of resistance to the threat to personal independence in an era of big data and algorithms. Only until the democratic value of the right to be forgotten is rightly appreciated would it be possible to properly strike a good balance between the right to know and the freedom of press on the one hand, and, on the other, the right to control one's personal data.

References

Mayer-Schönberger V (2009) Delete: the virtue of forgetting in the digital age. Princeton University Press, Princeton
Ministry of Justice (2015) Interpretation Letter No. 10403508020 (July 2, 2015)
Supreme Administrative Court (2017) Case 106 Pan Zi No. 54
Taiwan High Court (2013) Case 102 Shang Zi No. 915 (Civil Division)
Taiwan High Court (2016) Case 104 Shang Zi No. 389 (Civil Division)
Taoyuan District Court (2015) 104 Su Zi No. 985 (Civil Division)

The manufacturer's authorised representative in the EU is Springer Nature Customer Service Centre GmbH, Europaplatz 3, 69115 Heidelberg, Germany. If you have any concerns regarding our products, please contact ProductSafety@springernature.com

Printed and bound by CPI Group (UK) Ltd, Croydon, CR0 4YY

30/09/2026

02225602-0010